Advances in Intelligent Systems for Defence

Series on Innovative Intelligence

Editor: L. C. Jain *(University of South Australia)*

Published:

Vol. 1 Virtual Environments for Teaching and Learning
(*eds. L. C. Jain, R. J. Howlett, N. S. Ichalkaranje & G. Tonfoni*)

Forthcoming Titles:

Neural Networks for Intelligent Signal Processing
(A. Zaknich)

Complex-Valued Neural Networks: Theories and Applications
(ed. A. Hirose)

Biology and Logic-Based Applied Machine Intelligence: Theory and Applications
(A. Konar & L. C. Jain)

Levels of Evolutionary Adaptation for Fuzzy Agents
(G. Resconi & L. C. Jain)

Series on Innovative Intelligence – Vol. 2

Advances in Intelligent Systems for Defence

Editors

L. C. Jain
University of South Australia

N. S. Ichalkaranje
University of South Australia

G. Tonfoni
The George Washington University, USA

World Scientific
New Jersey • London • Singapore • Hong Kong

Published by

World Scientific Publishing Co. Pte. Ltd.

5 Toh Tuck Link, Singapore 596224

USA office: Suite 202, 1060 Main Street, River Edge, NJ 07661

UK office: 57 Shelton Street, Covent Garden, London WC2H 9HE

British Library Cataloguing-in-Publication Data
A catalogue record for this book is available from the British Library.

ADVANCES IN INTELLIGENT SYSTEMS FOR DEFENCE

ISBN 981-238-208-9

Printed in Singapore by World Scientific Printers (S) Pte Ltd

Foreword

When I look back over a twenty-five year research and development career, it is fascinating to note the change in emphasis in defense research. In the early years, the mainstream areas were in basic sciences and engineering. From the perspective of an electronics engineer, things were very different from today, signal processing was performed in (frequently analogue) dedicated hardware, and hardware implementation research was widely practiced. It is pertinent to remember that at that time, the first, low-powered, microprocessors were only just finding their way into products.

Over the years that followed, digital signal processing implementations progressively replaced their analogue predecessors. With the inexorable increase in processing power, programmable electronics then supplanted dedicated digital hardware in all but the most challenging applications. Thus by the late 1980s, the emphasis in information engineering had changed from hardware to software.

The early 1990s saw the maturation and many more technologies such as telecommunications, networking, and office automation applications, which quickly led to affordable, network-enabled enterprise information systems. Their advent has had a profound effect on military thinking and has given rise to what has become known as the revolution in military affairs: the paradigm shift in military operations made possible by ubiquitous military information systems. Consequently, this has led defense science research away from technology per se and towards gaining an understanding of how the fruits of technological progress can be employed to the greatest military effect.

The field of information systems is, however, a challenging research area because of its complexity and because of the dominance of the human rather than technical elements. Increasingly, researchers with hard science and engineering backgrounds have come to appreciate that the research philosophies, methodologies, tools and techniques that have served them so well are not effective in tackling problems at the level of complexity of information systems. Researchers are now turning to systems thinking, management science, multidisciplinary thinking, non-linear mathematics, and intelligent agents to provide the way forward.

This volume reflects the latest trends in military information systems research by bringing together international contributions that span from communications technology, through intelligent agents and large-scale modeling and simulation, to sociotechnical systems. Together they represent a good cross section of contemporary information systems research.

Stephen Cook

DSTO Professor of Systems Engineering
University of South Australia

Preface

'You see things and say "why?"
But I dream things that never were
and I say "why not?"'

George Bernard Shaw

The future of defense is the defense of our future. The world of today has so dramatically changed in such a short time that we need new concepts to rethink current problems. We also need to redefine obsolete terms and to create new terms to describe defense priorities. It is necessary to reinvent defense concepts, and in innovative ways.

The defense of the future needs to be redesigned as new priorities are emerging and becoming the new emergencies that must be handled on a daily basis. Traditional ways of thinking have proved to be ineffective as new scenarios have emerged.

It has always been mandatory for defense strategists to be capable of clear thought. It is also clear that in the future there is a need for sharper conceptual tools than in the past. It is now necessary to make sense of a huge amount of often conflicting pieces of information from which must be derived models for consistent action.

Fuzzy or contradictory flows of information if not identified can explode into conflicts of major proportions. Fuzzy reports are very likely to prevent defense strategists from having a clear vision of an increasingly complex situation.

More specifically, misunderstandings between individuals, from different cultures and countries, have always occurred, resulting in different ways of developing defense policies. There is no way to eliminate this diversity of interpretation, especially for concepts such as security and defense. There are however ways to contain the divergences.

Wars have always existed in the course of history. What is evident today, is that "amplification tools", such as technologies may turn potential local conflicts into global conflicts. Communication tools are also likely to turn into communication weapons and information opportunities may become conflict opportunities. Such incidents may be activated by a few inappropriate words.

If it is common belief and a widely accepted assumption that the world is controlled by some major players. The truth is that the world with its current structure, cannot be controlled by anyone. The complexity is far too great to allow any monitoring and controlling process of significance to occur.

It is now clear that a defense system, which has no depth of understanding of information warfare and communication strategies, will have no real future. Such an ill informed defense system will have no capacity to create an actual impact on or shape the future.

There may still exist the illusion of control, and this may easily change into disappointment when reality is confronted. Both synchronously and asynchronously evolving multiple chains of actions and reactions may be generated in ways never before considered. Events may also be triggered on a random basis.

In the highly interconnected world, as soon as an event occurs, news is immediately transported out of the originating context and turned into globally available information. This news is subject to all kinds of uncontrolled interpretation patterns, most of which rarely fit the originating scheme and structure.

The real meaning of some aspects of a certain problem may rapidly be lost as soon as an inconsistent representation has been assigned. In our highly interconnected world, we are continuously exposed to the illusion of being able to handle everything any time, or so we wish. The fact that we are able to access in real time a huge number of resources has hood winked us into believing that we are technologically advanced and are able at any time to have both a correct estimate and a complete control of any item, we elect.

A consistent interpretation of what happens in the world is considered to be just a matter of taking the time and effort to incorporate new events into consistently defined categories and pre-existing models of classification. Classification models are considered to have the inner power of generating reliable models of interpretation, from which immediate action may be derived. It is considered to be a fact that the world is run based upon a set of classification systems and stated assumptions. It is very likely that would not prove to be as effective as we may like to think.

Accuracy in interpretation of the available data is considered to be a most fundamental issue for the defense of the future. Consistent decision making as for which kind of strategies should be designed and then implemented, was much easier to perform and to control during the 'pre-high tech era'. In the past, problems were analyzed primarily within their own context as to be resolved. In this era problems such as improper reporting and interpretation of events are ignored.

Everything has changed in our world of global interconnectivity. Nothing may in fact be taken for granted in a continuously shifting scenario, where the information explosion has challenged the very crucial essence of any filtering mechanism. There is just no way to determine, which information is really accurate, and decision makers in defense must admit the possibility that they may be operating upon a set of false assumptions, based upon incomplete or inconsistent information. Paradoxically, this failure may be no one's fault.

Each institution devoted to information analysis may be struggling to comply but information problems have grown much too complex, to be handled in fully consistent ways. Tasks currently assigned are just impossible due to an explosive complexity, and advanced technology may still not provide any real solution.

Once again, this fact may not be charged to a 'sloppy' or even up to 'incompetent behavior' by the experts, advisors and information analysts. They may be just doing the best job, they can afford. There is really no one to be blamed or charged with incompetence or mischievous behavior.

We have seen a few examples of obvious failure lately as security predictions and defense strategies now need to be viewed and interpreted. This is a direct consequence of the overwhelming complexity of information sources and resources, which have grown exponentially and in uncontrolled ways. It is a fact that asymmetrical information sources are available all the time.

Information exchanges and communicative intercourses are currently asymmetrical, as a consequence of meaning transported over time and space at unprecedented speed. Everyone traveling overseas has experienced jetlag. Interpretive jetlag is not equally obvious.

The same event, interpreted by a certain culture when transported over a long distance, is subject to distortion phenomena. This is usually invisible, and must be analyzed using appropriate highly specialized set of tools (Tonfoni, 1998, 2002).

Up to now a traditionally accepted common assumption has been that defense starts when diplomacy fails. A war starts when conversations have not been effective or effective enough to prevent military action. An enemy is thought of as someone, who has been addressed and warned, but just does not seem to be willing or able to listen and understand. An enemy will probably make counterclaims that he/she was not listened to, and that he/she was not provided with the necessary tools to understand the final warning message.

Interpretation warfare is the major challenge for the present and for future defense. It must be undertaken in collaboration with information specialists. Information analysts and specialists will be in increasing demand in our society and their role will be of great value. It is desirable that there is an internally developed and maintained defense expertise for these advanced information systems.

New conceptual tools for the recognition of negative patterns resulting from mischievous actions are also required. Chains of actions, reactions, and the identification of disruptive patterns, all need to be analyzed both in depth and in breadth. A reliable system for drawing consistent analogies, designed to assist in providing consistent and effective defensive strategies, should be available and continually in service. These areas of expertise need to be incorporated within the currently existing framework of defense research and training.

Strategic Ecdotics is a specialized area of information science, which x-rays strategic information. This is done by reconstructing the originating communicative context of each document which is of relevance to any strategic decision making. When this technique is systematically applied to sensitive information, it may help provide a high precision conceptual platform. The intention is to produce an accurate step by step verification of defense knowledge.

Clusters of meaning and meaningful relationships depend upon the nature of the categorization scheme and the classification system in use. Sentences and paragraphs, extrapolated and taken out of their originating context, may explode like fragmentation bombs and prevent the recovery of authentical interpretive clues.

After disruption of the original meaning, a new meaning may be constructed. It will probably be unlike the original. Any model, which has not been fully verified is likely to develop its own irrelevant meaning. In the worst circumstances it may trigger and create precisely the conditions which was hoped to be avoided.

As in visual perception human beings use selective vision which in part is based upon what we expect to see and decide to see. Motives influence our perception of what we see in significant ways and the same thing can occur as expectations of actions. This also includes mistaken expectations. From the incoming information, analysis should be continually performed. Rushed and oversimplified models are likely to cause errors of various kinds. All kinds of exaggeration and distortion can have the most serious consequences.

We may think of language in terms of being analogous to light. That is light has a double nature in that it is both particle-like and wave-like. The particle-like nature of language is expressed in words and is the so called literal meaning of language sequences. This represents the denotational aspect of language. The wave-like nature of language is represented by the communicative intentions. It is also dependent on the communicative styles and communicative cultures (Tonfoni, 1996). If the literal interpretation only of a given expression is pursued, it can cause interpretation distortions and blurring effects to the extent that the original meaning of the information is lost. This is particularly true for some cultures. Any model resulting from a set of progressive simplification processes may develop an increasing number of inaccurate conclusions. This is the so called 'Snowball Effect'.

Wrong assumptions also cause divergency effects. These effects tend to move the process away from the originating context and area of consistent interpretation. Interpretive jetlag and continuously shifting contexts may also cause perceptive disorders for the information analysts. Difficulties such as these are likely to deeply affect the whole information analysis cycle in a way which allows no possibility of a return to the correct solution.

In order to avoid interpretive distortion and aberration, an accurate verification of each single document, used as a piece of evidence, must to be performed. Problems of this kind are of direct relevance to defense strategists, who must be able to justify their judgment.

Interpretation contamination in the interconnected world may be intentionally performed and is easily disseminated. Statistically information suggests that contamination is rarely a mean act of will, but it mostly occurs as the result of a whole set of processes, which occur on a random basis. When these incidents occur, they are immediately multiplied by the high speed carriers, and they may have an undesirable impact on current analysis.

Contamination of data may come in a number of different forms. It may be a sudden amalgamation of several sets of data into one single structure. These sources may have a very different nature and only show very superficial similarities. Contamination may also come as a random addition of an inconsistent structure, which does not match the existing structure of a certain event or of a chain of events.

The amalgamation of groups of data may result in further errors in interpretation. Inaccurate structures, once applied to a certain event or chain of events, tend to augment divergence effects. This effect may be described in terms of propagation waves, progressively moving away from the originating meaning cluster. Disentangling such contaminated structures, to avoid their disruptive potential, requires an extensive search, explicitly aimed toward identifying inconsistencies, gaps in the data and discrepancies of increasing magnitude.

Oversimplified configurations, at a reduced scale, need to be expanded back to the actual size model so that processing may proceed at a higher resolution. Only by doing this, will it be possible to identify precisely those patterns, which have been subjected to contamination. Contaminated patterns, if not identified and eliminated, may seriously challenge the overall interpretation assigned to a whole set of events and cases. New and highly consistent interpretive categories may have to be assigned as to produce trustworthy results. Reverting to the previous analogy between Ecodetics and light the following argument may be applied.

In optics, an echelon is "a specialized form of diffraction grating consisting of parallel glass plates of successively varying sizes, used to examine extremely fine structures." In strategic ecdotics, an echelon may be reframed in terms of a highly specialized information diffraction grating. This will consist of parallel interpretive schemes of various sizes, operating at various levels, and in parallel used to analyze highly complex and highly controversial meaning structures.

Accordingly we may find a whole set of cases, which show that an inaccurate interpretation model can cause the same events to happen, which were predicted by the same model but based upon inaccurate assumptions. It may be concluded that an inaccurate model, has the power to trigger those very facts and events, which were mistakenly predicted. Today's expectations are widely disseminated, and they may gain tremendous energy and speed through the media and they eventuate. A case of a self fulfilling prophecy.

One of the main assignments for future defense is that of prevention. More precisely it is about identifying and discarding those models, which have the potential to arouse expectations and to create conditions in which misplaced action is implemented.

Defense of the future is about keeping strategic decision makers from being caught in a risky loop, amplified by the pervasiveness of the media in our society. It is this loop, which feeds fear of virtual external enemies and which fuels negative predictions about what may happen next. This triggers negative expectations. Those negative expectations are immediately announced worldwide, and can result in devastating actions by these virtual external enemies, who can then turn into real enemies, as a consequence of this loop.

Defense of the future is also about defense from the 'enemy within' coming in the form of unverified misleading assumptions, which may be automatically accepted by defense strategists as reliable sources of specialized advice.

The defense of the future is about defending countries from 'the Hollywood-like syndrome', which means that war and attack plots are not only feared and foreseen, but also widely announced and illustrated in full detail. This creates the ideal conditions for those fears to materialize into real and devastating actions.

Defense of the future is also about preventing decision makers from falling into further kinds of interpretive trap, such as those caused by verbal expressions originally meant to be paradoxical or challenging. Once taken out of their originating context, these expressions may be interpreted as assertive, and are therefore likely to act as a trigger.

The making and the rewriting of a document, which deals with defense issues, needs to be based on the consistent identification of sound rhetorical strategies. We know with certainty that different kinds of rhetorics are the greatest barrier for intercultural exchanges. "A silly translation of a phrase in a diplomatic document might easily cause a war. It is only when first rate tools are put in the hands of first rate men that we avoid an expensive waste of effort." (Norbert Wiener, Inventions, 1954).

Rhetorics is the art of artificiality and today we live in a world where the artificial component has been expanding rapidly. In the past, rhetorics has been thought of as being in the realm of special effects in communication. Consequently it has appeared in literary texts and has legitimately formed a component of literary studies. Some rhetorical strategies have been recognized to be of direct relevance to diplomacy, and therefore have also been included as a component of diplomacy studies. Today rhetorics is at the core of defense as defense and diplomacy must be harmoniously merged.

Defense is a major concern for the future. It will involve the identification of the often invisible 'inner bug' and successful treatment of any problems arising from it. This aspect is of very real concern in high tech countries. We are dealing with the unpredicted and uncontrolled behavior as a consequence of a fast growth of information technologies and the ready availability of information today.

The world today is exposed to each individual and each individual may be exposed to the world. This development has modified ordinary behavior and the ways of thinking about conflicts. The increased visibility of conflicts, the implications and the global visibility, may well create conditions for a new kind of mechanics for conflicts and war schemes.

Defense conversations tend to be about some highly specific expressions. Such conversation is part of a delicate process which must be capable of accurate verification. Distortion and aberration effects, due to transference of the expressions through a variety of channels and media, may cause undesired modifications to occur.

Information Security is the science of making information secure. This area must also be able to ensure that information is accurate and reliable enough for decision makers to base decisions upon it.

Strategic ecdotics has the potential to provide substantial help to information analysts, as it is able to provide a highly consistent set of categories. This is intended to make qualitative reasoning possible based on the original information. Strategic ecdotics permits a reasoned analysis of the context sensitive document thus allowing the consistent classification and evaluation of information. The documents may come in a verbal or in a written format or both. The improvement is obtained by the application of context sensitive categories and the use of scientifically based principles.

Strategic ecdotics incorporates a methodology for document deconstruction. This is about "x-raying", in depth analysis and the evaluation of a document. By this means it is able to challenge the clarity and consistency of a document by disclosing hidden meanings, implications, contradictions, and alternative interpretations. Any military communication needs to be complemented by annotation. This requirement also applies to translation. Documents may be the result of the aggregation of a number of verbal sequences and may have originally come in the form of a random collection of quotations.

Documents, based upon this kind of information, are frequently found to be contaminated, as a particular kind of interpretation may interact with another. Deconstruction of entire lines and an in depth analysis, together with an accurate reclassification of each single transported sequence, may help reconstruct the intention and may resolve some potential conflicts in interpretation.

The tools used in the field of information security and strategic ecdotics are intended to make a specific set of mistakes surface. The technique involves the adoption of a certain scheme and the subsequent modification of this scheme. The tools provided are intended to avoid the problem of oversimplification. Information processing which is only based upon random recollection or casual observation of recursive speech patterns, is no longer adequate.

Selecting and consistently redesigning information to facilitate knowledge acquisition is a delicate task. It requires sensitivity and knowledge of the nature of the information material provided. This is essential if the material is to be converted into stabilized knowledge.

A distinction must be introduced between micro conflict resolution strategies which are concerned with specialized and context bound information, and macro conflict resolution strategies. These are concerned with the interplay of interpretive variables and their impact on decision making.

Macro conflict resolution strategies may be used to analyze different allocation and utilization of information resources. Choices will need to be made without the complete data spectrum being available. These decisions, concerning the quality and quantity of information are subject to revision and need to be made from projections. A comprehensive, flexible and dynamic framework for diverse, but compatible operations must be done on this information. This information is at the core of any defense long term policy.

Conversational exchanges occur as sequences and a set of recursive patterns may be recognized. This may provide a way in which cooperative attitudes can be separated from non cooperative misleading attitudes. Pre-allocation processes, which allocate the order and rhythm for a conversation is decided in advance by an external facilitator. These then regulate the space and time allowed for each participant.

Throughout history wars have had rituals. These have changed with different cultures and traditions. In the information era, wars are expected to involve the use of information and will need adequate information weaponry and equipment. To many people the speed at which new technology is altering their society is bewildering. This may even result in a wish to insert some kind of filter to weed out and slow down development. A kind of latter day Ludism where unwelcome development is impeded. This is of course an impractical and undesirable dream that may have roots in the Stone age.

Virtual reality environments which are intended to increase perception, may show disappointing results as demonstrated by many case studies. A continuous search for more intelligent technologies seems to have made some users into disturbed individuals, unable to take decisions for the fear of failure. By continuously trying to change system requirements, users may get confused and lose their traditional skills and they may be unable to master the new skills either. Under these conditions operators may make unpredictable mistakes which even the best technology cannot predict or correct.

Defense strategists, who as part of their work must engage in conversations, should be alert and have appropriate skills. Conversations intended for conflict resolution will have to be configured, according to each of the participants. In such discussions it is desirable that a strategist will be overall in charge of the entire process. The operators must be provided with accurate and effective equipment. This is necessary so that information from the conversations may be transferred between the parties to the conversation process.

Defense strategists need to see themselves as real participants to the process. It is important to be able to discriminate between participants, observers and bystanders, as the roles and outputs of each are substantially different.

The success of any conversational exchange, especially for conflict management issues, is the ability of the participants' to remain focused on the topics under discussion. This can be done by adopting a stance of multiple viewing, even if the participants intend to maintain their original opinions. By adopting a multiple viewing stance and having an accurate representation of controversial points, which may support the evidence required to obtain a new perspective, the participants may proceed towards forming a mutually agreeable solution.

Defense experts involved in such a collaborative interaction need to become joint narrators in a shared project. Many difficult conversational intercourses occur between partners belonging to different cultures. This is one of the major obstacles for a consistent interpretation of the respective aims.

According to Kluckhohn (1951) cultures imply "patterned ways of thinking, feeling and reacting. These are acquired and transmitted mainly by the use of symbols, constituting the distinctive achievements of human groups."

According to Rokeach (1968), values are "internalized standards for guiding action and enduring belief that a specific mode of conduct or end-state of existence is personally and socially preferable to alternative models of conduct or end-state of existence". Collaborative conversational interactions are meant to reside upon consensually agreed upon values so as to avoid disfunctionality, which may be derived from uncleared misunderstandings (Enninger, 1987).

There exist relevant spaces in conversations, which are points in the conversation when speakers may decide to take turns or not, or may do so after a delay so as to ensure the success of the collaborative interaction. Silences and changes in the expected turn on the side of one speaker may, in this case, represent constructive and protective strategies which ensure the final success of the conversational interaction.

According to Gumpertz (1982), there is also non communication, where no message is communicated. In this instance an unintended message may be communicated. To avoid possible miscommunication, silence or remedial strategies may be applied. The adoption of temporary solutions, designed as an interim step, may also be a constructive artifice.

The cooperative principle stated by Grice (1975), concerning conversational interactions, may be accepted as a fundamental discovery in the science of conversational analysis. This concept may have different kinds of implementation modes. The mode varies according to the contextual arrangement, the culture involved, and individual preferences and styles of operation.

A temporary silence will be part of the cooperative principle if this principle is understood by the conversational partners. Its use may help ensure the success of the communication process. If newspapers and media accounts are an information source for historical research in defense decision making, an accurate analysis is required. Careful attention must be given to both the making and the retraction of news in the press. Media amplification effects should also be recognized.

Creating news is a specific process and skills are required to analyze and interpret various kinds of information. It is necessary to derive a consistent interpretation of any data contained in the news if it is used for operational decision making.

Specific skills for enhancing interpretation abilities are necessary when confronting new information. Consistent comparison between the new knowledge and previously accumulated experience is required. Any information needs to be stabilized and verified before it can be accepted as reliable knowledge.

New metaphors and metaphorical environments must be developed (Tonfoni, 1998), in order to avoid undesirable effects resulting from poor metaphors and analogies. Models of reasoning, once established, are likely to persist.

Nelson (2001) states "a metaphor may be far from adequate but its shortcomings largely escape notice because its continuous conventional use has deadened our ears to the fact that it is but an untested metaphor." Wrong metaphors may adversely affect decision making in unpredictable ways, causing damage and undesirable effects. These errors will persist unless positive action is taken to correct the defect.

Reenstjerna (2001) states that "a choice of metaphor is critical, because thought processes tend to be limited by the constraints imposed by that metaphor. A metaphor can affect profoundly not only how we do things, but also what happens. No model arises from an ideological vacuum: it is based upon certain fundamental assumptions about how people imagine that a particular process operates."

Care must be taken when drawing analogies to discriminate consistently between different levels of likeness. Clarity, balance and full details must ensure that any comparison is not carried out beyond the intended bonds. Using analogies for explanatory purposes or for operational purposes requires an in depth understanding of the process.

In practice creating analogies means taking knowledge from one area and applying it to another area. It is based upon identification of structural and functional similarities of the knowledge. The use of metaphors and the practice of analogical reasoning are both essential in the expanding information technology area.

Thinking metaphorically and reasoning analogically are both processes, which need to be subjected to careful scrutiny. Again metaphors and analogies, once accepted end up shaping a certain representation of reality, and may affect reality in substantial ways. Once a certain interpretive framework has been widely adopted, it is hard to change or discard it, should it prove to be inadequate.

Defense strategists and decision makers need to master advanced and sophisticated tools, in order for them to identify misunderstandings and to fix interpretive mistakes. These may occur both in written documents and in recorded conversations.

As we have indicated, changes, in the information environment, may cause aberrant interpretation of the event. Aberration phenomena may come as a side effect of information overload and loss of context. They also come as a consequence of unpredicted and uncontrolled perceptual changes. These may result from fast expansion produced by new and pervasive information technologies. Some advanced technologies have not helped the end users and have resulted in confusion.

A light analogy may be useful when considering facts which must be analyzed for strategic reasons. Information may be contained in the part of the spectrum not visible to the human eye. It is necessary to use appropriate equipment to reveal this concealed information. In the same way information may not be immediately apparent on first examination and it is necessary to make it appear by the use of appropriate technology. Documents and even single paragraphs may contain additional valuable information. Turbulence generated within an information environment, may change the nature of the data. This may result in erroneous interpretation of the data.

Defense resources must provide services for many top priority tasks, such as:

(a) The assembly of information flows of many kinds and from many diverse sources;
(b) The monitoring, control, and verification of information;
(c) Filtering, identifying, and labeling of information;
(d) Classification, storing and declassifying of information;
(e) Retrieve, conversion and packaging of information;
(f) Repackaging, updating and upgrading of information;
(g) Selection, reconfiguring and delivering of information.

Information detection may be thought of as a kind of visualization enhancement process, which occurs through the visual apparatus and enable it to trace the communication patterns correctly.

There are different ways of perceiving time and this difference can be of considerable relevance. Time of perception in a high-tech society as compared to a low-tech society must be analyzed to determine any differences. The precise "representation of time" and "the time of representation" may vary significantly between cultures and also vary with the context. For example the time of reading an article is radically different to the time of writing. The mental time of a narrative allows the reader to build a set of different scenarios. Movie time is determined by the producer and is intended to convey a preset selection of scenarios and sequences to the viewer. In this case the movie producer dictates the rules for the perception time and time of perception.

Defense strategists currently make decisions based on representative models and on information compilations. These sources may include summaries incorporating clues for understanding. Defense strategists of the future will be both authors and producers of their operations — fully in control of the overall defense decision making process and responsible for both quality and the reliability of information.

Repositories of strategic defense information need to be built upon summaries. The summaries are in turn based upon a set of paragraphs labeled according to a set of categories. The categories will have been established as the result of an in-depth search and a consistent knowledge accumulation, following filtering and selection. Defense of the future should prevent polluted information from causing incorrect decisions and ensuring undesirable actions.

The present collection of chapters shows the wealth of research in advanced information systems and in technologies developed for defense purposes. Some chapters may become a component of advanced defense training programs. They provide an important overview of crucial research which has evolved over the last ten years.

Each chapter presented is supported by a solid research platform. Reference to a wide and specialized literature is also provided to support the main research topics covered. Readers may consider each chapter as a single piece of research. Any chapter may be easily reconfigured and adapted to handle information warfare problems of today. Some chapters may provide a new platform for combat communication in the information battlefield.

The high standards attained by research in advanced intelligent defense systems indicates that the knowledge and experience gathered for military purposes may also be used in other areas.

Chapter 1 provides the reader with a description of the challenges and vulnerabilities of Military Information Security. The strengths and weaknesses of the present situation are considered. An explanation is given why an automated command system, intended to increase a commander's situational awareness may contribute to uncertainty.

Chapter 2 presents a comprehensive overview and a well documented review of research in the delicate and crucial area of human and machine interaction. From an analysis of a variety of experiences and results, a new conceptual framework is provided for the design and development of human and machines teaming applications.

Chapter 3 presents an introduction to the field of Computer Generated Forces. Key concepts are outlined as well as the current research in this field. The debate concerning priorities in modeling choices and perspectives for the future are discussed. A wide bibliography is provided.

Chapter 4 examines possible future developments extending beyond the explicit topic. The use of automaton to rapidly illuminate tactical options is considered. Concepts of crucial relevance such as 'asymmetric warfare' and other factors arising from non conventional combat are examined. Possible future developments in this area are also considered.

Chapter 5 provides a presentation of the components and features of a Multi-User Wireless Link for real-time video transfer for defense applications. Derivative commercial applications are also envisioned and these are described in detail. Possible civilian uses such as monitoring in hazardous areas, surveillance, high quality video transmission and storage are considered along with other possible civilian uses.

Chapter 6 presents a balanced analysis of the Intelligent Agent Technology. It illustrates major gains achieved in productivity. The focus of attention is about agents of change and specifically addresses the impact of Intelligent Agent Technology on the analysis of air operations. The problems of obsolescence, maintenance, update and upgrade are mentioned and explored. Indications are also provided as a guide to possible use as technologies described suitable for other problems. Care is recommended when considering the use of novel technology.

Chapter 7 provides the reader with a documented illustration of a variety of systems models. This indicates the many ways in which agents may collaborate. Models of biologically inspired team formation for target observation are discussed. The behavioral implications for each is also discussed. A variety of biological systems is analyzed in depth in order to conceive sensors and to model their interactions.

Chapter 8 provides the reader with a description of the Disciple Approach. The model is expanded to indicate military applications of the Disciple Learning Agent, within an Artificial Intelligence Framework. Issues of major relevance such as the Strategic Center of Gravity, problem identification and analysis, Scenario Specification, Modeling, Task and Rule Learning, refinement and problem solving are also considered.

Chapter 9 presents a specialized technical description of a Parallel Associative Processing Bayesian algorithm for infrared dim point target detection. This has shown satisfactory detection performance where more usual algorithms fail. Background theory for image processing and mathematical morphology is also provided. Selected references to the broader fields are given for further research and applications.

Chapter 10 presents an overview of the concept of intelligent monitoring of complex systems. Attention is given to a selected core of current research in the area with a focus on an intelligent monitoring system which enables condition-based maintenance. Numerous opportunities for bi-directional technology transfer are seen in the ongoing research in data fusion and intelligent monitoring systems. Some of these technologies are likely to migrate to commercial systems and significantly affect our lives.

Ways need to be found to use some of the knowledge evolved in the defense service to make it applicable to daily life. It is possible that some studies carried out for defense purposes will help solve major civilian social behavioral problems.

Help to harmoniously merge old technology and new technology may be provided by defense specialists. Some users of high tech systems suffer from stress generated by the introduction of complex information tools.

Defense needs to be paid a lot of attention in order to ensure that real security for the future is implemented. It can only be based upon recognition and careful scrutiny of diversity and respect for the many and different dreams existing all around.

All of this may only happen, through a most accurate design of advanced information systems, and more fundamentally, by achieving the highest quality of conceptual equipment and context sensitivity ever reached by defense specialists in the history of defense. This is the real challenge posed today, to be continued into the future of defense, and to ensure the very future of mankind.

We are greatly indebted to the authors and reviewers for their contributions. We are grateful to Berend Jan van der Zwaag for his contribution. We thank the World Scientific Publishing Company for their interest in our work.

References

Enninger, W. (1987), "What interactants do with non-talk across cultures," in Knapp, K., Enninger, W., and Knapp-Pothoff, A. (eds.), *Analysing Intercultural Communication*, Narr, Tuebingen, pp. 269-302.

Goffman, E. (1983), "Replies and responses," in Goffman, E. (ed.), *Forms of Talk*, University of Pennsylvania Press, Philadelphia, pp. 5-77.

Grice, P. (1975), "Logic and conversation," in Cole, P. and Morgan, J. (eds.), *Syntax and Semantics 3: Speech Acts*, Academic Press, New York, pp. 41-58.

Gumpertz, J.J. (1982), *Discourse Strategies*, Cambridge University Press, Cambridge, U.K.

Kluckhohn, C. (1951), "The study of culture," in Learner, D. and Lasswell, H.D. (eds.), *The Policy of Sciences*, Stanford University Press, Palo Alto, pp. 86.

Nelson, C.K. (2001), "The haunting of communication research by dead metaphors: for reflexive analyses of the communication research literature," *Language and Communication*, vol. 21, pp. 245-272.

Reenstjerna, F.R. (2001), "Thinking about reference service paradigms and metaphors," *The Reference Librarian*, vol. 72, pp. 97-122.

Rokeach, M. (1968), *Beliefs Attitudes and Values*, Jossey-Bass, San Francisco.

Tonfoni, G. (1996), *Communication Patterns and Textual Forms*, Intellect, U.K.

Tonfoni, G. (1998), *Information Design: the Knowledge Architect's Toolkit*, Scarecrowpress, Lanham, MD.

Tonfoni, G. (2000), "Visualization of textual structures," in Yazdani, M. and Barker, P. (eds.), *Iconic Communication*, Intellect, U.K.

Tonfoni, G. (2002), *Changing Perceptions on Documentation Management*, Intellect, U.K. (in publication).

Wiener, N. (1954), *Inventions: the Care and Feeding of Ideas*, 1993 print by MIT Press, Cambridge, Mass.

Contents

Chapter 1.
Military information systems security: challenges and vulnerabilities
M.R. Frater and M.J. Ryan

1 Introduction 1
2 Security services 3
2.1 Confidentiality 4
2.2 Authentication 6
2.3 Integrity 9
2.4 Non-repudiation 10
2.5 Access control 10
2.6 Availability 11
3 Security architectures 12
3.1 System-high 12
3.2 Multi-level security 16
4 Challenges and vulnerabilities 19
4.1 Trade-offs in military systems 20
4.2 Vulnerabilities 21
4.2.1 More use of the spectrum 22
4.2.2 More points of attack 23
4.2.3 More types of attack 24
4.2.4 More complex systems 25
5 Conclusion 27
Endnotes 27

Chapter 2.
Teaming human and machine
P. Urlings and L. Jain

1 Introduction 29
2 Existing research 30

3 Initial framework 34
3.1 Level of machine assistance 35
3.2 Required characteristics 37
4 Functional classification 38
4.1 Hierarchical classification 40
4.2 Data processing classification 41
4.3 Cognitive classification 42
5 Task management and coordination 45
5.1 Sensor data management 45
5.2 Display and presentation management 46
5.3 Control management 46
5.4 Coordination management 47
6 Artificial intelligence and AIP techniques 48
6.1 Artificial intelligence paradigms 49
6.2 Requirements for AIP applications 51
6.3 Distributed problem solving 52
7 Conclusion and recommendations 54
References 55

Chapter 3.

Computer generated forces in distributed simulation

M.R. Stytz and S.B. Banks

1 Overview 62
2 Key concepts 65
3 Artificial intelligence technologies 70
3.1 Blackboard systems 70
3.2 Expert systems 71
3.3 Case-based reasoning 72
3.4 Evolutionary algorithms, genetic algorithms, and evolutionary computing 73
3.5 Bayesian networks / belief systems 76
3.6 Fuzzy logic and fuzzy systems 77
3.7 Neural networks 78
4 CGA requirements and their implications 80

5 Significant research projects and software architectures 83
5.1 Research projects 83
5.2 Software architectures 89
6 Major CGF systems and decision mechanisms 98
7 Human behavior modeling, skills modeling, decision modeling, and artificial intelligence technologies 107
8 Conclusions and suggestions for further research 120
References 128

Chapter 4.

Using automaton models to rapidly illuminate tactical options

M.K. Lauren and R.T. Stephen

1 Introduction 170
1.1 Models 171
1.2 Philosophy 176
1.3 Adding personality 178
2 Applications 181
2.1 Scenario 181
2.2 Control v squad mix 183
3 Parameter landscapes 186
4 Technical aspects of the MANA approach 188
4.1 Objects 189
4.1.1 Battlefield object 189
4.1.2 Squad and agent objects 189
4.2 Parameter set 191
4.3 Movement 193
4.4 The constraint parameters 197
4.4.1 Cluster constraint 197
4.4.2 Combat constraint 197
4.4.3 Advance constraint 198
5 Conclusions 198
References 199

Chapter 5.
Multi-user wireless link for real-time video transfer for defense applications
W.M. Daniels, S. Forbes, and L.C. Jain

1 Introduction 202
2 Background 204
3 DSTO concept demonstrator 205
4 Functional requirements 207
5 System requirements 208
6 Deployment scenario 211
7 General structure of the system 212
8 Communications system 215
8.1 Multiple-access techniques 215
8.2 Spread spectrum systems 215
8.3 Carrier sense multiple access/collision avoidance (CSMA/CA) 216
8.4 Time division multiple access (TDMA) 217
8.5 Frequency hopping spread spectrum (FHSS) 217
8.6 Code division multiple access (CDMA) 218
9 Video system 219
9.1 Real-time video compression techniques 219
9.2 Motion picture experts group (MPEG) 219
9.3 Wavelet compression 220
9.4 Knowledge-based techniques 220
10 Implementation conclusions and prototype 220
11 Conclusion 224
Acknowledgments 226
Acronyms, abbreviations and definitions 226
References 227

Chapter 6.
Agents of change: the impact of intelligent agent technology on the analysis of air operations

C. Heinze, M. Cross, S. Goss, T. Josefsson, I. Lloyd, G. Murray, M. Papasimeon, and M. Turner

1 A challenging application domain ... 229
1.1 Air operational analysis ... 231
1.2 Modeling and simulation ... 232
1.3 Intelligent agents ... 234
2 Innovative and successful application deployment ... 236
2.1 Fighter combat ... 236
2.2 SWARMM ... 237
2.3 BattleModel ... 240
2.4 F-111 stand-off weapon employment ... 241
2.5 The virtual air environment ... 244
2.6 AP3-C Orion operating procedure update ... 245
3 Leading edge agent technology development ... 248
3.1 dMARS and JACK ... 249
3.2 The agent architecture ... 251
4 Ongoing research and development ... 252
4.1 Perception and recognition of intention ... 253
4.2 Software engineering ... 254
4.3 Computer generated forces and other applications ... 256
4.4 Graphical interfaces for agent development ... 257
5 Discussion ... 258
5.1 Rapid development ... 258
5.2 Graphical interfaces ... 259
5.3 Appropriate abstraction ... 260
5.4 Enforced modularity ... 261
5.5 Distributed processing ... 261
5.6 Improved interactions ... 262
5.7 Software engineering ... 262
6 Conclusions ... 263
Acknowledgments ... 263
References ... 264

Chapter 7.
Agent-enabled sensor network for target observation
R.S. Bowyer and R.E. Bogner

1 Introduction 270
1.1 The target observation problem 270
1.2 Agent-enabled sensors 274
1.3 Centralized versus decentralized control in multisensor systems 277
1.4 Motivation for self-organizing sensor network 281
1.5 Motivation for team formation 283
1.6 Related work 285
2 System models 287
2.1 Characteristics and assumptions 287
2.2 Environment model 288
2.3 Sensor models 289
2.4 Target model 290
2.5 Inter-agent communication model 291
2.6 Social aspects of sensor networks 292
3 Biological systems 295
3.1 Morphogenesis – the emergence of form and function . 295
3.2 The agent phenotype (self) 296
3.3 Lessons learned from nature 299
4 A biologically inspired agent and the social potential 301
4.1 Agent goals 301
4.2 Agent capabilities 304
4.3 Agent behaviors 306
5 Agent architecture 306
6 Building relationships 308
6.1 Information 308
6.2 Calculating sensor relationships 309
7 Goal-directed social networking 313
7.1 Setting goals 313
7.2 Weight selection 314
7.3 Inter-agent relationships and how they are formed 315

8 Team formation ... 315
8.1 Motivation for team formation ... 315
8.2 What constitutes a team? ... 316
8.3 What triggers team formation? ... 317
8.4 A suitable organizational structure – who gets on the team? ... 319
8.5 What holds a team together? ... 320
8.6 When should a team disband? ... 321
9 Team forming algorithm ... 321
9.1 Algorithm mechanism ... 321
9.2 Corporate memory and the role of agent pheromone ... 322
9.3 The simulation environment ... 323
10 Simulations ... 325
10.1 Setup ... 325
10.2 Results and discussion ... 325
10.3 Concluding remarks ... 330
Acknowledgment ... 331
Glossary ... 332
References ... 334

Chapter 8.
Military applications of the Disciple learning agent
G. Tecuci and M. Boicu

1 Introduction ... 338
2 Center of gravity analysis ... 339
2.1 The center of gravity problem ... 339
2.2 A Disciple agent for COG analysis ... 340
2.3 Agent development with Disciple-RKF ... 346
3 Course of action critiquing ... 360
3.1 The course of action critiquing problem ... 361
3.2 The Disciple-COA critiquer ... 364
4 Workaround reasoning ... 367
4.1 The workaround reasoning challenge problem ... 368
4.2 The Disciple-workaround planner ... 370

5 Long-term research vision ... 372
Acknowledgments ... 373
References ... 374

Chapter 9.
A parallel associative processing Bayesian algorithm for IR dim point target detection

R.C. Warren

1 Introduction ... 378
2 Background theory ... 383
2.1 Associative processing ... 383
2.2 Mathematical morphology ... 385
2.2.1 Background ... 385
2.2.2 Mathematical morphology for ASM detection by IRST ... 387
2.3 Filtering and detection ... 389
2.4 Track before detect method ... 394
3 Algorithm assessment using measured target data in an uncluttered background ... 402
3.1 Spatial filters ... 403
3.1.1 Results ... 404
3.1.2 Discussion ... 411
3.2 Temporal-spatial filtering ... 413
3.3 Conclusions ... 416
4 Algorithm assessment using synthetically generated target in a cluttered background ... 416
4.1 Introduction ... 416
4.2 Detection and tracking method ... 419
4.3 Results ... 422
4.4 Conclusions ... 426
5 Conclusions ... 426
Acknowledgments ... 427
References ... 427

Chapter 10.
Intelligent monitoring of complex systems
D.L. Hall

1 Introduction 436
2 Motivation for intelligent monitoring systems 437
3 Concept of condition monitoring of complex systems 440
4 Challenges for intelligent monitoring systems 447
5 Multi-sensor data fusion 454
6 Summary 460
References 462

Index **469**

List of contributors **473**

Chapter 1

Military Information Systems Security: Challenges and Vulnerabilities

M.R. Frater and M.J. Ryan

It is widely recognized that Information-age technologies, such as networked information systems, have the potential to provide military commanders with greatly improved situational awareness, and through this to increase the capability of the forces they command. What is less widely discussed is the dependence on communications technology which results from this increase in capability, leading to an increased vulnerability to information-based attacks. Small Information-age forces are often seen as having the same capability as much larger Industrial-age forces. It might be said that an Information-age force is made stronger by Information-age technology, but is probably also made more brittle by it. This chapter discusses issues in information systems security, examining both the sources of vulnerability and measures that can be taken to protect information systems from attack. The important conclusion is that military information security architectures should be looked at in the context of the six security services: authentication, confidentiality, integrity, non-repudiation, access control and availability.

1 Introduction

While there have been a number of major technological revolutions in warfare, perhaps the most significant will be in the role of information and knowledge, and in particular, in the degree of situational awareness made possible by the increasing number of communications and information systems supporting combat forces. Purveyors of information-age technologies promise ubiquitous battlefield networks through which all automated battlefield entities are seamlessly integrated so that real-time data can be shared, as and when required. To modern commanders the allure of this prospect is irresistible. It has been several hundred years since a commander has had the ability from a convenient hilltop to survey personally the disposition of all friendly and adversary forces. Now, in the Information Age, the modern commander, with senses enhanced by electronic sensors and modern communications and information systems, is promised the ability to stand on an electronic hilltop and once again 'see' whatever portion of the battlefield is desired at whatever detail is appropriate.

While there is little doubt that the Information Age has produced technologies that provide great promise of decisive advantage on the modern battlefield to the commander who can gather and exploit information most effectively, there is a significant dark side to the information revolution. As communications and information systems become vital to military and civilian society, they can become major targets in war and can also serve as a major means for conducting offensive operations. Consequently, military adoption of information technology creates a new vulnerability—the same information technology that provides the fuel for the networks that support modern commanders also provides one of the major means for their destruction. An increased reliance on communications and information systems increases this vulnerability. So, while automated command systems increase a commander's situational awareness, they can also be turned against him and used to contribute to his uncertainty. It is the very increase in capability brought about by information-age technology that leads to this increase in vulnerability, perhaps a case of increasing strength leading to brittleness.

In many ways, the evolution of the network, both for communications and information, is the key to the benefits and the vulnerabilities associated with information-age technology. It is the network that enables the sharing of information; it is the network that potentially exposes the whole force to attacks from a single source. This chapter examines the provision of security services for military information systems, concentrating on the network.

2 Security Services

The aim of security services in a networked information system is to make available the resources (including information) of the network to legitimate users without degradation in the quality of service due to attacks, and to deny these resources to unauthorized users. Security services aim, therefore, to minimize the impact of attacks on the network. They do this first by protecting the network against the attack, that is, by making the task of an attacker more difficult. In many cases, it is not possible to provide guarantees that protection measures will be successful, so the ability to detect attacks is an important second aspect of security services. Having detected a successful attack, a variety of responses may be possible.

There are six information or network security services [1]:

- *confidentiality*, aimed at preventing disclosure of information to unauthorized parties;
- *authentication*, allowing parties to a transaction to be confident of the identity of other parties;
- *integrity*, which seeks to prevent modification of data or insertion of counterfeit data;
- *non-repudiation*, allowing parties to a transaction to be confident that the other party cannot deny taking part;
- *access control*, which seeks to prevent an adversary from gaining advantage by use of friendly systems; and
- *availability*, aiming to ensure that the quality of service in

friendly information systems and networks is not degraded by attacks.

Each of these security services is provided by a combination of technologies (that is, automation) and procedures (that is, human involvement). These security services are an important part of the total force protection capability of a modern force.

Provision of effective network security can only be achieved if these security services are implemented in the context of a security architecture, which should include procedures as well as technological measures. This security architecture is itself part of the network architecture, and is driven by a security policy for the network. The security policy for the network also does not exist in isolation; it is part of a military force's overall force protection plan, which also includes physical protection.

2.1 Confidentiality

In commercial systems, confidentiality is mostly concerned with protecting information from disclosure. This protection is provided by encrypting data for transmission (or sometimes for storage) [2], as illustrated in Figure 1. Before being transmitted on a channel, data is passed through an encoding process (called *encryption*), which converts the original data into a form that cannot be read without being decoded (*decrypted*) by a device with access to some form of secret information (usually called a *key*). Encryption systems should be designed so that it is not necessary to keep the basic algorithms secret, but only the secret information that modifies the behaviour of these algorithms. The difficulty of restoring the original data without access to the key is a measure of the strength of the encryption.

In secret-key (or symmetric) encryption, the same key is used for encryption and decryption. In public-key (or asymmetric) systems, two different keys are used. One is used for encryption, the other for decryption. Either key may be made public without compromising the value of the other.

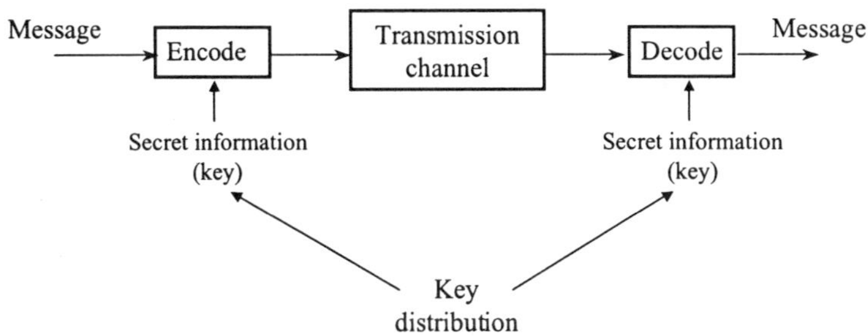

Figure 1. Structure of an encryption system.

Traditional military encryption systems are based on secret-key technology. Public key encryption systems are often used in commercial applications for authentication and exchange of a randomly generated secret key at the beginning of a transaction, and are beginning to find their way into similar military applications. This helps to overcome one of the most significant overheads of secret-key encryption systems: the management and distribution of keys. Secret-key encryption systems tend be less computationally intensive that public-key systems, favoring their use for exchanges of large amounts of data.

In military systems, not only is protecting the data carried by a network important, but it is also often necessary to protect against traffic analysis, which may require protection of both the addresses of the sender and recipient of the data, or even the existence of the transmission itself. Four levels of protection of a communications channel can be described:

- *Protection of data*. The data that is carried by a communication channel may be protected by encryption of the bodies of messages, while leaving the message (or packet) headers unencrypted. This has the advantage of simplified switching, and not requiring encryption systems to exist in switching equipment.

This approach provides no protection against traffic analysis, which seeks patterns in data such as message headers from which to identify likely future actions of a force.

- *Protection of data and source and destination addresses.* This is usually achieved by encryption of both message bodies and their headers. This leads to increased protection against traffic analysis because the data on source and destination addresses is no longer available in plain text, but increases the complexity of a network because encryption systems are required in at least some switching equipment. Limited traffic analysis may still be possible on channels that do not transmit continuously. Virtual private networks in commercial systems use this approach to confidentiality.
- *Protection of a whole link.* Where transmission on a communications channel is continuous, all data may be encrypted, allowing the whole structure of transmissions to be concealed. This protects against interception and analysis of transmissions, but not against detection or direction finding.
- *Protection of the existence of link.* By using the techniques of electronic protection, including directional antennas and spread spectrum techniques, the very existence of a communications channel may be obscured, protecting against all of an adversary's electronic support systems, including detection, intercept, direction finding and analysis.

2.2 Authentication

Authentication allows one party (Alice) to confirm the identity of another party (Bob). This is mostly useful as a prelude to a transaction that Alice wishes to carry out only with a particular party (say Bob), or with a party whose identity is known. For example, a consultant who is paid to write a report for a particular customer may need to confirm the customer's identity before sending the report. The customer, on the other hand, may employ many consultants, and be happy to receive a report from any one of them. The customer may use authentication to reliably identify a particular consultant involved in a transaction.

In order to be authenticated by Alice, Bob must demonstrate possession of a *verifiable secret*. Alice does not necessarily need to possess the verifiable secret, only to be able to verify that Bob has it. For example, Alice may send a challenge to Bob, with Bob returning the challenge encrypted with his private key (which is unknown to Alice). So long as Alice possesses the public key corresponding to Bob's private key, she is able to decrypt Bob's response and verify that this response was encrypted with Bob's private key. So long as Bob maintains the security of his private key, Alice can be confident that any data encrypted with it was encrypted by Bob.

Hash functions are often used in the authentication process. A hash function is a one-way function (that is, the computational cost of calculating its inverse is prohibitive) which produces an output of fixed length, regardless of the number of bits used at its input. The value at the output of a hash function is a summary of the data at its input, and is often called a *digest*.

Authentication is usually accomplished through an exchange of dedicated authentication data. If there are two parties to a transaction, a separate authentication is required for each to authenticate the other. Example authentication techniques include:

- *Password*. Alice authenticates Bob by requiring Bob to provide a secret shared between them. Typically, to allow the authentication of multiple parties, each is assigned two authentication parameters: one public (that uniquely identifies the individual) and the other secret (that, in combination with the public parameter, authenticates the individual). Most computer systems use a combination of (public) account name and (secret) password for authentication. Having supplied a valid combination of account name and password, the user is granted access to the system.

- *Challenge/response.* The authenticating party (Alice) issues a challenge to the other party (Bob), who generates a response, which is returned to Alice. Bob's response is based on a secret transform of the challenge, which has the property that it can be verified by Alice. Because both the challenge and response are transmitted in plain text, it is necessary to ensure that the same challenge is not repeated. This may be achieved either by randomly generating the challenge for each authentication, or by using a challenge whose value changes with time. Unlike password-based authentication, challenge/response authentication systems are designed so that no secret information is passed across the channel in plain text, making them more secure in an environment where this information might be intercepted by a potential intruder. An example of the use of challenge/response authentication is by the Post Office Protocol (POP), used to access a mail server. When a session is established by a client, the server responds with a time stamp. The client may then authenticate itself by returning an MD5 hash digest of this timestamp and a shared secret string (the password). Other systems generate the response by encrypting the challenge, using either secret-key or public-key encryption.*Certificate.* Bob can identify himself to Alice by providing a digital certificate, signed by a certificate authority that Alice trusts. So long as Alice possesses the public key of the certificate authority, she is able to verify the authenticity of the certificate, from which she can obtain Bob's public key. If Bob's public key is used to encrypt a randomly generated session key, only Bob is able to obtain the session key. If Alice uses this session key to encrypt data transmitted to Bob and insists that Bob use the session key to transmit data to her, she can be confident that only she and Bob can read that data. The certificate method of authentication is commonly used by web browsers on the Internet, and supports the exchange of a session key as part of the authentication process.

In practical systems, authentication needs to be followed by methods implementing other security services (such as access control) that can secure one or more transactions.

2.3 Integrity

Integrity is a security service that protects against the modification of data (including temporally displaced reinsertion of data) and the insertion of counterfeit data. It therefore protects against modification and fabrication attacks. Simply encrypting data may provide some level of integrity, preventing an adversary from inserting a predictable object, that is, any data inserted will be randomised by decryption. Protection against reinsertion of recorded data into the network usually requires that some form of reliable timestamp be attached to all transmitted data.

The key minimum requirement for integrity is the availability of a trusted guarantor for each message, block of data or time stamp. This guarantor may sometimes be the originator or sender of a message or a trusted third party. In other cases, especially for where a time stamp is involved, a third-party guarantor (or reliable timestamp generator) may be required.

Integrity is usually provided using digital signatures for a document or block of data. As a minimum, a digital signature contains an encrypted summary (which is usually the digest generated by a hash function) of the document. The digest is usually generated by a hash function [3], and is encrypted using the guarantor's private key. Verification of the integrity of a message is achieved by comparing a digest generated by the receiver with the decrypted digest transmitted with the message.

For military systems, it is often desirable to combine a secure time stamp with a digital signature. This time stamp should be secure in both its value and its binding to a particular message. A time stamp provided by a trusted server (which signs a digital signature for the time stamp) is secure in its value. Binding to a particular message can be achieved by combining the signed time stamp with the message before signing by the message guarantor.

2.4 Non-repudiation

For non-repudiation, technical means are required to give parties to a transaction confidence that the transaction cannot be denied by other parties.

These technical means are only effective in the context of procedures for their use, including definition of circumstances under which repudiation is allowed. Credit cards are a good example of a service that allows limited repudiation where unauthorised use of the card occurs, especially for transactions occurring over the internet and telephone. There are significant legal implications for both military and commercial networks in the definition of circumstances under which repudiation is allowed.

The non-repudiation service in military networks has traditionally been provided for text message traffic by operators in communications centres, who manually log all messages transmitted and received. This service is not provided by many commercial messaging and data transmission systems, but will be required in the future for all military communications systems, especially those used for command and control.

2.5 Access Control

The purpose of access control is to prevent an adversary from making use of the capacity of friendly networks and information systems. Without appropriate access control, an adversary may be able to steal capacity or degrade friendly quality of service. In commercial networks, such theft has always been a major issue for both telephone and electric power distribution systems. Theft of capacity from satellite transponders is also relatively easily achieved. For military systems, protection is required against both the degradation in friendly quality of service and the increase in adversity capability.

While authentication tends to be its public face, access control must also provide ongoing protection against unauthorised access for the whole period during which services are provided an authorised party. In other words, the responsibility associated with access control continues long after the initial authentication. Examples of measures taken to provide this ongoing access control include:

- restriction of privileges to a particular, authorised connection, which is unlikely to be reliable in modern, packet-oriented network;
- encryption of a link or data on a link; and
- repeated authentication during a connection.

In a network with effective access control, a user is rarely granted unfettered access. Therefore, even for authenticated users, access control limits access to the specific areas that have been granted. An important means by which an adversary may gain advantage by the use of friendly communications or information systems resources is to launch a secondary attack against another part of the network. Prevention of such attacks is therefore a key task of access control, and requires that an adversary not be able to gain unfettered access to the whole network from a successful attack in one part of the system. An important consequence of this is that no user of the network should be able to be granted access to all parts of the network.

2.6 Availability

Availability is the security service that guarantees that information resources (including communications capacity, computing capacity and information itself) are made available to legitimate users as required. It protects against unauthorised actions that lead to the degradation of quality of service.

Traditionally, a variety of electronic protection techniques, including passive techniques (such as directional antennas) and active techniques (such as spread spectrum transmission) have been used to minimise the impact of jamming on friendly radio nets. These techniques, which tend to be based in the physical layer, remain relevant for all radio links in the tactical network, and are an important means of restricting an adversary's access to the network. The ability of an adversary to launch an attack in one part of the network whose impact is in a completely different (possibly distant) part of the network, however, requires that a range of different types of protection are provided at the network layer and higher layers.

3 Security Architectures

In this section, we describe two generic security architectures, known as *system-high* and *multi-level security*.

3.1 System-high

System-high is the security architecture used in most existing military communications and information systems. The security classification given to the system is that of the most sensitive data that it carries. Physical means are used to enforce an access control policy at nodes of the network, which ensures that only individuals who are cleared for access to this most sensitive information are granted physical access. The level of protection provided to a particular communications or information system can then be determined by the requirements of its most sensitive data.

When data flows between parts of a system-high network with different security classifications, it always passes through an encryption device (Figure 2). This means that there is no possibility of direct electronic transfer of data between parts of the system with different security classifications. This means that leakage of sensitive data into less protected parts of the system cannot occur.

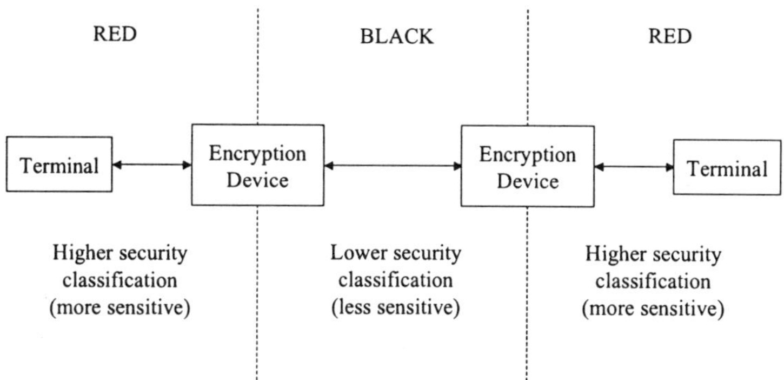

Figure 2. Interface between two parts of a system-high communications or information system separated by a system with lower security classification.

The terms *red* and *black* are used to refer to those parts of the network in which the data is available in its original and encrypted forms respectively. It is usual that the encryption device providing the red-black interface is physically separate from other equipment to avoid the possibility of accidental paths between red and black parts of the system. Other equipment resides in either the red or black part of the system; nothing apart from encryption devices is permitted to be connected to both.

The use of encryption shown in Figure 2 can be seen as allowing a higher-classification system to tunnel through a lower-classification network. Its purpose is therefore very similar to the concept of a *virtual private network* in commercial networks.

Figure 2 shows the use of encryption devices where two parts of a system-high system are connected via a point-to-point communications link. The more general case, in which a network connects a number of classified parts of the system, is shown in Figure 3. In this more general case, protection against traffic analysis may not be provided, because information on source and destination addresses of data is often available in the black network, and used for delivery of data.

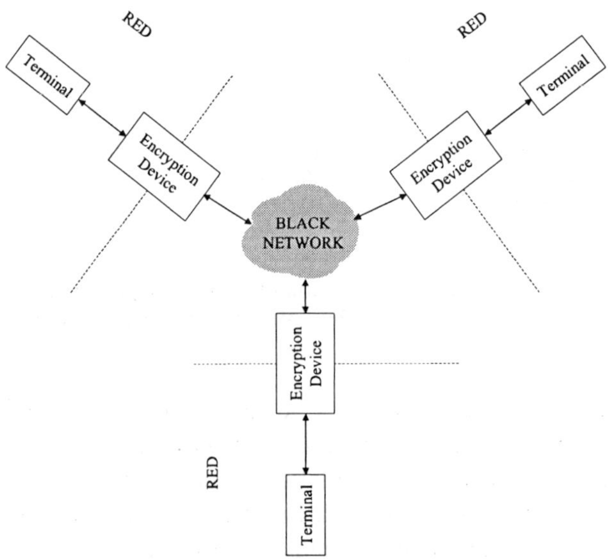

Figure 3. System-high security using a lower-classification network to carry data.

A system-high communications or information system provides the six security services as follows:

- *Confidentiality.* By encrypting all data leaving a sensitive system, data is only available to those with physical access to parts of the system where the data is available in its unencrypted form.
- *Authentication.* The use of encryption means that authentication is provided by the physical access control provided at locations

where data is available in its unencrypted form. While this does not support authentication of an individual, it does permit authentication as a member of the community among whom information is exchanged.

- *Integrity*. Modification of data and insertion of counterfeit data can only be achieved at locations where data is not encrypted. The use of physical access control at these locations supports the provision of the integrity service.
- *Non-repudiation*. Separate means are usually used to provide non-repudiation. Such means include logs in communications centers and using public key technology to return a secure, electronic receipt for a received message.
- *Access control*. Physical access control to all areas where data is not encrypted is vital to the success of system-high security. The use of encryption on all data flowing outside these protected parts of the system limits electronic access to those with access to these encryption keys. In other words, physical and personnel security are central to any successful implementation of system-high security.
- *Availability*. The use of encryption at the red-black interface protects against many information systems attacks. It is implicitly assumed that the black network provides its own electronic protection to guarantee availability of its communications resources. Such electronic protection might protect against attacks such as jamming.

System-high security has the advantage that the technical foundation of its security is the strength of the encryption used. Because encryption of data is rigidly enforced at all boundaries where the security classification changes, leakage of sensitive data into lower-security parts of the network should not be possible. Use of system-high security with other physical and electronic access control mechanisms can increase security by providing compartmentalization of the network and its information. The major disadvantages of system-high security are that it does not provide for the electronic transfer of data between parts of the system that have different security classifications, and there are only two levels of access provided: full access and no access. Flexible support for finer control of access is not available. Finally, if system-high security is implemented in a large network, security is reduced by the large number of users that have access.

3.2 Multi-level Security

Multi-level security is often seen as the "holy grail" of security architectures for military communications and information systems. It aims to allow information to move between parts of an information systems with different security classifications. This means that the use of encryption at all boundaries where changes of security classification occur is not necessarily required, but that information systems must ensure that classified information does not pass into parts of the network that are not sufficiently protected.

The US Department of Defense has run a major program to develop multi-level security solutions for a number of years. Agencies involved in this program include the Joint Interoperability and Engineering Organization and Center for Information Systems Security of the Defense Information Systems Agency, the Joint Staff, the Defense Intelligence Agency and the National Security Agency.

In multi-level security, each user is assigned a security level, which is the highest security classification of data to which access is permitted. Each time that the user logs on to a system, a security level is set for the session, which can be any level up to the highest permitted for that user. Multi-level security therefore requires technology that can enforce different levels of access for different users, and in many systems requires that this different access be enforced reliably by computer software. Multi-level security does not remove the need for personnel and physical security. In fact, the need for both is as great as for system-high security.

Multi-level security systems must provide a *guard* to control the flow of information across each security boundary [4]. Each guard filters data flowing across the boundary in accordance with a security policy. A guard may permit data to flow in only one direction (either low to high, or high to low), or may permit data to flow in both directions. While a one-way guard may be simpler to implement, it has the disadvantage of not allowing acknowledgements of the receipt of data to be returned, preventing the use of many protocols to ensure the reliable delivery of data.

The concept of the guard is similar to the commercial concept of a *firewall*. Like firewalls, a guard may be implemented as a device that is physically separate from other components of the network or information system. Alternatively, the guard may live within an information system, and be implemented in software. The difficulty of providing guarantees of security for systems using guards can be judged by the high rate of software updates required for firewalls in the Internet.

A low-high guard allows data to move from a part of an information system or network with low security classification to one with higher security classification (in accordance with a security policy). The advantage of allowing this to occur is that a user of the higher security system has direct, electronic access to data from a broader range of sources. The major risk in allowing this to occur is to the integrity of data: it is necessary to ensure that incorrect or corrupted data from a less protected system cannot migrate to a highly protected system. For example, the security policy for a low-high guard may allow emails to be passed after checking that they are virus-free. Emails in which a virus is detected, and some types of attachments, such as executable files, might be judged to be too high a risk, and may be blocked. A low-high guard may also seek to block data containing forged identifiers (such as a sender's address in an email), or data that is suspected of being part of a denial of service attack (such as flooding a network with data). The ability to ensure the integrity of data flowing between its source and the guard is therefore critical; in this sense it is not always true that a lower level of protection is acceptable for data in a network with a lower security classification.

The major issue for a high-low guard is whether or not particular data can be released to a lower-security system. This task requires that the guard is able to determine the classification of both the data and the system to which it is to be released. Simple ways in which the classification of the data might be checked include defined words that are used to describe the security classifications, such as "confidential" or "secret". In general, a high-low guard has a much more complex task than a low-high guard.

A guard may also perform sanitization of data to allow its security classification to be downgraded. In some automated sensor systems, this might be done by applying processes to reduce the accuracy or resolution of data. In other cases, such as free-form text in electronic mail, human intervention would be required because no currently available automated system is able to interpret the data.

The guards at security boundaries are the single most important contributor to the security services provided by a multi-level secure communications or information system:

- *Confidentiality*. A high-low guard will release data only to systems that have a sufficiently high security classification. This means that data is not made available, except to individuals who are authorized at the required level.
- *Authentication*. Authentication of individuals to a local terminal is provided by access control. Unlike system-high security, multi-level security systems usually perform authentication of individuals.
- *Integrity*. Integrity is provided by the use of authentication and physical access control to systems, and by the security policy implemented by low-high guards. A variety of other authentica-

tion options, discussed in Section 2.2, may also be employed.

- *Non-repudiation.* Separate procedures are required for non-repudiation. These may be the same as those used for system-high security.
- *Access control.* Physical access control to all systems is required. Individual electronic access control is usually provided by means of passwords.
- *Availability.* A multi-level security system relies partly on protection of the underlying networks. Some additional protection is provided by the security policies implemented by low-high guards.

Multi-level security has the potential to offer similar levels of protection to system-high security, with much greater flexibility. The major risk of multi-level security comes from this flexibility: the complexity of the system, and the possibility of transferring data across security boundaries, make possible a large variety of security breaches that simply cannot occur in system-high security.

4 Challenges and Vulnerabilities

There is an overriding presumption in the expansion of the role of military communications and information systems that it is desirable to have a high degree of meshed networking between command, sensor and weapon elements. It is certainly true that there are considerable advantages to be accrued to the commander who can make use of the power of the network, but it does not necessarily follow that a higher degree of connectivity is better. This is not unique to networks since, in military systems, it is often true that increasing strength can lead to brittleness. If small, highly networked forces are used in place of large, traditional forces, this brittleness is exacerbated. The small force may be less likely to be defeated. When it is, however, it is likely to be less able to recover than a larger force.

For example, consider the options open to the designer of a main battle tank. It is desirable to have a high degree of protection, which implies very thick armor. Increasing armor thickness increases the vehicles weight, however, and the vehicle becomes less agile and therefore an easier target. Similarly, the tank can be made a more powerful weapon system by increasing the caliber of the gun, however the overall vehicle weight is again increased, which further reduces the vehicle's capability. In short, armored vehicle design is a compromise; vehicle survivability requires a balance between firepower, mobility and protection. Increased vulnerability results if one aspect is arbitrarily increased at the expense of the others. There is a lesson here for military networks; increased connectivity can lead to increased vulnerability.

This section explores some of the trade-offs and vulnerabilities associated with increasing military use of communications and information systems.

4.1 Trade-offs in Military Systems

The design of most military systems involves a range of trade-offs between competing capabilities. For an armored vehicle, we have already discussed the trade-off between firepower, mobility and protection. A soldier would be better protected against bullets by wearing full body armor. Due to the weight and bulk of the armor, however, such a soldier is unlikely to be able to perform tasks such as patrolling. Almost every military system, including aircraft, ships and ground vehicles, experiences such trade-off.

In the specific context of land communications systems, there is a trade-off between capacity, mobility and range [5]. The operational use of different types of communications systems leads to different capabilities, some emphasizing capacity, others emphasizing mobility or range. All three capabilities cannot be increased simultaneously.

The provision of military communications and information systems is no different to other services: trade-offs must be made between the increases in capability created by new systems and factors such as their cost, weight, mobility and security. Requirements for security often conflict with the goal of global connectivity usually associated with communications networks.

Network management is an important example of this conflict. Network management in commercial networks is usually most efficiently provided by a single central authority. This authority is able both to define policy and implement it, providing a high degree of coordination and consistency. Centralized implementation of network management conflicts with many requirements of military forces, including security and modularity. The security requirements of a military force mean that a successful attack in one part of the network should be limited to that part of the network. Centralized network control provides many possibilities for crippling attacks to be launched from a single location, once the security has been breeched at that location. Modularity requires that a part of a military force be able to be detached for a specific task, and have at least a limited capacity for self-sustainment. For a force using networked communications and information systems, this capacity for self-sustainment includes being able to manage its own network.

4.2 Vulnerabilities

Networked communications and information systems are seen by military forces as significant force multipliers, allowing small forces to provide capabilities previously only available from much larger forces. What is not often discussed is the vulnerabilities that are a direct consequence of these systems, arising from a greater use of and dependence on the electromagnetic spectrum, increasing the number of points in the communications system from which attacks can be launched and increasing the number of types of attack that can be used by an adversary. Without effective security, which implies a multi-layer defense, the impact of these vulnerabilities may be catastrophic.

4.2.1 More Use of the Spectrum

An attack on the communications and information systems that underpin command and control is a direct attack on the commander's C2 ability. Modern C2 systems rely heavily on the use of the electromagnetic spectrum, from the sensors involved in surveillance and target acquisition to the communications and information systems involved in the passage and processing of information. This increased reliance on the electromagnetic spectrum is an increased vulnerability that can be exploited by an adversary.

EW opportunities are very limited when sensors are scarce, communications systems are rudimentary, networks are sparse, and command systems are manual. In a future networked force that utilizes automated, interconnected command systems, there are many more targets for EW. Not only are there several orders of magnitude more sensors and communications nodes, but each command node also depends on sophisticated information systems that are interconnected by a high-capacity local area network, often based on wireless technology. Each of these communicating and processing elements relies on the electromagnetic spectrum for its operation. This increased use of the spectrum presents many more opportunities to adversary EW systems, both ES and EA. While the utility of a command system has been increased dramatically by automation, its vulnerability has also correspondingly increased.

It follows, therefore, that an automated, networked C2 system is inherently more vulnerable than a manual one. It also follows that, to reduce this vulnerability, dominance of the spectrum is an essential tenet of operation for a networked force. While an industrial-age commander may have been able to ignore the potential impact of EW on manual systems, the information-age commander is hamstrung if denied the ability to operate within the electromagnetic spectrum. The corollary, of course, is military success in the Information Age is predicated upon an ability to conduct electronic combat; the commander who dominates the electromagnetic spectrum dominates the battlefield.

4.2.2 More Points of Attack

Access to current command and control information requires an adversary EW effort to intercept or attack communications links. This is relatively easy with combat net radio (CNR) nets, which employ radios that radiate omnidirectionally and are deployed relatively close to adversary EW assets. CNR nets are also single-frequency, half-duplex nets, which means that both sides of the conversation can be monitored. Trunk networks are much harder to target because they tend to use highly directional antennas that are generally pointed away from EW assets. Trunk links are also duplex, which means that the EW asset has to be located between the antennas to monitor both sides of the conversation. Tactical CNR nets also tend to be hierarchical, so that an attack is limited in its scope to the net currently under attack—that is, the effect is not propagated through the network.

An information-age force has far more points of attack because there are many more communications systems that must radiate omnidirectionally if they are to form a reliable, survivable meshed network across which information can be transported. Additionally, these networks will strive to provide a universal service, whereby any two terminals can communicate to each other without having to be identical. Attacking the network is now easier because there are many more points of attack. More importantly, however, the universal service provided by the network supports an attack from a point that may be well away from the node under attack. For example, a command-post computer could be attacked by an EW element that is well beyond line-of-sight. In fact a tactical unit could be attacked by a strategic information warfare element that is many hundreds of kilometers away and may even be back in the adversary's home country.

The use of system-high security with small networks greatly limits the range of locations from which an attack can be launched. The use of large networks greatly increases the options to an attacker. Only by limiting the connectivity by the use of security policies can this vulnerability be managed. Paradoxically, limiting the connectivity of a network to essential command control links returns us to the hierarchical communications structure eschewed in concepts such as network-centric warfare [6].

4.2.3 More Types of Attack

In traditional hierarchical networks, electronic warfare has been able to target communications systems by listening to their content, by jamming to prevent their correct operation, or by using the adversary's communication systems to create confusion. In networked forces, electronic warfare opportunities (command and control vulnerabilities) are dramatically increased because there are more different types of attacks.

Electronic Support. In the networked world, there are many more transmitting sources, which provides much more opportunity for electronic support. Certainly, the networked force would be wise to encrypt all transmissions, but some traffic analysis can still occur and units announce their presence by continuous radiation of situational awareness. As an example, a networked section may well be empowered by the ability to share situational awareness data while on patrol, but their presence is announced by their radiation some 5–10 km in front of them. The section has an advantage by being networked, but they also offer an advantage to an adversary that is scanning the RF spectrum as well as the skyline.

Jamming. Jamming networked radios is considerably easier than jamming a voice radio net because the jammer can simply join the network as a radiating station that will significantly disrupt the multiple-access technique employed by the net. Additionally, new sorts of attacks, such as denial of service attacks and distribution of viruses are possible because of the existence of a network that offers universal service.

Deception. It is quite difficult to deceive operators in a voice net because the deception asset needs to be able to mimic a human operator in procedures as well as language and accent. In a high-speed data network, much more data is flowing around and it is hard for a target terminal to weed the good from the bad, to detect packet spoofing, and so on. The potential for confusion is increased directly by the increasing use of information.

Neutralization. Traditionally, neutralization requires so much power to be transmitted onto a target that the target's electronics are destroyed or disabled. Neutralization has rarely been used in practice because the power levels are difficult to produce. RF directed-energy weapons (RF DEW) [7] offer a new sort of weapon system that is only practicable because there are so many electronic systems as targets.

4.2.4 More Complex Systems

Simplicity is an important principle of military communications; simple plans and systems are much more likely to survive the fog of war. The provision of a secure, mobile, battlefield-wide network is a complex venture that requires sophisticated planning and network management. Additionally, to ensure that commanders are served continuously by their command systems, considerable redundancy is required, which further complicates planning.

The existence of the network also complicates operations. From the preceding discussion, it follows that, if the adversary has an EW capability, they will be able to obtain dominance of the spectrum, even if it is only for short periods. Commanders therefore need to be able to operate without the support of their communications and automated systems, that is they will need to revert to manual procedures. The first rule of automation, however, is to re-engineer business processes before automation, so it is unlikely that the manual procedures are the same as the automated procedures. The loss of automated systems, even for a short period, therefore not only represents a sudden significant drop in capability, but it requires additional complex doctrine if the transition is to be made relatively smoothly.

Standard operating procedures (SOPs) now need to describe two sets of procedures (automated and manual) and need to address the manner in which units and subunits transition between the two different modes of operation. The latter requirement may sound somewhat trivial, but in fact introduces a considerable degree of complexity. For example, SOPs need to address: when transfer is made; who makes the decision; how is the decision promulgated (given that the need for the decision is predicated upon loss of communications); how does an automated headquarters control subordinate units, some of which may be using manual procedures (do all drop back to manual, or is there a mechanism by which automated database transfers can be made to humans); once communications are restored, how are databases re-synchronized; and so on.

In addition, networks must be encrypted, which when combined with the much wider distribution of communicating entities, exacerbates key-management and increases the risk of compromise. The much larger numbers of communications and information-processing nodes deployed also markedly increases the repair and maintenance effort as well as the logistics effort required to resupply batteries and so on. Public key encryption technology partly solves this problem, but introduces new problems of certificate distribution and revocation. In all the implementation of Information-Age technologies contributes significantly to a considerable increase on complexity of battlefield systems.

Complexity in the Internet is perhaps the primary reason that security systems require constant software updates. Developers of security systems are always responding to successful attacks. Unlike the other vulnerabilities discussed above, the potential impact of unforeseen, successful attacks made possible by system complexity is difficult to judge, which may prevent effective management of the risk.

5 Conclusion

Networked communications and information systems are seen as providing significant force multipliers in for future military forces. Like other military systems, however, no single design goal (such as connectivity) can be pursued exclusively without limiting the overall capability of the system due to a degradation of other important design goals, such as security. The creation of an effective military capability in this area can only be achieved by effective management of such trade-offs, and the vulnerabilities that are inherent in the increased communications connectivity and automation provided.

The authors would like to thank Mr. Bill Blair for his insightful comments on a draft of this chapter.

Endnotes

1. Discussion of security services (using a slightly different taxonomy) can be found in: W. Ford and M.S. Baum, *Secure Electronic Commerce*, Upper Saddle River, NJ: Prentice Hall, 2nd Edition, 2001.

2. Discussion of encryption can be found in (for example):
Denning, D.E., *Cryptography and Data Security*, Reading, MA: Addison Wesley, 1983;
Schneier, B., *Applied Cryptography: Protocols, Algorithms and Source Code in C*, New York: Wiley, 1994;
Singh, S., *The code book : the science of secrecy from ancient Egypt to quantum cryptography*, New York : Anchor Books, 1999;
Sinkov, A., *Elementary Cryptanalysis, a Mathematical Approach*, The Mathematical Association of America, 1966;
Stallings, S., *Network and Internetwork Security*, 2nd Edition, Englewood Cliffs, NJ: Prentice Hall, 1995; and
Torrieri, D. J., *Principles of Secure Communications*, Norwood, MA: Artech House, 1985.

3. See, for example, Schneier, B., *Applied Cryptography: Protocols, Algorithms and Source Code in C*, New York: Wiley, 1994.

4. "Multilevel Security in the Department Of Defense: The Basics", Washington DC: US Department of Defense, March 1995.

5. M. Ryan and M. Frater, *Tactical Communications for the Digitized Battlefield*, Norwood, MA: Artech House, 2001.

6. Alberts, D., J. Gartska, and F. Stein, *Network Centric Warfare*, CCRP Publication Series, Washington, DC: US Department of Defense, 1999.

7. M. Frater and M. Ryan, *Electronic Warfare for the Digitized Battlefield*, Norwood, MA: Artech House, 2001.

Chapter 2

Teaming Human and Machine

P. Urlings and L. Jain

Advances in automation and especially artificial intelligence have enabled the formation of rather unique teams with human and (electronic) machine members. This chapter reviews existing research in this area and proposes a conceptual framework for teaming human and machine. The basis of this framework will be the introduction of the machine into the traditional situation where the human is solely responsible for managing, control and execution of all activities. Focus will be on the identification and classification of activities to be allocated to the machine. Task management and coordination between human and machine will be identified as a specific area of research and design concern.

1 Introduction

During recent decades we have witnessed not only the introduction of automation into the work environment but we have also seen a dramatic change in how automation has influenced the conditions of work. While some 30 years ago the addition of a computer was considered only for routine and boring tasks in support of human, the balance has dramatically shifted to the computer being able to perform almost any task the human is willing to delegate. The very fast pace of change in processor and information technology has been the main driving force behind this development.

Advances in automation and especially Artificial Intelligence (AI) have also enabled the formation of a rather unique team with human and electronic members. This has led in military aviation to the development of concepts under the header of "pilot associate" or "crew assistant". The team is still led by the human (pilot or crew) with the machine as a subordinate associate or assistant, sharing responsibility, authority and autonomy over many cockpit or mission system tasks. As these systems become more complex, the automation that the human has to interact with is becoming increasingly intelligent and capable. The requirements for teaming human and machine in a highly dynamic, unpredictable task environment has led to impressive achievements in many supporting technologies. These include methods for system analysis, design and engineering and in particular for information processing, for cognitive and complex knowledge engineering (Taylor and Reising 1998).

Task allocation between human and machine has become an increasingly difficult design challenge especially when flexibility is needed to handle rapidly changing operational situations and to avoid the risk that decisions are locked into a specific automation design. Earlier mistakes in automation have emphasised the importance of a correct balance between the control and monitoring tasks especially for the human member of the team. In order to maintain overall efficiency and safety, it has been suggested that the human should stay involved in all critical control tasks. Automation should therefore allow for intensive management and coordination between the different members of the human-machine team.

2 Existing Research

This chapter proposes a conceptual framework for analysis and design of complex human-machine teaming applications. It will be based on a review of existing research that is well documented in the public domain. Pioneer programs are the US "Pilot's Associate" (LaPuma and Marlin 1993), the British "Mission Management Aid" (Gibson and Garret 1989) and the French "Copilote Electronique" (Champigneux 1993). These early programs all aim at the automation of routine tasks and the provision of effective aids in the single-pilot cockpit environment of a military fighter aircraft. Follow-on programs continue to target this demanding operational environment. Examples are the British "Cognitive Cockpit" (Taylor *et al.* 2000), the German "Tactical Mission Management" system (Schulte 2001) and the Netherlands "Pilot Oriented Workload Evaluation and Redistribution" project (Hesselink *et al.* 2001). The German "Crew Assistant Military Aircraft" (Walsdorf and Onken 2000) supports the multi-member cockpit crew during the tactical mission of a military transport aircraft.

Similar developments in civil aviation can be found in the French "Electronic Centralised Aircraft Monitoring" system for the Airbus A300 series (Dekker 1983). The development of this system was driven by the commercial need to reduce the number of cockpit team members and to support two-pilot cockpit operations. Other examples are the German "Assistant for Single Pilot Instrument Rules Operation" and its follow-on "Cockpit Assistant System" (Gerlach and Onken 1995, Onken 1997). Both programs aim to improve safety under instrument flight conditions by the introduction of an artificial intelligence system, capable of monitoring flight-plan and airfield approaches. Non-aviation applications have been promoted in the mid-1990s by the European EUREKA program. Several research projects under its Prometheus umbrella aimed to support the drivers of next generation cars. Examples are the French "Prolab-2" which optimised the behaviour of the machine as co-driver (Lefort-Piat 1994) and the German "Driver Assisting System" which has been developed as a monitoring and warning aid for the human driver in longitudinal and lateral control on motorways (Onken 1994).

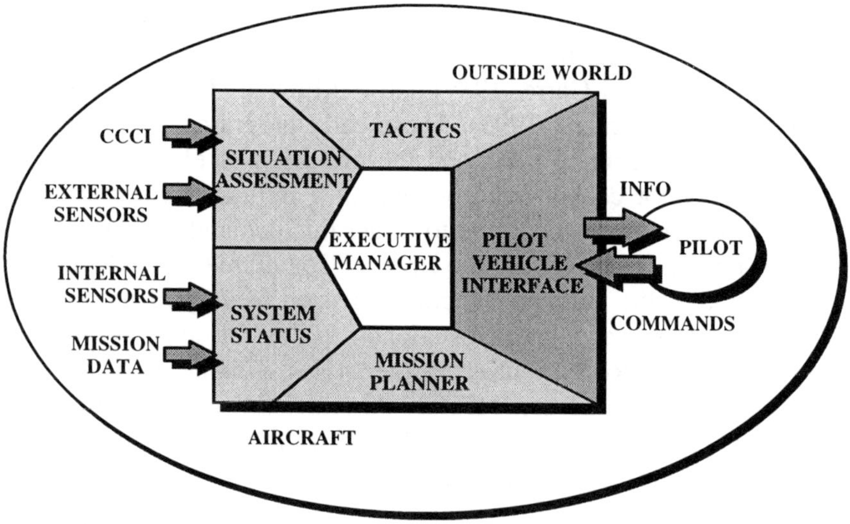

Figure 1. Architecture of the "Pilot's Associate".

Although all these programs go by different names and aim at different applications, their design and functional architectures have many elements in common (Hunt 1992, Urlings *et al.* 1995). The high-level architectures of the "Pilot's Associate" and the "Copilote Electronique" are given here as representative examples, see Figures 1 and 2. The common elements found are:

- the aim and objectives of these programs are, without exception, the demonstration and evaluation of the potential benefits of AI for the purpose of enhancing operational effectiveness and improvement of human operator workload;
- the functional architectures all implement a highly modular structure in which the functions to be automated directly correspond with single operator tasks;
- dedicated functional modules are allocated to manage and coordinate between the human operator and machine assistant; these functions are additions to the non-automated situation and are often called execution manager, information manager, task manager or mission manager;

- the research programs employ a multitude of Advanced Information Processing (AIP) techniques, methods and tools; most frequently used are knowledge based (rule and case based) systems, planning (script based and plan-goal graphs), fuzzy logic and (distributed) blackboard systems; and
- the size of knowledge bases of prototype implementations is substantial; in the case of the "Pilot Associate" there are 828 rules for the pilot-vehicle interface and 990 rules in total for the other modules; the four main modules of the "Copilote Electronique" roughly consist each of 400 rules and 3000 objects.

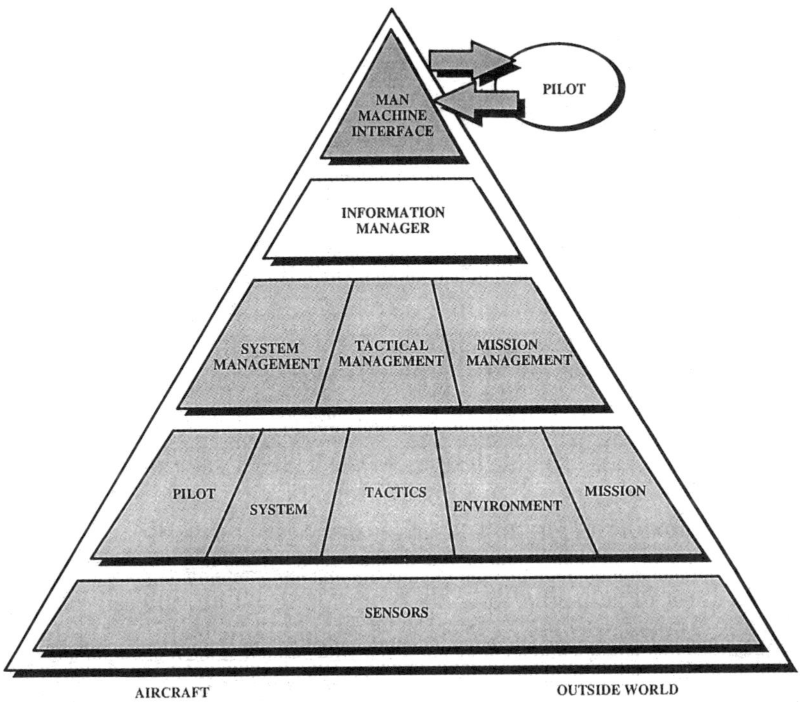

Figure 2. Architecture of the "Copilote Electronique".

3 Initial Framework

The previous programs all have in common the provision of effective aids to the human in problem solving and the automation of routine tasks traditionally performed by the human. The architectures developed in particular in the early programs ("Pilot Associate", "Copilote Electronique" and "Mission Management Aid") also suggested a more generic approach. Such an approach and generic framework has been proposed by the European Coordination for the Long term in Defence (EUCLID) study on "Crew Assistant" (Urlings *et al.* 1995). Although the EUCLID study aimed at the operation of a single-pilot military aircraft, its findings are equally valid for non-aviation and multi-operator situations.

The "Crew Assistant" proposes an automated system on-board an aircraft that supports the human operator (pilot or crew) in performing his/her flying and operational tasks. In the traditional manual situation, each aircraft system and sensor will interface directly with the operator through dedicated controls and displays. The operator has to interpret multiple displays and operate multiple controls simultaneously in order to perform the functions that are related with his main task. In the non-assisted manual situation, the interpretation of all sensor information and the control of all systems remain with the human operator. This manual situation is illustrated in Figure 3a, where the upward arrows show the information flow from the sensors to the displays, and downward arrows show the control flow to the systems. The cockpit elements are divided into displays (inputs from sensors to the operator only) and controls (output from the operator to the systems only). The aircraft elements are divided into sensors (output to displays only) and actuators (input from controls only). In practice, most cockpit and aircraft systems will integrate these functional elements.

Figure 3b illustrates the situation when a part of the operator task is automated and assigned to an automated assistant. The original task is then split into (sub)tasks delegated to the assistant and (sub)tasks that remain with the operator. Depending on how much of the original tasks are delegated to the assistant, this will result in a change in the amount of information offered to the operator and in a change in the amount of control required from the operator. Transfer of all operator tasks to the assistant, results in the automated situation of Figure 3c where the operator is at most only informed of the status and results of tasks delegated to the assistant. Full control authority in that case is given to the machine-assistant.

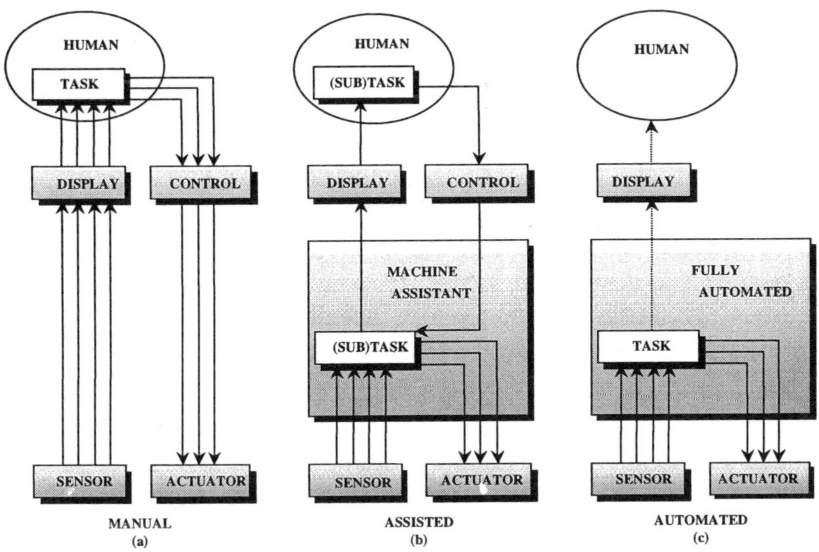

Figure 3. Human-machine teaming environment.

3.1 Level of Machine Assistance

Only the operator-assisted situation of Figure 3b is of interest for the subject of this chapter. The manual and fully automated situations rather describe the boundary situation of the problem. The operator-assisted situation can be expanded to be consistent with the many possible levels of assistance ranging from manual, manual augmented, cooperative, automatic manual sanctioned, to fully autonomous automatic as formulated by others. Different classes of operator have been proposed, based on an increasing level of intelligence of the tasks to be supported and on an increasing level of complexity of the information produced (Nguyen *et al.* 1997):

- ***Routine-performing assistant*** – A system in this mode is aimed at assisting the operator by providing computations to maximise the efficiency of human information processing. The system performs some routine tasks, which are pre-programmed with step-by-step procedures. Regardless of the complexity of the task, the system is deterministically static in the sense that the problem domain is carefully articulated and a matching solution domain is anticipated.
- ***Advisory assistant*** – A system in this mode is characterised as an expert who can provide some specific advice to operators on request. It is also equipped with procedures for monitoring the operational environment with the purpose of detecting any abnormal conditions and, hence, recommending appropriate solutions. Thus the system requires specific knowledge about the tasks the system is expected to carry out.
- ***Decision assistant*** – A system in this mode is aimed at assisting the operator by providing computations to increase the effectiveness of the human decision-making process. Therefore the system has, at the least, a reasoning capability to draw conclusions from data that the system perceives. Also, the system must be able to handle certain, uncertain, and incomplete data, which may come from a dynamic environment.
- ***Cooperative assistant*** – A system in this mode is targeted to augment the operator by performing a part of some control activities. The system should be able to work in a team environment with other systems and under the operator's leadership. The system should have a set of goals and a set of plans to achieve these goals. The system requires not only a reasoning capability as for the decision assistant, but also communication capability to interact with other systems and operators in order

to fulfil its goals. It may possess a learning capability.

- ***Associate assistance*** – At this highest intelligence level, a system is considered as the operator's associate and can replace the operator in some special or designated (emergency) circumstances. In order to fulfil its role, the system must have all the intelligent capabilities of the cooperative assistant, plus the ability to make decisions related to its self-operation in the absence of the operator's direction. Once authorised the system acts independently but should be able to be overruled by the operator at any time.

3.2 Required Characteristics

The framework of the operator assisted system environment (Figure 3b) is generic and is applicable to various domains exhibiting certain human-machine teaming characteristics. The required characteristics of real-world problems that make this framework applicable are (Urlings and Zuidgeest 1997):

- ***Observability*** – Sensors should provide information about the states or parameters of the system and its environment. The system must be observable and sensor measured information must be available to the operator or operator assistant. It should be possible to estimate or determine the states, parameters and modes of the system and its environment from these measurements.
- ***Controllability*** – The operator or operator assistant should be able to control the system and to transition the system from any initial state to any specified state in time. The system should also offer adaptive displays and controls, adapting to different situations and task requirements by transfer authority backward and forward between the operator and the operator assistant, and the allocation of functions between them.
- ***Operator interfaces*** – Interaction between operator and an operator assistant depends heavily on the available display and control interfaces. Contemporary operator interfaces combine a

blend of display and control technologies, ranging from conventional electro-mechanical dials to flat-panel colour displays and from mechanical switches to voice-controlled input devices. By far the greatest majority of displays use vision although audio signals are used to provide alerts in danger or failure situations. The great majority of controls are still manual and they can be located anywhere provided the operator can reach them. The size of the (sub)task delegated to the operator assistant will determine the change in the amount (and type) of information offered to the operator and the change in the amount (and type) of control required from the operator. Interfaces need to be adaptable to different levels of task automation.

- ***Operator task decomposition*** – In order for the proposed framework to be applicable to a real-world problem, it is important that the operator task can be decomposed and split between the human operator and the automated operator assistant. The aim of automation is to provide the operator with an improved system and situation awareness and to enable him/her to make the best possible decisions in any situation.

4 Functional Classification

The next step in the development of the framework is the modular decomposition of the tasks allocated to the machine assistant. The assistant can be seen as a collection of relatively independent functions that assist the human operator in separate tasks and hence will require different capabilities. The machine assistant can also be seen as a data processing unit that processes low-level sensor data in several stages to easy-to-access information to be accessed by the human operator (see Figure 4). When analysing candidate operator tasks to be allocated to the machine assistant, it is attractive to classify these tasks into several levels of priority, complexity or workload. There are several schemes for task classification that support this approach.

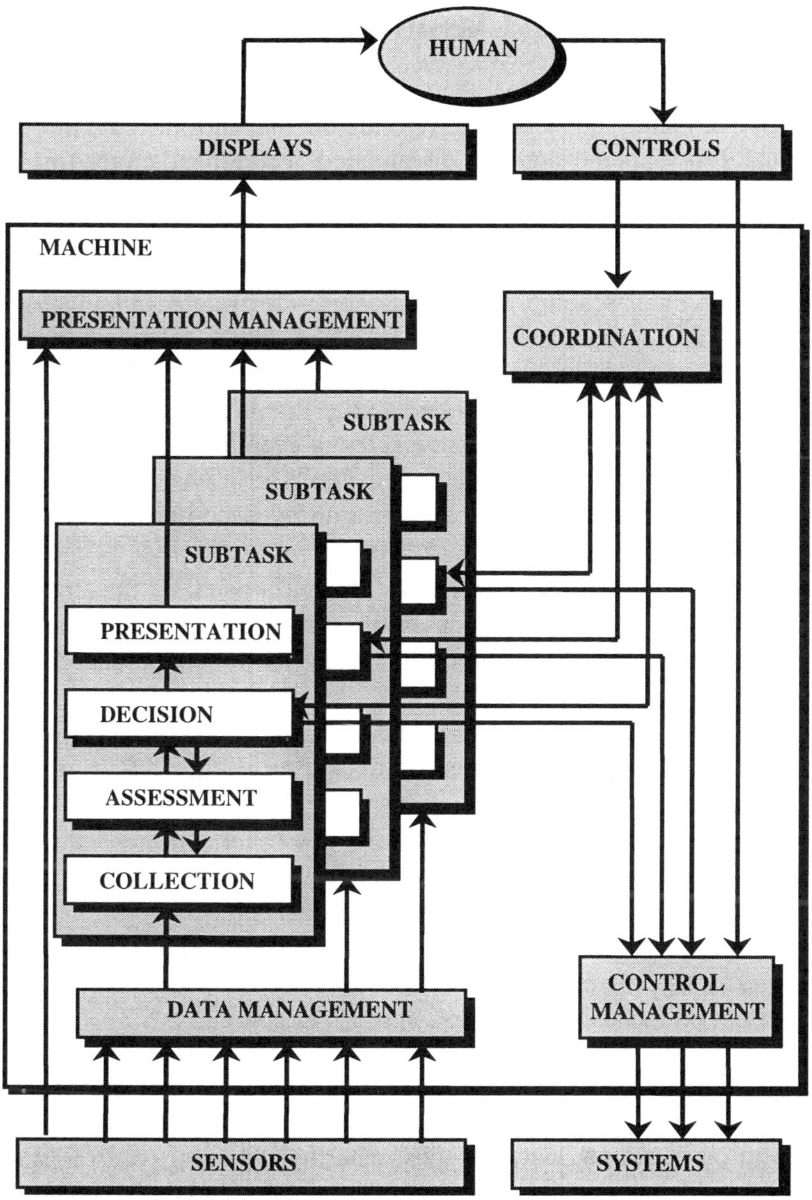

Figure 4. Framework for Human-Machine teaming environment.

4.1 Hierarchical Classification

An Advisory Group for Aerospace Research & Development (AGARD) study on "Cockpit Automation" has introduced a hierarchical task scheme and has formulated recommendations for its automation. The tasks in this original scheme range from routine (flight control) to complex (command and control) and include system management (Hollister 1986). Others have formulated a similar hierarchical scheme of task categories for related applications (Abott 1993, Abott and Rogers 1993). A major relationship between these levels is that the human will only pay full attention to the next level once all tasks allocated to the previous one are handled adequately. This reflects the rules of thumb often used by human pilots in prioritising their tasks. The "Crew Assistant" project illustrates such hierarchical classification by decomposing the operator tasks into the following categories:

- ***Aviate*** – which includes all tasks related to handling the aircraft, to basic flying and manoeuvring, to monitoring system health and status, and to encountering system malfunctions and emergencies;
- ***Navigate*** – which includes all tasks that keep the aircraft on the intended (navigational) mission or authorised (air traffic) flight plan;
- ***Communicate*** – which includes the tasks that communicate and coordinate with all friendly elements that contribute to or may interfere with mission or flight intentions; and
- ***Operate*** – which includes the tasks that deal with all unfriendly and non-cooperative entities that directly interact or will have effect on the successful mission or flight completion.

The hierarchy between these categories should be that the lowest level represents routine and skilled (control) based tasks and their execution is almost instantaneous. The highest level requires specific knowledge and expertise from the operator and their execution is less time critical. This follows very closely the well-known classification proposed by Rasmussen, who breaks down human's cognitive behaviour into skill-based, rule-based and knowledge-based. This also links the hierarchical structure to a cognitive classification.

4.2 Data Processing Classification

Another classification scheme is based on an analysis of data processing and sensor data fusion. For each function allocated to the machine assistant, the basic flow of data is from the sensors to the cockpit displays and it is the goal of the assistant to direct this flow by processing sensor data into information for display. The main objective is to provide the human operator with concise and relevant information. In this process, a number of steps can be distinguished, each representing a processing level at which data are combined with information, knowledge and procedures and interpreted into information for a next step (Zuidgeest and Urlings 1996). Four processing levels can be distinguished (see Figure 4):

- ***Collection*** – At the collection level, data are collected and prepared for further assessment. This includes the collection of data from sensors and other input devices, the transformation of these data into a format that can be read by the assessment level, the execution of complex operations in which data from different sensors are integrated into a standard data format (e.g., by sensor data fusion), and the preliminary filtering of data by rejecting irrelevant data or by giving priority to data that are urgently needed by higher processing levels.
- ***Assessment*** – At the assessment level, the collected data are assessed on normal or abnormal properties. This includes the comparison of data from the collection level, mutually or by comparison with reference data (e.g., threshold values), the execution of complex processing (e.g., the analysis of system trends by examining a range of chronological data values and the prediction of values in order to anticipate future problems),

and the assessment of the operational environment on the basis of sensor data.

- ***Decision*** – At the decision level it is decided what has to be presented to the human operator on the basis of inputs from the assessment level and possibly provide autonomous control. This includes the filtering of data from the assessment level in order to prevent saturation of the human's cognitive resources, the generation of advice on handling abnormal situations, and, if authorised, the execution of autonomous action by the machine assistant.
- ***Presentation*** – At the presentation it is decided how the information from the decision level is presented to the human operator. This includes an assessment of the available man-machine interface resources and human preferences, and the presentation of information in such a way that the human is directly cued and able to process the information efficiently and effectively.

These different levels of data processing proposed by the "Crew Assistant" show similarity with the hierarchical model and processing levels proposed for data fusion in command and control applications (White and Llineas 1990). The main difference is that the data fusion process specifically supports situation and threat assessment within a command and control application while the "Crew Assistant" process will support a variety of crew tasks, including situation and threat assessment.

4.3 Cognitive Classification

The cognitive classification has been proposed by Rasmussen as a promising automation approach to overcome the deficiencies of traditional automation (Rasmussen 1986). Traditional automation is considered not being capable to compensate for the natural human cognitive deficiencies, which results in an automation concept in which the human either becomes overloaded or is kept out-of-the-loop. The core element of the cognitive approach is the natural human information processing loop, which has the following elements: monitoring of the situation (perception and interpretation), diagnosis of the situation (including goal activation and evaluation), planning and decision-making, and plan execution.

The cognitive classification forms the basis of the "Copilote Electronique" and is illustrated by its pyramid-approach which decomposes tasks into the following vertical assistance domains (Champigneux 1993):

- ***System planning*** – planning the avionics system (re)configuration, scheduling of actions and resources, and assessing efficiency and feasibility of the plan;
- ***System evaluation*** – monitoring discrete events, monitoring continuous signals, assessing real avionics system status and dependability;
- ***Tactical planning*** – planning tactics according to threats, scheduling actions and resources according to the tactics, handling conflicts among proposed tactics;
- ***Tactical assessment*** – analysis of friendly and opposite forces, elaboration of forecast evolutions, assessment of risk and efficiency according to present plan;
- ***Mission planning*** – selecting routing options according to the mission context, planning new routes, monitoring possible routes with quality estimates, selecting current route; and
- ***Mission assessment*** – mapping of pre-mission meteorological information onto possible routes, mapping of pre-mission geographical data onto possible routes.

The "Crew Assistant" also illustrates the cognitive classification. Similar tasks as above were identified during extensive interviews with 33 military pilots from four different Air Forces, flying several aircraft types. Key criteria for task identification were their operational relevance, their impact on pilot workload and mission effectiveness and the expected applicability of AIP technologies

(Urlings *et al.* 1995). This resulted in tasks focusing on system, mission and tactical planning:

- ***System planning*** – addresses monitoring of normal system performance (and in particular engine performance), trend analysis, and reporting of information on system status;
- ***System evaluation*** – relates to analysis of anomalies, to presentation of appropriate warnings, to (checklist) assistance in countering malfunctions, and (when authorised) to automatic execution of corrective actions;
- ***Mission planning*** – includes the capability to monitor mission/flight progress, to evaluate the impact of environmental entities (adverse weather and enemy threats) on this plan and, if needed, to assist in or to perform an automatic (re)planning;
- ***Tactical planning*** – addresses management of self protection systems, assessment of sensor information, selection of available countermeasure options, and (automatic) execution of the selected tactics; and
- ***Situation awareness*** – relates to the capability to combine and interpret all available environmental data in order to derive an easy to assess situation picture of this environment; situation awareness may be limited to navigational information but, for military applications, includes all relevant strategic and tactical information;

The cognitive approach has been adopted by most current automation research in military aviation and other application areas. Examples are the "Cognitive Cockpit" (Taylor *et al.* 2000), the cognitive man-machine cooperation approach of the "Crew Assistant Military Aircraft" (Walsdorf and Onken 2000) and the cognitive concepts as applied in the "Tactical Mission Management" system (Schulte 2001). Major reason is that the human problem solving strategy and the cognitive process as initially proposed by Rasmussen have been developed into a practical system engineering framework that has the potential to identify and solve current shortfalls in automation (Rasmussen *et al.* 1994, Vincente 1999).

5 Task Management and Coordination

Task management and coordination has been initially suggested by Abott to reflect supervision and organisation (monitoring and scheduling) of the other tasks. Task management is extremely important in dealing especially with abnormal situations. In abnormal situations a correct implementation of task management may be the only difference between an accident and an incident (Schutte and Trujillo 1994). The challenge will be (to classify and) to incorporate this management and coordination layer into a multi-task structure of a human-machine team. Such structure is not only subject to the specifics of the machine environment but also has to deal with the individual differences and personal style of the human operator.

The proposed conceptual framework in Figure 4 identifies the need for extensive task coordination between the human and machine and between the different tasks of the machine itself. This has been reflected in four functional modules or interface capabilities to organise the corresponding data, information and control flow. Different aspects of task managament and coordination will be discussed below.

5.1 Sensor Data Management

Importing sensor data – Data management is responsible for importing and filtering all sensor data as required by the active machine assistant functions. One function might require data from multiple sensors while several functions might require data from the same sensor. Data management is responsible for correlation of filtered data with the assistant internal data. Sensor data management and data collection will be closely integrated in the system design of a machine assistant.

Sensor data fusion – Data management is closely related to sensor data fusion, but the overall fusion problem might be resolved outside the machine assistant. The functional architecture assumes responsibility for correct data to remain with each sensor individually and the responsibility for correctly fused data with the involved sensors collectively.

5.2 Display and Presentation Management

Limited display resources – The machine assistant will be operational in an environment that is expected to rely heavily upon multi-function display technology. This implies that conflicting requirements in the presentation of information are likely to emerge when multiple assistant functions simultaneously require access to the same display. Solving these conflicts has to match the way the human would solve them. Remaining conflicts should be prioritised and, when required, additional guidance should be requested from the human operator.

The machine assistant may also be in conflict with a sensor not included in the assistant if both require access to the same display interface. Since such a conflict emerges by the introduction of a machine assistant, the assistant should solve this. The conflict could also be solved by displays that are dedicated only to the machine assistant.

5.3 Control Management

Overruling by the operator – For each function that is delegated to the machine assistant, the human operator shall be able to overrule the assistant. Overruling may cause sensors and systems to receive control inputs from both the human operator and machine assistant, which may be conflicting. This conflict is prevented within the design of the machine assistant by routing all control inputs through a control management module. Note that overruling of system control is basically different from de-selecting the machine assistant.

Conflicting system control – A conflict in system control exists when the same sensor or system is employed simultaneously both by the human and machine assistant while each performs a different task. Control management prioritises and solves such conflicts and, when required, informs the operator and requests additional guidance. When multiple functions are assigned to the assistant, these may also conflict in controlling the same systems. Solving these conflicts has to match the way the human would solve them.

Operator requested input – Occasionally, the machine assistant may not be able to collect all data required to perform a function, e.g., because a sensor is malfunctioning or because there is no sensor available. Such data can be obtained by requesting the human operator to provide them. Loading mission data via an operator inserted data cartridge is part of this capability.

Sensor management – When data collection requires activation or redirection of a sensor, this control is subject to operator authorisation and does not differ from control of other systems. Control management, therefore, should include sensor management.

5.4 Coordination Management

Machine authority – By delegating a task to the machine assistant, the human operator inevitably has to specify the nature of its interaction and the authorisation for presentation and control. This delegated authority can be expressed in standard levels of automation (e.g., stand-by, manual, semi-automatic and automatic). Full automation is outside the scope of the machine assistant and in the "automatic" mode the assistant should at least inform the operator on the status of its activities and should instantaneously accept a reset by the operator at any time. The human should have a correct and complete understanding of the functioning of the assistant in all modes, in order to allow a smooth transition between different modes and to maintain consistency with manual operations. The human remains in the loop and may regain control at any time.

Coordination between functions – When several tasks are delegated to machine assistant, interactions will take place, which requires coordination between the corresponding functional modules. This includes:

- translation and decomposition of the request for assistance by the operator into the activation of all needed functions within the assistant;
- prioritisation between assistant functions when simultaneous execution of machine assistant functions results in conflicts that are related to the human operator (limited cognitive capabilities), to system resources (limited available displays or supporting sensors) or to other resources limitations (computer memory, processing power or throughput capability); and
- cooperation between assistant functions when some functions need specific results from another; this control (or request for data) is performed at the decision level though the actual exchange of data may remain at the assessment level.

6 Artificial Intelligence and AIP Techniques

The presented conceptual framework for human-machine applications shows a modular approach in which various functional elements can be marked as knowledge intensive. It also shows that the machine-assistant interactions are complex and that these interactions should show intelligent behaviour in order to remain transparent to the human-operator at all times. Artificial Intelligence (AI) and Advanced Information Processing (AIP) are generally expected to be capable to handle this complexity and to support a sophisticated human-machine interaction by minimising the cognitive gap between human and machine.

6.1 Artificial Intelligence Paradigms

As is typical with a new field of scientific research, there is no precise definition for AI. Generally speaking, however, so called AI-based data and information processing techniques are those that are inspired by an understanding of information processing of humans, or more general in biological systems. In some cases an attempt is made to mimic some aspects of intelligent behaviour. When this is the case the process will include an element of adaptive or evolutionary behaviour similar to biological systems, and like the biological model there will be a very high level of connectivity between distributed processing elements (Jain 1995, Jain and Allen 1995, Jain 1997a, Jain and Jain 1998).

Intelligence is also not easy to define. However, we can say that a system is intelligent if it is able to improve its performance or maintain an acceptable level of performance in the presence of uncertainty. The main attributes of intelligence are learning, adaptation, fault tolerance and self-organisation. Data and information processing paradigms that exhibit these attributes can be referred to as members of the family of techniques that make up the knowledge-based engineering area. Researchers are trying to develop AI systems that are capable of performing, in a limited sense, "like a human being." (Jain and de Silva 1998, Jain *et al.* 1998). The main AI paradigms are:

- ***Artificial Neural Networks*** – Artificial Neural Networks (Jain and Martin 1997, Jain and Vemuri 1998) mimic biological information processing mechanisms. They are typically designed to perform a non-linear mapping from a set of inputs to a set of outputs. Artificial neural networks are developed to try to achieve biological system type performance using a dense interconnection of simple processing elements analogous to biological neurons. Artificial neural networks are information driven rather than data driven. They are non-programmed adaptive information processing systems that can autonomously develop operational capabilities in response to an information environment. Artificial neural networks learn from experience and generalise from previous examples. They modify their behaviour in response to the environment, and are ideal in cases where the required mapping algorithm is not known and tolerance to faulty input information is required (Jain 1997b, Jain and Jain 1997, Narasimhan and Jain 1996). The underlying reason for using an artificial neural network in preference to other likely methods of solution is that there is an expectation that it will be able to provide a rapid solution to a non-trivial problem. Depending on the type of problem being considered there are often satisfactory alternative proven methods capable of providing a fast assessment of the situation.

- ***Evolutionary Computing*** – Evolutionary computation is the name given to a collection of algorithms based on the evolution of a population towards a solution of a certain problem. (Vonk *et al.* 1997, Van Rooij *et al.* 1996). These algorithms can be used successfully in many applications requiring the optimisation of a certain multi-dimensional function. The population of possible solutions evolves from one generation to the next, ultimately arriving at a satisfactory solution to the problem. These algorithms differ in the way a new population is generated from the present one, and in the way the members are represented within the algorithm. Three types of evolutionary computing techniques have been widely reported recently. These are Genetic Algorithms, Genetic Programming and Evolutionary Algorithms. The Evolutionary Algorithms can be divided into Evolutionary Strategies and Evolutionary Programming.
- ***Fuzzy Logic*** – Fuzzy logic was developed in the mid-1960s for representing uncertain and imprecise knowledge (Sato *et al.* 1997, Jain and Vemuri 1998). It provides an approximate but effective means of describing the behaviour of systems that are too complex, ill-defined, or not easily analysed mathematically. Fuzzy variables are processed using a system called a fuzzy logic controller. It involves fuzzification, fuzzy inference, and

defuzzification. The fuzzification process converts a crisp input value to a fuzzy value. The fuzzy inference is responsible for drawing conclusions from the knowledge base. The defuzzification process converts the fuzzy control actions into a crisp control action.

- ***Hybrid Systems*** – The modern trend is to fuse these novel AI paradigms for offsetting the demerits of one paradigm by the merit of another. For example, it will be useful to use neural network to determine the membership function of fuzzy system automatically.

6.2 Requirements for AIP Applications

Despite being a relative new field of scientific research, Artificial Intelligence has already contributed greatly to the development of a range Advanced Information Processing (AIP) technologies. Candidate technologies for implementation in the proposed framework are knowledge-based systems, natural language and speech understanding, perception (including advanced sensor data processing and fusion), planning, learning, and distributed problem solving. AIP technologies that will be applied in order to develop the conceptual framework into an operational system are expected to have features that satisfy the following design requirements (Zuidgeest and Urlings 1996):

- ***Modularity*** – The assistant shall be based on technologies that allow logical decomposition of the system into smaller components (modules) with well-defined interfaces. Modularity facilitates development, enables future upgrades and reduces life-cycle costs by improved maintenance.
- ***Real-time performance*** – The assistant shall have guaranteed response times in a highly dynamic environment. It may be better to provide an acceptable response in time than to provide a response that is best, but too late. This can be extended with the requirement for a response being not too complex. Although a complex response is in time, its contents might be difficult to

understand. Real-time performance is a critical factor in acceptance of the machine-assistant by its human-operator.

- ***Reliability*** – The assistant shall have built-in hardware and software elements that are designed to reduce the risk of a complete system failure. The applied technologies should allow for graceful performance degradation in case of failure.
- ***Integration*** – The assistant includes many diverse functions needing different implementation methods and techniques. The technology used should support integration with conventional as well as advanced methodologies preserving modularity.
- ***System engineering*** – The assistant shall be developed and maintained by a well-defined and widely accepted system engineering methodology. The technology used should support such a methodology in order to reduce development and life-cycle costs.
- ***Maturity*** – The assistant shall be based on mature and proven implementation technologies. This is expressed by the availability of tools, successful prototypes and operational applications.

6.3 Distributed Problem Solving

An emerging candidate technology for realisation of the proposed human-machine framework is Distributed Problem Solving (DPS) (Gasser 1992). This technology provides a natural transition from the functional framework to a system architecture where the inherent distribution and modularity of functions is preserved in distributed problem solving modules. DPS technology considers two main approaches: distributed knowledge sources, often referred to as blackboard systems, and multi-agent systems. Both consist of multiple agents but they differ in structure at global architecture level and at agent level.

In a blackboard system, the agents are knowledge sources interacting through a shared memory: the blackboard (Nii 1986). Here, only knowledge is distributed, but data, information and control are central as compared to multi-agent systems. A common shared data structure for a complex operator-assistant system with heterogeneous knowledge, data and functions is not likely to be obtained. A central blackboard system control will also be a bottleneck for real-time performance. Therefore the application of a blackboard system seems to be limited to single operator assistant functions only.

A multi-agent system normally consists of heterogeneous agents that have a range of expertise or functionality. These agents have the potential to function stand-alone but are also able to cooperate with other agents (Decker *et al.* 1989). This provides the technology basis for the overall system architecture, which will be based on multiple cooperative agents, where each agent implements an assistant function. Each agent has its own local data, information, knowledge, operations and control that are relevant for the problem or task domain of the function. This encapsulation increases modularity and reduces complexity.

The agent capabilities and their potential to cooperate and achieve goals beyond the capabilities of a single agent determine the total system functionality. Cooperation in particular is the key to a sound human-machine team architecture that is compliant with basic requirements such as modularity, reliability, real-time performance and comprehensible, predictable system behaviour. It directly addresses the key problem in the operator-assistant: how to manage interaction between human and machine, both being agents on their own.

To implement successfully DPS applications, it is necessary to model and simulate their functional and system architectures very early in the design stage. Tools such as the Simulator of Heterogeneous Architectures, Agents and Active Resources (SAHARA) have been specifically developed to describe the architecture of a human-machine assistant at different levels (Champigneux 1993). At the functional level the system is viewed globally and the designer can verify that the desired behaviour has been achieved and the proper information is available. At the structural level the designer can express and analyse the system organisation. Implementation tools such as the Knowledge-based Operating System (KOS) have been specifically designed to meet the demanding real-time performance requirements. Performance of executing more then 1000 rules per second have been claimed (Heudin 1991). The distributed system and multiple agent approach enable researchers and system designers also to employ commercially available tools. The Common Object Request Broker Architecture (CORBA) in particular supports a distributed problem solving system design and allows multi-agent applications to communicate with each other no matter where they are located or who has designed them (Walsdorf and Onken 2000).

7 Conclusion and Recommendations

This chapter provides a conceptual framework for the description and development of human-machine teaming applications. The focus is on the introduction of the machine into the traditional situation where the human is solely responsible for all management and control activities. Essential in this framework are the inclusion and definition of what is often called the "task manager", "executive manager", "information manager" or "mission manager". Such manager is involved in monitoring, scheduling and allocation of task and tasks resources, and represents the main functional interface between the human and machine members in a team.

Another essential component in such framework is the decomposition of the primary tasks into a cognitive structure with several levels of priority and complexity. The task manager and the hierarchical task structure are closely linked together, they are critical in the proposed framework and they are the main areas for further research. An emerging candidate technology for realisation of such framework is Distributed Problem Solving and in particular its multi-agent approach.

By taking a teaming approach in the design of a human-machine application, further research will address the specific requirements that can be derived from the expectation that both human and machine behave as members of this team. Authority, task delegation, self-learning, performance monitoring, and assertive behaviour are expected to be the main components of these requirements. Authority, task delegation and team member assertive behaviour are also critical elements in a training concept called Crew Resource Management (CRM) (Weiner *et al.* 1993). Due to its original success in aviation, CRM has been adopted for a number of teaming situations in many other professions. Further research is proposed into the translation of the principles of CRM to the teaming environment of human and machine. It is expected that the principles of CRM are directly relevant for the management and coordination task. It is also expected that the results will not be limited to the requirements and recommendations for the machine member of the team only.

References

Abott, T. (1993), *Functional Categories for future Flight Deck Designs*, NASA TM 109005, Langley Research Center, Hampton VA.

Abott, T. and Rogers, W.H. (1993), "Functional categories for human-centered flight deck designs," *Proceedings of the 12th Digital Avionics Systems Conference*, New York, AIAA/IEEE.

Champigneux, G. (1993), "In-flight mission planning in the Copilote Electronique," *AGARD Lecture Series on New Advances in Mission Planning and Rehearsal Systems*, October.

Decker, K.S., *et al.* (1989), "Evaluating research in cooperative distributed problem solving," in Gasser, L. and Huhns, M.N. (eds.), *Distributed Artificial Intelligence Vol II*, London Pitman/Morgan Kaufmann.

Dekker, G.J. (1983), *Certification Aspects of the Avionics Software of the Airbus Industries A310 Aircraft,* NLR 83101 C.

Gasser, L. (1992), *An Overview of Distributed Artificial Intelligence, Theory and Praxis*, Eds. Gasser, Avouris, Boston Kluwer Academic.

Gerlach, M. and Onken, R. (1995), *CASSY – The Electronic Part of a Human-Electronic Crew*, WL-TR-96-3039, Wright Patterson AFB, December, pp. 140-145.

Gibson, C.P. and Garret, A.J. (1989), "Towards a future cockpit – the prototyping and integration of the Mission Management Aid (MMA)," *AGARD Symposium on Situation Awareness in Aerospace Operations.*

Hesselink, H.H. *et al.* (2001), *On-board Decision Support through the Integration of Advanced Information Processing and Human Factors Techniques: the POWER Project*, RTO Lecture Series 227, Amsterdam, RTO-ENP-019, November.

Heudin, J.-C. (1991), *KOS: a Knowledge-Based Operating System for Real-Time On-board Applications of Artificial Intelligence*, Sodima, IEEE.

Hollister, W. (1986), *Improved Guidance and Control Automation at the Man-Machine Interface*, AGARD Advisory Report No 228, December. (Chairman AGARD GCP Working Group 07.)

Hunt, G. (1992), *Mission Planning Systems for Tactical Aircraft (Pre-flight and In-Flight)*, AGARD Advisory Report 313, December. (Chairman AGARD Joint Working Group 15.)

Jain, L.C. (ed.) (1995), *Electronic Technology Directions towards 2000*, ETD2000, IEEE Computer Society Press, USA.

Jain, L.C. and Allen, G.N. (1995), "Introduction to artificial neural networks," in (Jain 1995), pp. 35-62.

Jain, L.C. (ed.) (1997a), *Proceedings of the First International Conference on Knowledge-Based Intelligent Engineering Systems*, Volumes 1 and 2, IEEE Press, USA.

Jain, L.C. (ed.) (1997b), *Soft Computing Techniques in Knowledge-Based Intelligent Engineering Systems*, Springer-Verlag, Germany.

Jain, L.C. and Jain, R.K. (eds.) (1997), *Hybrid Intelligent Engineering Systems*, World Scientific Publishing Company, Singapore.

Jain, L.C. and Martin, N.M. (eds.) (1997), *Fusion of Neural Networks, Fuzzy Systems and Evolutionary Computing Techniques: Industrial Applications*, CRC Press , USA.

Jain, L.C. and de Silva, C.W. (eds.) (1998), *Intelligent Adaptive Control: Industrial Applications*, CRC Press, USA.

Jain, L.C. and Jain, R.K. (eds.) (1998), *Proceedings of the Second International Conference on Knowledge-Based Intelligent Engineering Systems*, Volumes 1 and 2, IEEE Press, USA.

Jain, L.C., Johnson, R.P., Takefuji, Y., and Zadeh, L.A. (eds.) (1998), *Computational Intelligence Techniques in Industry*, CRC Press, USA.

Jain, L.C. and Vemuri, R. (eds.) (1998), *Industrial Applications of Neural Networks*, CRC Press, USA.

LaPuma, A.A. and Marlin C.A. (1993), *Pilot's Associate: a Synergistic System Reaches Maturity*, AIAA-93-4665-CP, pp 1131-1141.

Lefort-Piat, N., Morizet-Mahoudeaux, P. and Berge-Cherfaoui ,V. (1994), *Intelligent Techniques for Vehicle Driving Aid*, University of Compiegne, France.

Narasimhan, V.L. and Jain, L.C. (eds.) (1996), *The Proceedings of the Australian and New Zealand Conference on Intelligent Information Systems*, IEEE Press, USA.

Nguyen, X.T., Urlings, P.J.M., and Qureshi, Z.H. (1997), "A Framework for Automation with Human-in-the-Loop," *Proceedings 3rd Australian Workshop on Intelligent Decision Support Systems*, Melbourne, October.

Nii, H.P. (1986), "Blackboard systems: the blackboard model of problem solving and the evolution of blackboard architectures," *AI Magazine*, Summer.

Onken, R. (1994), "DAISY, an adaptive, knowledge-based driver monitoring and warning system," *Vehicle Navigation and Information Systems Conference Proceedings*, IEEE.

Onken, R. (1997), "Functional development and field test of CASSY – a knowledge based cockpit assistant system," *Knowledge-Based Functions in Aerospace Systems*, AGARD LS 200, AGARD, Neuilly-sur-Seine.

Rasmussen, J. (1986), *Information Processing and Human-Machine Interaction: an Approach to Cognitive Engineering*, New York: North Holland.

Rasmussen, J., Mark Pejtersen, A., and Goodstein, L.P. (1994), *Cognitive Systems Engineering*, Wiley Series in System Engineering.

Sato, M., Sato, Y., and Jain, L.C. (1997), *Fuzzy Clustering Models and Applications*, Springer-Verlag, Germany.

Schulte, A. (2001), "Mission management and crew assistance for military aircraft – cognitive concepts and prototype evaluation," *RTO Lecture Series 227*, Amsterdam, RTO-ENP-019, November.

Schutte, P.C. and Trujillo, A.C. (1994), *Flight Crew Task Management in Non-Normal Situations*, NASA Langley Research Center, Hampton VA.

Taylor, R.M. and Reising, J. (1998), "The human-electronic crew; human-computer collaborative teamworking," *RTO HFM Symposium on Collaborative Crew Performance in Complex Operational Systems*, Edinburgh, RTO MP-4, April.

Taylor, R.M. *et al.* (2000), "Cognitive cockpit engineering: coupling functional state assessment, task knowledge management and decision support for context sensitive aiding," *TTCP Technical Panel 7, Human Factors in Aircraft Environments*, Workshop on Cognitive Engineering, Dayton OH, May.

Urlings, P.J.M., Brants, J.J., Zuidgeest, R.G., and Eertink, B.J. (1995), "Crew assistant architecture," *RTP 6.5 Crew Assistant*, EUCLID CEPA-6, NLR Amsterdam, February.

Urlings, P.J.M. and Zuidgeest, R.G. (1997), "A generic architecture for crew assistant systems," *Conference Proceedings of the AGARD MSP Symposium on Advanced Architectures for Aerospace Mission Systems*, Istanbul, CP-581, July.

Urlings, P.J.M., Qureshi, Z.H., and Nguyen, X.T. (1998), "Automation in operator assistant systems," *Proceedings of the 2nd KES International Conference on Knowledge-Based Intelligent Electronic Systems*, Adelaide, April.

Van Rooij, A., Jain, L.C., and Johnson, R.P. (1996), *Neural Network Training Using Genetic Algorithms*, World Scientific Publishing Company, Singapore.

Vicente, K.J. (1999), *Cognitive Work Analysis: Toward Safe, Productive, and Healthy Computer-Based Work*, Lawrence Erlbaum Associates.

Vonk, E., Jain, L.C., and Johnson, R.P. (1997), *Automatic Generation of Neural Networks Architecture Using Evolutionary Computing*, World Scientific Publishing Company, Singapore.

Walsdorf, A. and Onken, R. (2000), *Cognitive Man-Machine Cooperation: Modeling Operator's General Objectives and Its Role within a Cockpit Assistant System*, IEEE.

Weiner, E., Kanki, B., and Helmreich, R. (1993), *Cockpit Resource Management*, San Diego, Academic Press.

White, F.E. and Llineas, J. (1990), "Data fusion: the process of C3I," *Defense Electronics*, vol. 22, no. 6, pp. 77-83, June.

Zuidgeest, R.G. and Urlings, P.J.M. (1996), "A generic architecture for in-flight crew assistant systems based on advanced information processing technology," *20th ICAS Conference*, Sorrento-Italy, September.

Chapter 3

Computer Generated Forces in Distributed Simulation

M.R. Stytz and S.B. Banks

The development of computer-generated synthetic environments relies heavily upon computer-generated forces, also called computer-generated actors (CGAs), that provide accurate behaviors at reasonable cost so that the synthetic environments are useful, affordable, complex, and realistic. Unfortunately, the pace of synthetic environment development and the level of desired CGF performance continue to rise at a much faster rate than CGF capability improvements. This insatiable demand for realism in computer-generated forces arises from the growing understanding of the significant role that modeling and simulation can play in a variety of uses in military and civilian application areas. These uses include training, analysis, procurement, mission rehearsal, doctrine development, force level training and task level training, information assurance, cyberwarfare, force structure analysis, sustainability analysis, life cycle costs analysis, material management, and infrastructure analysis and more. In these and other uses of military synthetic environments, computer-generated forces play a central role because they are the technology with the potential to significantly increase the realism of the environment while also reducing the cost of operating the environment. The progress that must be made to realize effective and realistic CGFs for simulation environments and the technical areas that should be the focus of future work are the subject of this chapter. The chapter introduces the technologies and progress made in the construction and use of CGFs and provides the technical foundation needed to comprehend the field and current research.

In this chapter, we will provide a comprehensive introduction to the field of computer generated forces. The chapter will open with a discussion of the field, its uses, necessary definitions, key concepts, and major challenges, both in research and implementation. The second section will provide an introduction to the state of the art, a review of significant previous research and projects, and the main CGF software architectures. Section three will contain a discussion of the basic requirements for CGF and the implications of these requirements for CGF architecture, design, and implementation. Section four will contain a discussion of the major CGF systems, how they operate, their capabilities, and the decision mechanisms they use. Section five will contain a discussion of human behavior modeling, skills modeling, decision modeling, and artificial intelligence technologies as they relate to CGFs. Section six will contain a discussion of knowledge base assembly and the role of subject matter experts. Section seven will contain a discussion of CGF software architectures, knowledge base architectures, and CGF development environments. Section eight, the concluding section, will contain a brief review of the chapter, suggestions for research and further reading, and, and a discussion of the areas most in need of work in order to make the CGFs that we need.

1 Overview

The development of computer-generated distributed synthetic environments (DSEs) and distributed virtual environments (DVEs) relies heavily upon computer-generated forces (CGFs), also called computer-generated actors (CGAs) to achieve a complex, accurate, and useful environment. To fulfill this role, CGAs must exhibit accurate behaviors at reasonable cost so that the synthetic environments are useful, affordable, complex, and realistic. The current state of CGA technologies is one where our ability to compute has far outstripped our ability to model; it is much easier to increase the computational power at a host than it is to improve any aspect of a human behavior model or other component of a CGA.

The most recurrent criticism of current CGAs is their brittleness and rigidity, ie., they are unable to respond in a realistic manner to situations that were not explicitly entered into their knowledge base. Therefore, an important objective is to develop CGAs that are adaptable to a variety of circumstances and that are robust in their response to situations. The new class of CGAs must exhibit realistic unpredictability and variability in behaviors, and there is a need for realistic behavior of adversaries and teammates, which requires representations for intent, deception, and adaptability in adversarial CGAs. However, addressing the knowledge acquisition issue is just the first step. Extensive knowledge bases must be coupled with adaptable decision making and behaviors that overcome the brittleness and rigidity of current systems. The objective of this new class of CGAs is improved CGA fidelity and performance coupled with a reduction in development and maintenance costs. The creation of a comprehensive learning capability that supports all types of decision making systems is also needed.

Unfortunately, the level of desired CGF performance is rising at a much faster rate than CGF capabilities. While there are many reasons for this increasing importance of CGFs, they can be distilled into two dominant themes, namely the desire for reduced cost and the impracticality of dedicating large numbers of human operators to distributed simulation operations. This increasing importance also arises from the increasing awareness of the significant role that modeling and simulation can play in a variety of uses in military and civilian application areas. These uses include training, analysis, procurement, mission rehearsal, doctrine development, force level training and task level training, information assurance, cyberwarfare, force structure analysis, sustainability analysis, life cycle costs analysis, material management, and infrastructure analysis. In these and other uses of synthetic environments, computer-generated forces play a central role because they are the technology with the potential to significantly increase the realism of the environment while also reducing the cost of operating the environment. Nevertheless, their quality remains a matter of issue because of the complexity of the decision-making that must be performed, the difficulties in assembling the knowledge base for a CGF, the challenges inherent in human behavior and group behavior modeling, the need for accurate physical models, and the computational cost of achieving high fidelity operation. The progress that must be made to realize effective and realistic CGFs for simulation environments and the technical areas that should be the focus of future work are the subject of this chapter. The chapter introduces the technologies and progress made in the construction and use of CGFs and provides the technical foundation needed to comprehend the field and current research.

In this chapter, we provide an introduction to the field of computer generated forces. The remainder of this section contains a discussion of the key concepts and decision-making technologies used in CGFs. Section two contains a discussion of the basic requirements for CGF systems and the implications of these requirements. The third section provides an introduction to the state of the art, a review of significant previous research and projects, and the significant CGF software architectures. Section four holds a discussion of the major CGF systems, how they operate, their capabilities, and the decision mechanisms they use. Section five presents a discussion of human behavior modeling, skills modeling, and decision modeling as they relate to CGFs. Section six, the concluding section, presents a brief review of the chapter, contains suggestions for research, and provides a discussion of the areas most in need of attention in order to achieve the CGFs of the future.

2 Key Concepts

In the CGF research and development field, there are a number of central ideas and concepts. Chief among these is the need for the incorporation and use of the proper doctrine for the CGAs as the basis for insuring proper behavior. Given the incorporation of doctrine, it is crucial that artificial intelligence technologies be used for decision making, with the selection of technology being based upon the tasks that the CGA must perform. The CGAs should be capable of autonomous operation, that is, they should be able to operate without continuous or even intermittent human supervision. The CGAs should be interoperable with other human and computer controlled actors; that it, they should have representations of the DVE/DSE and of the entities and actors in it that are identical to those used by the other participants in the DVE/DSE.

Within a CGA, there are nine major logical components. These logical components are the decision-making component, the knowledge base, the human behavior modeling and skills component, the dynamics component, the sensors component, a model of the physical world, weapons models, command and control, and group behaviors. We will return to these components in detail later in this chapter, but at this point we will briefly discuss each component.

The decision-making component of a CGA is responsible for using information in the knowledge base and the status of the DSE/DVE to determine a course of action for the CGA, which may involve activities ranging from long-term planning to immediate response to a life-and-death situation. The knowledge base contains all of the static information available to the CGA to support its decision-making activities. The human behavior model is a computable representation of human behavior and is used to moderate the CGA's activities so that they appear to be more human-like, allow the CGA to respond to circumstances in a human-like manner, and, in conjunction with the skill level, allow the CGA's performance to correlate with a desired level of skill and experience. The human behavior model should have as a subcomponent a combat psychology model. The dynamics component of a CGA contains a computable model for the motion of the CGA and thereby determines how a CGA travels through the DVE/DSE. The sensors component contains a computable model for each of the sensors available to the CGA, which can can include as examples a model of human vision or a model of microwave radar. The model for the physical world contains a representation for the terrain, buildings, manmade features, weather, Sun position, and all other aspects of the DVE/DSE not related to an individual actor or entity; the world model provides the backdrop context through which the actors and entities move. The weapons model contains a computable representation of the motion and capabilities as well as destructive power of each weapon available to the CGA. The command and control model is a representation for how information and directives are transmitted between CGAs and also designates a chain of command and a commander. A group behavior model is much like an individual human behavior model, except that it provides a computable description of how a group of CGAs will respond to circumstances within the DVE/DSE.

The key element of any successful CGA is its behavior. Correct behavior is achieved using adaptive decision mechanisms and comprehensive, expandable human behavior models. In a CGA, the decision mechanism must be structured so that the amount of information considered when making the decision can be adaptively varied and so that additional behavioral possibilities can be considered as time and circumstances permit. However, the most difficult and expensive challenge is modeling human behavior. A human behavior model, or representation, is a computer-based, computable model whose outputs faithfully mimic the behavior of a single human or a group of humans executing a task or set of tasks. Human behavior modeling addresses the need for making the behavior and reactions of a CGA seem realistic by developing models that yield an acceptable analog of the output of the human decision-making process. Human behavior modeling requires the broad-based acquisition of domain-specific knowledge that humans use in the decision-making process. In a DSE/DVE, human behavior models are needed for individuals, small organizations, individual vehicles, groups, command and control elements, and very large organizations. Finally, acceptable fidelity for a human behavior model can only be attained by incorporating skill level representations, stress factor representations, and situational psychological representations. Incorporation of rich, flexible, and comprehensive human behavior models into CGAs will lead to more accurate behaviors for CGAs and result in a more accurate performance on the part of the CGAs. CGA behaviors are founded upon knowledge acquisition activities related to the environment to be simulated. Artificial intelligence techniques are used in conjunction with the knowledge to analyze the situation that confronts a CGA and to select an appropriate response to the situation in light of the CGA's role, goals, mission, and skills.

In CGFs for distributed simulation environments, fidelity is a paramount consideration. Fidelity is defined as the accuracy of a representation (where a representation is a simulation model in execution) as compared to its counterpart in the real world. CGA-related types of fidelity are: 1) sensory fidelity, 2) physics fidelity, 3) modeling fidelity, 4) actor behavior fidelity, 5) time, latency, and lag fidelity, 6) information fidelity, 7) system fidelity, 8) command and control fidelity, 9) phenomenology[1] fidelity, and 10) threat fidelity.

Sensory fidelity addresses the need for information acquired by a CGA's simulated sensor systems to replicate the accuracy and vol-

---

[1] The term phenomenology refers to weather (sun, clouds, rain, snow, mud, fog, etc.) as well as emissions into the environment, such as radar, radio, and microwave.

ume of information produced by the actual sensor in the real world. *Physics* fidelity addresses portrayal of the effects of gravity, electromagnetic spectrum emissions and propagation, motion (including friction), and energy consumption and conservation on the CGA. A third form of fidelity is *modeling fidelity*. Modeling fidelity addresses two related issues. The first is insuring that models have correct relative proportion (scale) and move at correct relative velocities within the simulation environment. The second issue is maintaining fidelity of model portrayals between different participants, which may require coordinate system transformations as well as asymmetric scaling transformations.

A fourth fidelity concern is *actor behavior fidelity*. In simulations that contain a mixture of human-controlled actors and computer-generated actors, the behavior of the computer-generated actors should replicate the behavior that a human-controlled actor would exhibit in similar circumstances. The human behavior modeling component is typically the focus of efforts to validate the accuracy of the performance of the CGA. Human behavior modeling requires acquisition domain-specific knowledge about the human mental models, the information brought to the decision-making process, and the key factors in the decision-making process as understood by people at different levels of skill in the domain(s) of interest. For military simulation purposes, this knowledge acquisition process involves incorporation of doctrine and training knowledge into the CGA's knowledge base and/or decision processes.

Information fidelity addresses the volume and timeliness of the information available to a CGA within the DSE/DVE. This area of fidelity addresses the type of information available, the quantity of information to be assessed, the arrival rate at the locus of decision-making in the computer-controlled actor, and the veracity of information available. A system with high information fidelity insures that the information available to the CGA is the same as in the real-world. The seventh area of fidelity is *system fidelity*. System fidelity addresses the activities and response of collections of actors and requires the capability for coordinating and directing the collections by humans or other CGAs. The level of *command and control fidelity* for a system is determined by the capability to employ the command and control systems used within the real-world within the DSE/DVE. *Phenomenology fidelity* is the correspondence between a simulation of weather, radars, jamming systems, and other spectral emissions within a DSE/DVE and the measurable performance and effects of their real-world counterparts. *Threat fidelity* is the correspondence between a portrayal of the threats within a DSE/DVE and the actual behavior of the same threats in the real-world.

In addition to consideration of the types of fidelity discussed above, there are three areas of capability that must be considered. These areas are *adaptive decision-making*, *learning*, and *actor migration*. *Adaptive decision-making* requires that the CGA's decision mechanisms robustly deal with incomplete information and uncertainty and permit the CGA to maintain credible behavior at run-time under a variety of circumstances, time pressures, and lack of information. While a primitive CGA operating in a primitive environment may not require adaptivity, a CGA operating in a complex environments such as a military simulation does need adaptivity in CGA systems. The requirement for *automated incorporation of CGA experiences* into a knowledge base and decision mechanism increases the complexity of the CGA because it must be able to modify these two components during execution. The interplay between the knowledge bases and decision mechanisms further complicates the task of implementing a system that can learn. Learning is necessary because it is impossible to assemble a knowledge base that can address every eventuality in a simulation if the CGA will operate in a complex environment and if the CGA must exhibit realistic behaviors.

Actor migration is the transfer of computer or human control of a CGF from one computer to another across a network. Migration is required for two reasons, one is that a CGA must be able to migrate into and out of human control so that humans can provide mission re-planning, critical activity support, and other decision capabilities that are beyond the abilities of CGAs. The other reason is to satisfy latency and fidelity requirements. To migrate a CGA information related to actor state, fidelity, and knowledge base must be transmitted. There are three types of actor migration, which are differentiated by the starting and ending type of control for the actor. The three cases are where an actor moves from one host in a computer-controlled situation to another host in a computer-controlled situation (CGA-CGA); where an actor moves from a computer-controlled situation to a human-controlled situation (CGA-HCA); and where an actor moves from a human-controlled situation to a computer-controlled situation (HCA-CGA). Each type of migration has specific latency and data transfer requirements and can be either a planned migration or an unplanned migration.

3 Artificial Intelligence Technologies

3.1 Blackboard Systems

Blackboard systems are an architectural approach to distributing data to a number of intelligent agents, they are not strictly for decision making. Blackboard systems were initially developed to support natural language processing and have since been used for a wide array of purposes, including situation awareness. In the blackboard approach to data distribution, an environment is decomposed into one or more hierarchical containers of data, symbolized as panels of information. When using a blackboard system, knowledge sources encode data and insert it into the appropriate location(s) in the blackboard. Other knowledge entities reason about the information that has been posted and make inferences about future situations or states and then post their conclusions back into the blackboard structure. The collection of information on the blackboard constitutes the CGA's awareness of the situation in the world and of its own decision state. Once a knowledge entity has its information, any form of reasoning can be used to reason on the data; the blackboard imposes no restrictions in this regard.

3.2 Expert Systems

Expert systems are knowledge-based systems that use expert level knowledge to achieve expert level performance. Generally, expert systems have mechanisms for representing knowledge, interacting with existing databases, learning, performing inference, making decisions, and making recommendations. The term expert system derives from the fact that the knowledge in the system is derived from interviews with experts in the domain of operation for the system as well as from the desired level of performance. There are three main components of an expert system, the rulebase, the factbase, and the inference engine. The rulebase is the set of rules that encode specific knowledge relevant to the domain of operation. The factbase contains a set of asserted values of the properties of objects and events in the domain, the factbase encodes the current state of the domain. The inference engine is a formal implementation of one or more reasoning mechanisms. The inference engine operates on the factbase to produce new facts and on the rulebase to produce new rules. The general operation of an expert system is usually characterized by a match-select-apply cycle, using either forward-chaining or backward chaining. Forward chaining is used to derive the consequences of a new fact being added to the database, backward chaining is used to solve a particular problem and to plan.

3.3 Case-Based Reasoning

Case-based reasoning is based upon the concept of using explicit, documented experiences, called *cases*, to solve new problems. In case-based reasoning, the decision making process retrieves an appropriate case from a larger set of documented cases, where the similarities between the present problem and the retrieved case are the basis for the case selection. A case-based reasoning system consists of three basic components: (1) a library of historical cases, (2) a means of using the key elements of the present problem to find and retrieve the most similar case(s) from the case library, and (3) a means for making modifications to the proposed solution when the case on which the solution is based is not identical with the current problem. The construction and organization of a library of cases is a crucial task. The effect of retrieving an improperly matched case is often more computationally expensive than selecting the wrong rule to execute in a rule-based system. Inefficient searches through the case library due to poor organization can result in unacceptable performance. Therefore, the cases in the library must be indexed to provide efficient search of the library. How the indices are expressed is a key issue in case-based reasoning. If the indices are defined too broadly, too many cases may be retrieved in a situation because they are classified as being similar to the problem. Conversely, if the indices are expressed too narrowly, no case may be retrieved in a situation because no case is deemed similar. Cases can be developed by an expert based on experiences, from documented sources, or directly through task execution or work simulation.

To problem solve using a case-based system, a new problem (target case) is presented to the system and is compared with all the cases in the case library. The comparison is accomplished using a similarity metric defined over a subset of the case features (index features), which returns a case or set of cases that most closely resemble the new problem at hand. In general, the returned case(s) will not match the new problem exactly and, therefore, a case-based system usually includes a mechanism for adapting or tailoring the specifics of a selected case to match those of the new situation. These adaptation or generalization mechanisms are highly domain specific and construction of a case-based system within a new domain generally requires a completely new knowledge engineering effort to assemble the case modification mechanism. Therefore, although case-based reasoning has many advantages, primarily due to the fact that the process is patterned after human reasoning and lends itself to analogical reasoning, there are also tremendous computational burdens incurred by the problem solving technique. Library organization, case feature selection and indexing, similarity metric choice and evaluation, and case feature modification are all very computationally expensive processes. The quality of the cases contained in the library and the proper number of cases to use are also important development issues.

3.4 Evolutionary Algorithms, Genetic Algorithms, and Evolutionary Computing

The term *evolutionary algorithm* is a generic term used to describe computer- based problem solving systems that use computational models of evolutionary processes to find solutions in complex search spaces. An evolutionary algorithm employs a population of structures (generally initialized randomly), that evolve according to rules of selection, recombination, mutation and survival, that are referred to as genetic operations. The environment for the algorithm determines the fitness of each individual structure (or chromosome) in the population. The set of possible solutions for the optimization problem is considered as a population of individual chromosomes. The degree of adaptation of an individual chromosome to its environment is specified by its fitness. In each evolutionary cycle, each individual chromosome in each generation of the population is scored on its fitness for the environment. Reproduction then focuses attention on high fitness individuals. Recombination and mutation perturb those individuals, providing the means for generating subsequent populations of solutions that have members with higher fitness function values. The fittest individuals/chromosomes are more likely to be selected for reproduction (retention or duplication) into subsequent generations. Recombination and mutation modify the properties (genome) of selected individuals, yielding potentially superior genomes. The four major types of evolutionary algorithms are: 1) genetic algorithms, 2) evolutionary programming, 3) evolution strategies, and 4) classifier systems. These four approaches share the conceptual base of simulating the evolution of individual structures via processes of selection, mutation, and reproduction. The evolutionary algorithm approaches have some advantages over traditional algorithms, chief among them being that they are adaptable to changing circumstances and that they can generate acceptable and actionable solutions rapidly enough to be of use.

A *genetic algorithm* derives its behavior from a metaphor of the processes of evolution in nature. A genetic algorithm is a heuristic optimization method that operates through determined, randomized search. The evolutionary process is emulated in a genetic algorithm by the creation of a population of individual solutions represented by chromosomes. A gene is a subsection of a chromosome that encodes the value of a single parameter being optimized by the genetic algorithm. The individual chromosomes in the population evolve by combining or mutating chromosomes to breed new chromosomes in the subsequent generation. A genetic algorithm operates using the following cycle: 1) evaluate the fitness of the individual chromosomes in the current population; 2) create a new population by performing operations such as crossover, reproduction, and mutation on the "fittest" chromosomes; 3) discard the old population; 4) iterate using the new population. One complete cycle of this loop is a generation. The first generation (generation 0) of this process operates on randomly generated chromosomes. Thereafter, the genetic operations and fitness measure operate to improve the population.

Evolutionary programming is similar to genetic algorithms but places its emphasis on the behavioral linkage between parents and offspring, rather than seeking to emulate specific genetic operators. Evolutionary programming is similar to evolutionary strategies. In evolutionary programming, as in genetic algorithms, there is an underlying assumption that the fitness of a solution can be characterized in genetic terms, and that there are one or more optimum solutions. The basic evolutionary programming method has three steps: 1) choose an initial population of trial solutions at random; 2) replicate each solution into a new population; and 3) mutate each solution using a distribution function for mutation and each offspring solution is assessed by computing its fitness.

An *evolutionary strategy* is generally more successful than other iterative solution methods on a large number of test problems. An evolutionary strategy is adaptable to most problems in optimization because an evolutionary strategy needs little information about the problem. A single chromosome in an evolutionary strategy population consists of a genotype, which represents a solution in the search space and real-valued object variables. In a two-membered evolutionary strategy one parent generates one offspring per generation by applying normally distributed mutations to their chromosomes until a child performs better than its ancestor and takes its place. For an evolutionary strategy to be successfully employed, the chromosomal population has to be fairly large and a reproduction selection algorithm must be used that allows a wide variety of parents to reproduce.

The general concept of a *classifier system* is to start without knowledge and use a randomly generated classifier population as the baseline classifier and then let the system learn by induction. For learning to occur, the input data stream must contain recurrent input patterns. The most straightforward approach to classifier systems is the animat (animal + robot = animat), which is also sometimes called artificial life. The general concept for an animat is that the animat is a cognitive system that can classify the changes in state in its environment and then react to these changes. An animat system requires: 1) an environment; 2) receptors/sensors for the environment; 3) effectors to manipulate the environment; and (4) the decision system itself. A simple animat would use a production system as its decision system and classifier. The key to success for this approach is choosing a knowledge representation that makes it easy to manipulate the rules.

3.5 Bayesian Networks / Belief Systems

The basis of Bayesian networks is Bayes' Theorem. Bayes' theorem relates phenomena observed in the present (the evidence) to phenomena known to have occurred in the past (the prior) and to theories about what is going to happen in the future (the model). The Bayesian formula is expressed as:

$$P(t \mid y) = \frac{P(y \mid t) P(t)}{P(y)} \qquad (1)$$

where P(t) is the probability of event t occurring; P(y|t) means the probability that y will occur given that t occurred; P(t|y) means the probability that t will occur given that y occurred; and P(y) is the probability of y occurring. Bayes' theorem has proven to be effective at many different pattern-recognition tasks such as fingerprint identification, voice identification, facial matching, and handwriting analysis. One advantage of the Bayesian approach is that the patterns that occur naturally "teach" the computer about the relationships between the elements of the pattern. A Bayesian approach is appropriate in an environment of uncertainty, conflicting data, and/or incomplete information.

3.6 Fuzzy Logic and Fuzzy Systems

Fuzzy logic deals with fuzzy propositions, which are propositions that contain fuzzy variables and fuzzy values. In fuzzy logic, an assertion may have a degree of both truth and falseness. In fuzzy logic, an object can have a partial membership in one or more sets. The degree of membership in a given set for a data item is determined by the characteristic function, or membership function, for the set. The membership function is used to compute the degree to which a value in the fuzzy rule's domain is contained within the range of the fuzzy rule. The values of the membership function are in the range [0, 1], the value returned by the membership function for a piece of data is called the membership degree. A linguistic variable, such as temperature, is used in fuzzy logic to describe a quantity or an idea that is best represented by fuzzy sets, also called term sets. The value of the linguistic variable can be assigned to one of these term sets. A powerful technique in fuzzy logic is to "fuzzify" a crisp value for a linguistic variable and then determine the term set(s) to which the crisp value belongs, thereby allowing the linguistic variable to be valuated as the union of its fuzzy sets.

In fuzzy logic, fuzzy rules specify the relationship between input membership function values and the output membership function value. A fuzzy rule can have one or more input variables and one or more output variables. The fuzzy rules deal with fuzzy values, which are usually represented by their membership functions. Fuzzy inference is an inference method that employs fuzzy implication relations, fuzzy composition operators, and an operator to use the fuzzy rules.

Fuzzy systems are expert systems that employ fuzzy rules and fuzzy inference to perform decision making. A fuzzy logic system is defined by three components; 1) fuzzy input and output variables, 2) fuzzy rules and membership functions, and 3) a fuzzy inference mechanism (decision engine). The fuzzy rules and their membership functions comprise the system's knowledge base. The input data for the system can either be exact or fuzzy. The fuzzy inference mechanism uses fuzzy inference methods and activates all of the satisfied fuzzy rules during each evaluation cycle. When using a fuzzy expert system, there can be a partial match between the input data and the fuzzy condition elements in the antecedents for the fuzzy rules, which means that each rule can, in some sense, be satisfied to some degree and therefore fired each processing cycle. Generally, a measure of the degree of matching (satisfaction) is calculated for each rule and the rule fires only if the antecedent fuzzy value exceeds a specified threshold.

3.7 Neural Networks

An artificial neural network is a biologically inspired computational model that consists of a set of processing elements (neurons), connections between the neurons, weights assigned to each connection, and training and recall algorithms. A neural network is a computational model that uses parallel pathways of computation that mimic the high degree of parallelism and interconnectivity of the human nervous system to perform tasks such as pattern recognition. A neuron is defined by inputs (which have weights assigned to them), an input function that computes the net input signal based on the weights (and possibly the neuron's previous state), a function to compute the neuron's activation level, and the output signal strength.

The neural network consists of processing elements and connections between them that are based upon a model of an artificial neuron. Weights are assigned to the connections and these weights capture the "intelligence" for the system. The commonly used artificial neuronal model is based upon four elements: 1) input connections, 2) an input function, 3) a signal function, and 4) an output function. Each neuron can have one or more input connections, each of which has a weight assigned to it. The input function calculates the net input signal to the neuronal computational element and is usually a simple summation function. The signal function calculates the activation level (excitation level) of the neuron. The output function calculates the signal value emitted through the axon of the simulated neuron, generally the output signal is assigned the value computed by the activation (signal) function.

Because neural networks are robust and employ approximate reasoning (much like fuzzy logic) they can deal with uncertain, noisy, or ambiguous data and can compute an answer even with only sparse data sets. Using neural networks for problem solving is a two-phase process. The first phase is the training phase, in this phase known data sets are presented to the network and the weights assigned to the connections between the neurons are computed. In the second phase, called the recall phase, data are fed into the network and the network calculates a result.

Learning in a neural network is accomplished by applying inputs to the network until the output of the network matches the desired values. The set of inputs used to train the network is called the training set, the elements of the training set are called training examples. The training process causes the connection weights to change, thereby forcing the network to learn. In general, a network improves its performance/response to a given stimulus in direct proportion to the number of times it is exposed to the stimulus. Learning in a neural network is not a property of a single neuron, rather it is a property of the network.

Having completed a short review of the background material for CGAs, we will now turn to a discussion of CGA performance requirements. This discussion is presented in the next section.

4 CGA Requirements and Their Implications

In this section we briefly examine the set of CGA requirements and the implications of these requirements on the structure of a CGA and upon its assembly. The following requirements outline the capabilities desired for an ideal CGA: 1) a CGA must support ease of modifiability and re-use throughout the software system, 2) a CGA must provide high fidelity representations of motion and human behavior, 3) a CGA must support adaptable decision mechanisms, 4) a CGA must have hardware independence, 5) a CGA must be able to operate at varying levels of fidelity, 6) a CGA must be able to operate at a number of skills levels, 7) a CGA must have a terrain reasoning capability, 9) a CGA must have data fidelity, and 10) a CGA must be able to perform actor migration. The next few paragraphs elaborate upon a few of these requirements and their implications for the CGA architecture, design and software development.

Modifiability. Modifiability is the ability to expand upon existing CGA capabilities and requires several capabilities: 1) the ability to expand a domain-specific knowledge base, 2) a flexible software architecture, and 3) the capability to operate on a variety of hardware. The requirement for knowledge base expandability will permit the CGA to incorporate new strategies, tactics, and maneuvers as necessary. This requirement also addresses the need to maintain CGAs , re-use software and knowledge bases, and support improvement of the system's performance by permitting well-encapsulated and localized changes to the knowledge base. Furthermore, a CGA must be as independent as practicable of any software encumbrances that would impede porting it to a new hardware platform or operating system.

High fidelity representations. High fidelity representations are achieved by using accurate: 1) world representations, 2) dynamics for vehicle motion, 3) sensor and weapons models, and 4) models of human behavior. The world representations are based upon Earth-surface representations composed from primitive data elements organized within a hierarchical representation of the terrain data. Additionally, the world representation must have an equivalent fidelity counterpart for manned systems and for other CGA systems. The implementation of correct vehicle dynamics is another aspect of achieving a high fidelity representation. Correct vehicle dynamics insures that the vehicle only moves according to its capabilities and does not achieve a level of performance that is unrealistic given the terrain, weather, and atmospheric conditions. Likewise, the weapons and sensor models must use the same sensitivity, field of view, and range as their real-world counterparts. This level of fidelity must span the variety of sensors from the eyesight of the operator of the CGA to the radar and infrared sensing systems of the CGA.

Adaptable decision mechanisms and a Variety of Skill Levels. Adaptable decision mechanisms allow the CGA to exhibit a degree of flexibility in dealing with situations that occur in the DSE/DVE Adaptable decision mechanisms permit the system to maintain robust, credible behavior for CGA at run-time under a variety of external circumstances and at different levels of simulated operator skill. The sub-requirement for robust, credible behavior is necessary so that the CGA can act and react even when confronted by conflicting or incomplete information. If the CGA does not exhibit robust behavior, then it will fail or have a predictable pattern of behavior. Multiple skill levels can be achieved by varying manual skills, by varying the range of options available to the decision making component, and by changing the combat psychology model.

Data fidelity. The requirement for data fidelity arises from the need to provide an accurate environment wherein CGAs behave in a manner that is analogous to the manner in which they operate in the real world. The degree of data fidelity for a system is determined by the correspondence between the system's simulation of the sensors and sensor displays and the sensors and sensor displays in the real-world system.

Support for migration. The requirement for support for migration arises from the necessity of reducing latency between source and recipient of data across the network. In many circumstances in training environments the time between friendly action and opponent reaction is crucial to a successful training experience. Currently, the best way to insure that accurate latencies are achieved is to migrate one or more of the participants in an interaction to a host computer system whose interaction latency with the remaining, unmigrated, participants is acceptable.

These requirements parallel the requirements discussed by a number of other authors. For example, the work reported by Webber and Badler (1993), parallels these requirements for CGAs but also emphasizes the graphical appearance. Webber and Badler claim that CGAs require an interactive graphical model, biomechanical properties, human-like behavior, perception, task performance and guidance, and a capability for interaction with other agents. Therefore, they argue, a CGA must have knowledge of their own skills and the conditions under which they apply. CGAs must also understand requests for tasks, they must be able to arbitrate and integrate multiple concurrent task actions, they must engage naturally with their environment, they must be able to adapt their behavior to new situations, and they must act in real-time.

5 Significant Research Projects and Software Architectures

There are a number of open issues and research problems that must be addressed in order to develop CGAs with the qualities that are needed. In this section we will briefly review these efforts, and later in this chapter discuss some of these efforts in more detail. In the first subsection, we present a brief overview of the significant research efforts and categorize them by their main research thrust. This subsection also provides some indication of the topics of CGA research that have received considerable attention and those fields where research has been performed sparingly, if at all. In the second subsection, we examine some of the research in CGA software architectures. In light of the lack of a single dominant, widely usable software architecture for CGAs, several research groups have addressed some of the research challenges associated with the development of a broadly applicable CGA software architecture. Our goal here is to present an indication of the breadth and scope of the research that has been performed and also indicate topics where further research is required.

5.1 Research Projects

The run-time challenges for a CGA arise from the need to compute human-like behaviors and reactions to a complex dynamic environment at a human-scale rate of time. Additionally, the CGA behavior must be realistic and accurate enough so that other CGAs and human participants react to its outputs as though it were operated by a human. Therefore, the capability to construct large, complex reasoning systems and the development of comprehensive knowledge bases for use by the decision machinery are needed to enable the implementation of CGAs of acceptable fidelity. A large body of work has been developed that addresses these and other issues that must be addressed when assembling a CGA. To date, many aspects of the process of defining a CGA, its behavior, architecture, and reasoning systems have been reported. Unfortunately, no consensus has emerged concerning the best means for implementing these capabilities and the literature provides little insight into the difficulties involved and approaches that have proven useful.

The requirements for CGAs have been addressed infrequently in the literature (Banks and Stytz 1999, Berkowitz 1996, Brasse *et al.* 1999, Chandrasekaran and Josephson 1999, Cox 1998, Foss and Robinson 1999, Franceschini and Cox 1998, George *et al.* 1999, Gillis 1998, Hoagland *et al.* 1998, Jones *et al.* 1996, Laird *et al.* 1998b, Lorenzo *et al.* 2000, McIntyre and Middleton 1995, McIlroy and Heinze 1996, Pratt 1996, 1998, Rosenbloom *et al.* 1994, Stone and Pettitt 1998, Stytz *et al.* 1996b, 1998, Surdu *et al.* 1998, Tyler *et al.* 1998, Webber and Badler 1993), but this may be because general-purpose requirements are difficult to enumerate at this stage of the development of the field. Lessons learned and system specification have been discussed even more sparingly (Balzer 1996, Cisneros *et al.* 1996, Pratt *et al.* 2000, Rigg *et al.* 2000).

As might be expected for a software system, architectural aspects for the CGA have been addressed often and a wide number of approaches have been reported (Adkins 1996, Allen and Kalphat-Lopez 1998, Becket and Badler 1993, Bergenthal *et al.* 2000, Bokma and Slade 1993, Bowden *et al.* 1997, Bowden and Davies 1998, Brasse *et al.* 1999, Busetta *et al.* 1999, Butler 1998, Calder *et al.* 1996, Calder and Drummey 1999, Carey and Kleiner 1998, Ceranowicz 1994, 1998, Chaib-draa and Paquet 1993, Coradeschi *et al.* 1996, Courtemanche and Ceranowicz 1995, Courtemanche 1999, Courtemanche and Burch 2000, Cox 1998, Craft and Karr 1996, Cusack and Hoare 1999, Franceschini *et al.* 1998, 1999, Gagnon *et al.* 1999, Ge *et al.* 1995, George *et al.* 1998, Gonzalez and Ahlers 1995, Jackson and Pearman 1999, Jones *et al.* 1993, Jones and Laird 1994, Hartzog and Salisbury 1996, Howard and

Lee 1998, Kocabas and Oztemel 1998, Kuijpers *et al.* 1998, Kuokka 1993, Laird *et al.* 1987, 1995, Lankester 1995, Laughery and Dahn 1999, LaVine *et al.* 1999a,b, Marshall *et al.* 1996, McIlroy and Heinze 1996, Murray *et al.* 1999, Oztemel and Kocabas 1996, Parsons 1994, Patrone and Nardo 2000, Pongracic *et al.* 1997a, Pratt *et al.* 1994, Reece and Kelly 1996, Roberts and Crawford 2000, Sale *et al.* 1999, Sherman 1994, Siksik 2000, Steuart and Murray 1996, Stytz and Banks 2000, 2001, Takvorian 1998, Webber and Badler 1993, Vrablik *et al.* 1998, Vrablik and Longfritz 1999). The architectural approaches range from modular software libraries and interacting processes to closely coupled objects and data-flow architectures. Given the low cost of gaming systems, they have recently become popular infrastructures for CGA systems and for CGA experimentation, as discussed by a number of researchers (Laird and vanLent 2001, Laird 2002, Jacobson and Hwang 2002, Kaminka *et al.* 2002).

The crucial aspect of any CGA, as has been proved in many demonstrations, is its reasoning mechanism. The reasoning mechanism should offer acceptable performance coupled with robustness and suitability to the problem domain. A wide variety of reasoning systems have been proposed (Adamson 1996, Adamson and Joshi 1996, Adkins 1996, Becket and Badler 1993, Bimson *et al.* 1994, Bowden *et al.* 1995, Braudaway 1993, Burke 1998, Carter *et al.* 1993, Cisneros *et al.* 1996, Coradeschi *et al.* 1996, Craft and Karr 1996, Davies and Gabrisch 1995, Ehrhart and Bigbee 1998, Gelenbe 1999, Gonzalez and Ahlers 1995, Hepplewhite and Baxter 1996a,b, Jones *et al.* 1993, Karr *et al.* 1995, Katz 1993, Keirsey *et al.* 1994, Kenny *et al.* 1996, Kocabas and Oztemel 1998, Koenig and Liu 1999, Laird *et al.* 1994, 1995, Laird and vanLent 2001, Landweer 1993, McIlroy *et al.* 1997, McKenzie *et al.* 2001, Page and Widdowson 1999, Parsons 1994, Penney 1996, Perry 1997, Pratt 2001, Rajput and Karr 1996, Rajput *et al.* 1996, Rosenbloom *et al.* 1994, Schwamb *et al.* 1994, Siksik 1993, Tambe and Rosenbloom 1994, VonderLippe and Courtemanche 1999, Vrablik and

Longfritz 1999, Weaver and Mullen 1994), but to date no consensus has been reached. Results to date indicate that the problem domain and operational environment more than any other factors determine the suitability of a reasoning system. No guidelines have been proposed that would allow a system developer to choose a reasoning mechanism based upon a characterization of the problem domain or operational environment.

Associated with the issue of choice of reasoning system are the issues of knowledge acquisition and representation as has been discussed by many researchers (Adamson 1996, Adamson and Joshi 1996, Adkins 1996, Allsop *et al.* 2001, Aronson 1994, Banks *et al.* 1998a, Banks and Stytz 1999b, 2000, Baxter and Horn 1999, Becket and Badler 1993, Bimson *et al.* 1994, Bokma and Slade 1993, Booker *et al.* 1998, Booker *et al.* 1999, Bowden and Davies 1998, Briggs and White 1997, Calder 1998, Calder *et al.* 1999b, Carter *et al.* 1993, Chaib-draa and Paquet 1993, Chandrasekaran and Josephson 1999, Cheek and Eamon 1999, Cisneros *et al.* 1996, Coradeschi *et al.* 1996, Dompke *et al.* 1999a, Dubas *et al.* 1999, Eggleston *et al.* 2001, Edwards and Stytz 1996, Fogel *et al.* 1996, Funge 1999, Galanis *et al.* 1997, Gamma *et al.* 1995, Gelenbe 1999, Goldman 1996, Gonzalez and Ahlers 1995, Goss *et al.* 1999, Gratch 1996, Gratch *et al.* 1999, Harmon 1998, Hepplewhite and Baxter 1996a,b, Hicinbothom and Lyons 1999, Hieb *et al.* 1995, 1996, Hille *et al.* 1994, Hill *et al.* 1998, Hoagland *et al.* 1998, Jackson 1996, Johnson 1994, Jones *et al.* 1993, 1996, 1999, Jones 1999, Keirsey *et al.* 1994, Kocabas *et al.* 1995, Kocabas and Oztemel 1998, Kuokka 1993, Laird *et al.* 1995, 1998b, Lannon *et al.* 2001, Laughery and Dahn 1999, LaVine *et al.* 1999b, McCoramack and von der Lippe 1999, McIlroy and Heinze 1996, Murray *et al.* 1999, Naikar *et al.* 1999, Pongracic *et al.* 1997a, Powell 1998, Pratt *et al.* 1994, 1999, Psotka 1998, Puckett *et al.* 1999, Rajput *et al.* 1996, Rosenbloom *et al.* 1994, Rubinoff and Lehman 1994, Sale *et al.* 1999, Schwamb *et al.* 1994, Severinghaus *et al.* 1998, Tolk 1999, Tuttle *et al.* 1999, Tyler *et al.* 1998, Vrablik *et al.* 1998, Weaver and Mullen 1994, Williams *et al.* 1998a,b, Zachary *et al.* 1998). Behavior modeling is also an important issue and has been addressed often (Adamson and Joshi 1996, Adkins 1996, Archer *et al.* 1999, Allsop *et al.* 2001, Banks *et al.* 1998a,b, Banks and Stytz 1999b, 2000, Baxter and Horn 1999, Belyavin 1999, Bimson *et al.* 1994, Booker *et al.* 1998, 1999, Bowden *et al.* 1995, Burke 1998, Calder *et al.* 1993, Calder 1998, Calder and Drummey 1999, Carr and Roberts 1998, Ceranowicz *et al.* 1994, 2000, Chandra 1997, Chandrasekaran and Josephson 1999, Cisneros *et al.* 1996, Coradeschi *et al.* 1996, Courtemanche and Monday 1994, Courtemanche and Ceranowicz 1995, Courtemanche *et al.* 1995, Cunningham 1994, Devro and Gonzalez 2001, Dompke *et al.* 1999a,b, Deutsch 1993, Eggleston *et al.* 2001, Ehrhart and Bigbee 1998, Feinerman *et al.* 2001, Fogel *et al.* 1996, Franceschini and Cox 1998, Franceschini *et al.* 1998, Funge 1999, Gagne 1995, Galanis *et al.* 1997, Gamma *et al.* 1995, Garcia-Crespo *et al.* 1997, Gillis 1998, Gillis and Hursh 1999, Glenn *et al.* 1998, Gonzalez and Murillo 1999, Goss *et al.* 1999, Gratch 1996, 2000, Gratch *et al.* 1999, Harmon 1998, Heinze *et al.* 1999, Hepplewhite and Baxter 1996a,b, 1998, Hicinbothom and Lyons 1999, Hicinbothom 2001, Hille *et al.* 1994, Hill *et al.* 1998, 2000, Holmes 1996, Hudlicka and Billingsley 1999, Ioerger *et al.* 2000, Jones *et al.* 1994, 1996, 1998, Jones and Laird 1994, Karr *et al.* 1995, Karr and Holbrook 1999, Katz 1993, Keirsey *et al.* 1994, Kenny *et al.* 1996, Kim and Hill 2000, Kocabas and Oztemel 1998, Koenig and Liu 1999, Kraus *et al.* 1996, Lacy and Dugone 2001, Laird *et al.* 1987, 1994, 1998a, Landweer 1993, Lannon *et al.* 2001, LaVine *et al.* 1993, 1999b, McCauley *et al.* 1998, McCoramack and von der Lippe 1999, McIntyre and Middleton 1995, McIlroy and Heinze 1996, McKenzie *et al.* 2001, Meliza and Vaden 1995, Might and Cannon 1999, Naikar *et al.* 1999, Page and Widdowson 1999, Parsons 1994, Pongracic *et al.* 1997b, Powell 1998, Psotka 1998, Rajput and Karr 1996, Rajput *et al.* 1996, Roberts *et al.* 1999, Roberts and Crawford 2000, Rosenbloom *et al.* 1994, Sapaty 1996, Schwamb *et al.* 1994, Steuart and Murray 1996, Takvorian 1998, Tambe and

Rosenbloom 1994, Tambe *et al.* 1995a,b, Tambe 1996, Tolk 1999, Tyler *et al.* 1998, Tyler 2000, VonderLippe and Courtemanche 1998, 1999, von der Lippe *et al.* 2000, Vrablik *et al.* 1998, Wallace *et al.* 1999, Weaver 1993, Weaver and Mullen 1994, Williams *et al.* 1998, Wu and Yoshikawa 1997, Yufik *et al.* 1997, Zachary *et al.* 1998, Zhang and Hill 2000, Zimm 1999). In knowledge acquisition and representation, as in the case of reasoning mechanisms, no general consensus has been developed regarding solutions to the issues that are involved, but this may be due to the nature of the application-specific solutions proposed and to the stringent demands of the problem domains that were investigated. Because it is difficult to estimate the performance of a human behavior model or decision system before it is implemented, a number of projects have been initiated to compare and evaluate human behavior models and decision systems (Tenney and Spector 2001, Gluck and Pew 2001a,b, Feinerman *et al.* 2001, Banks and Stytz 2001, Rigg *et al.* 2000).

To aid in the specification of behaviors and scenarios, a number of computer languages and interfaces have been proposed (Aronson 1994, Balzer 1996, Berkowitz 1996, Coulter and Laird 1996, Cox and Flaglien 1999, Goldschen *et al.* 1998, Jones 1999, Jones *et al.* 1999, Lattimore and Reicken 1993, Ogata *et al.* 1999, Pittman *et al.* 1996, Pratt *et al.* 2000, Smith 1993, Towne 1999). Several teams have also addressed the difficulties and issues encountered when validating human behavior models and satisfying fidelity requirements (Cross and Ronnquist 1999, Dompke *et al.* 1999b, Gonzalez and Murillo 1999, Harmon and Youngblood 1999, Laird and vanLent 2001, Laird 2002, Kaminka *et al.* 2002, Pace 1999, Roza *et al.* 1999) and their reports indicate the difficulties and uncertainties involved. A number of researchers have addressed issues related to threat system generation as well (Allen and Kalphat-Lopez 1998, Carver *et al.* 1996, Carr and Roberts 1998, Coradeschi *et al.* 1996, Courtemanche *et al.* 1995, Fogel *et al.* 1996, Kraus *et al.* 1996, Oakes 1999). A related issue is the modeling of the military command and control processes that the CGAs must operate within (Baxter and Horn 1999, Bimson *et al.* 1994, Bokma and Slade 1993, Bouche *et al.* 1999, Burke 1998, Calder 1998, Calder *et al.* 1999a, Carr and Roberts 1998, Cunningham 1994, Davies and Gabrisch 1995, Dompke *et al.* 1999, Ehrhart and Bigbee 1998, Foss and Robinson 1999, Ge *et al.* 1995, George *et al.* 1998, 1999, Gillis 1998, Gillis and Hursh 1999, Gratch 1996, Grama *et al.* 1998, Grant *et al.* 1999, Harmon *et al.* 1994, Hartzog and Salisbury 1996, Hicinbothom and Lyons 1999, Hill *et al.* 1998, 2000, Howard and Lee 1998, Karr and Holbrook 1999, Kenny *et al.* 1996, Lankester 1995, Liang *et al.* 2001, McCauley *et al.* 1998, Mall *et al.* 1995, McKenzie *et al.* 2001, Parsons 1994, Salisbury *et al.* 1995, Sapaty 1996, Sherman 1994, Tambe 1996, Tolk 1999, Tyler 2000, von der Lippe *et al.* 2000, Vrablik *et al.* 1998, Vrablik and Longfritz 1999, Weaver and Mullen 1994, Williams *et al.* 1998, Zachary *et al.* 1998). Here again, a wide variety of approaches have been examined for architectures, approaches, reasoning systems, models of military hierarchy, order generation and dissemination, intelligence gathering, and command and control. Planning and inter-CGA coordination are important issues for CGAs, unfortunately little work that has been reported (Allen and Kalphat-Lopez 1998, Burke 1998, Carey and Kleiner 1998, Eggleston *et al.* 2001, Ehrhart and Bigbee 1998, Gillis and Hursh 1999, Hepplewhite and Baxter 1996a,b, Hill *et al.* 1998, Kocabas and Oztemel 1998, Oztemel and Kocabas 1996, Pongracic *et al.* 1997a,b, Steuart and Murray 1996, Tambe and Rosenbloom 1994). As in most areas of CGA technology, no consensus has emerged but there is clearly no superior approach.

5.2 Software Architectures

A software architecture describes the parts, data and functions of a system or application, including the components, objects, composition, control flow(s), data flow(s), interconnections, and functionality of a system. To address the challenges associated with development of CGA and to support hosting of a variety of decision mechanisms, a number of software architectures have been developed. In this section of the chapter we discuss some of the architectural approaches that have been pursued.

Two early CGA architectures are described by Braudaway (1993) and by Bokma and Slade (1993). Both of these systems use a blackboard paradigm. The significant difference between the two systems is that Braudaway's system uses a single blackboard for all of the CGAs hosted on a single computer host whereas the Bokma and Slade system uses a distributed blackboard system wherein each CGA on a host has its own blackboard. In Braudaway's system, CGAs are expert systems and they perform both human-like decision-making and algorithmic processing. In Braudaway's approach, the decision-making component of the system is based on a task-oriented hierarchical behavioral model and has a limited capacity for simulating the collective behavior of a number of humans. The decision-making component monitors the simulation environment, makes decisions, determines actions based upon the decisions, and issues orders. The blackboard component of the system is used to coordinate all of these activities and coordinate the agents that work on each activity. The blackboard has three components, the blackboard data structure, knowledge sources, and a controller. Each knowledge source is a self-contained specialist for a particular sub-problem and operates as an intelligent agent. The agents use the blackboard to communicate with the other agents. The agents operate independently, thereby permitting additional agents to be added to the system without changing existing knowledge sources. A knowledge source in the blackboard can be implemented as either a rule-based program, a procedural program, a logic program, or a functional program.

The Consensus Method, described by Bokma and Slade (1993), also takes an agent and blackboard based approach to implementing CGAs. Their architecture is designed to operate across a network and consists of knowledge sources, local databases, local controllers, and communication channels at each host computer. At each host computer, there are a number of independent intelligent agents, called knowledge sources, that share a local blackboard. The execution of the intelligent agents at each computer host is managed by a local controller at each computer host. The local controller also manages communication between hosts. Communication occurs using pre-defined communication channels, ad-hoc communication between agents is not allowed. Like the system described by Braudaway, the Consensus approach is oriented toward the simulation of units (groups such as platoons or battalions) and not toward the simulation of individuals.

The ITEMS (Interactive Tactical Environment Management System) system described by Siksik (1993), (Labbiento 2000) models systems and intelligence for actors within a DSE/DVE and was designed to simulate Army company-level activities within a battlespace. The ITEMS approach to achieving realistic behavior employs expert system technology. In ITEMS, actors are assembled from libraries that specify sensors, weapons, communications, dynamics, and knowledge. To perform decision-making, ITEMS uses rule-based expert systems in conjunction with frames and the system partitions the decision process into four types; 1) mission, 2) opponent selection, 3) player action, and 4) command and control. Within ITEMS, knowledge is encapsulated in frames, with each frame responsible for representing facts about a specific type of knowledge. To reduce the knowledge representation problem, inheritance is used to permit specialization of knowledge from a frame that represents general knowledge.

Calder *et al.* (1993), discuss ModSAF (Modular Semi-Autonomous Forces), which is composed of a set of extensible software modules that were identified and defined using object-oriented techniques. The ModSAF hierarchical architecture has six main functional areas; 1) simulation world state maintenance, 2) persistent object database, 3) simulation modules to define entities, 4) terrain database, 5) graphical user interface, and 6) support modules. The persistent object database is used to share command and control information between CGAs and, in conjunction with the persistent object protocol, to send command and control information to actors at other computers on the network. The ModSAF architecture is designed to enable the accomplishment of tasks. Tasks are defined as any behavior performed by a unit or individual on a battlefield. There are five types of tasks in ModSAF; 1) unit tasks, 2) individual tasks, 3) reactive tasks, 4) enabling tasks, and 5) arbitration tasks. In ModSAF, tasks are accomplished using asynchronous finite state machines. A task is represented by a finite state machine based task model that defines the specific task states plus a state to define processing to perform when a task completes and a state to define operations to perform when a task is suspended. The task model uses four data structures, called system parameters, task parameters, shared task state, and local task state, that contain all of the information needed to define a specific task, its execution, and its ongoing state. Because the ModSAF architects wanted to be able to compose tasks of arbitrary complexity, task frames are used in ModSAF to group related tasks that execute at the same time. In ModSAF, a mission for a CGA is a collection of task frames and each consecutive task in the collection is started by the completion of a specific task within its previous task frame. Because tasks and task frames can be complex and a task may be suspended so that a task of higher priority can be performed, ModSAF provides task frame stacks for each ModSAF actor (unit or individual).

The Distributed Mission Training Integrated Threat Environment (DMTITE) Project (Banks *et al.* 1998a,b, Banks and Stytz 1999a,b,c,d, Stytz and Banks 1998a,b, 1999a,b,c, 2001, Stytz *et al.* 1998) was undertaken to develop aircraft threats for military DSEs/DVEs and that interact with human-operated and computer-controlled actors in a realistic fashion. The DMTITE architecture is a data-handling architecture that uses object-oriented techniques, component software[2] (Beugnard *et al.* 1999, Brown and Wallnau 1998, Cicalese and Rotenstreich 1999, Digre 1998, Meyer and Mingins 1999, Wang *et al.* 1999) object frameworks[3], containerization (Stytz *et al.* 1996a), and design patterns[4] (Bohrer *et al.* 1998, Cooper 1998, Gamma *et al.* 1995, Keepence and Mannion 1999) technologies. The architecture supports the incremental growth of the software system and knowledge bases as requirements and the design evolve (Coplien 1999, Donohoe 1999, Herbsleb and Grinter 1999, Shaw and Garlan 1996, Woodward 1999). Actor data processing is accomplished in two stages, computations of the physical world state in the Physical Representation Component (PRC) and reasoning upon the state in the Cognitive Representation Component (CRC). The reasoning outputs are then used to control the actor and to generate outputs for the DSE/DVE. Components define the major parts of the DMTITE architecture, objects flesh out the specification, design, and implementation of the components. Frameworks knit the components of the architecture together. Two levels of frameworks are used. One framework is at the highest level of the architecture (Common Object DataBase (CODB), sub-CODBs, and containers) and the second level framework supports the individual actors. The highest level framework provides the information routing and data management services required by the major system components. The second level framework provides a set of services that all CGAs require. For decision-making, each individual DMTITE CGA accesses a knowledge base that was assembled specifically for its actor type. The information accessed by an actor is determined by the fidelity level and skill level specified for the actor.

---

[2] A software component can be defined as being a nontrivial, nearly independent, and replaceable part of a system that fulfills a clear function in a well-defined architecture.

[3] A framework provides a support infrastructure, interface standards, and execution scaffolding for implemented components and objects.

[4] Design patterns are catalogues of useful, tested methods for writing effective object-oriented programs.

Mall *et al.* (1995) present a computer-generated command entity that can operate autonomously within a DSE/DVE and can produce judgmental behaviors. Their system uses an architecture that supports the integration of heterogeneous reasoning systems. The system's knowledge base interconnects two representations of the battlespace, one is a semantic net and the other is a spatial representation of the battlespace. The semantic net represents the symbolic relationships between objects in the battlespace and the map captures the spatial relationships and tactical influence factors. Reasoning in the system is accomplished using intelligent agents that are specialized to accomplish different reasoning tasks but use the same knowledge base. The components for each agent consist of a reasoning mechanism, specialized knowledge, communication facilities, and knowledge base access methods. Communication between agents is supported by the semantic net and direct one-to-one communication. The reasoning system used for each intelligent agent is selected based upon its assigned area of expertise.

Calder *et al.* (1996), discuss a software architecture for a command forces (CFOR) command entity (CE) system hosted on ModSAF that simulates company level and above commanders by performing the planning, executing, tracking, and re-planning activities performed by real-world commanders. The CFOR exists as a separate process on its host computer and assumes that higher-level commanders and subordinates exist as separate processes as well. The main software components of their system are the Situation Awareness class, the Terrain class, the Planner class, the Mission Tracker class, and the Event Processor class. All communication between the command entity and higher and lower-level forces is accomplished using the Command and Control Simulation Interface Language (CCSIL). In their architecture, general knowledge is contained in the generic base classes and domain-specific knowledge is contained in the derived classes. Execution within the CE software is event driven. Knowledge is organized into frames, which they call events. Events are managed by the Event Processor.

The Event base class defines a generic interface to an event and allows the CE to determine if an event has occurred, what is expected to be true about the state of the world when the event occurs, and the actions to take when the event occurs. The Situation Awareness class receives all incoming CCSIL messages and extracts the relevant information from them to update the world representation. The Terrain class is used to compute mobility corridors, battle positions, and routes. The Planner class generates, evaluates, and selects courses of action using the knowledge contained in Constraint Sets. The system has two types of Constraint Sets (CS), component CS and composite CS. A component CS is used to evaluate alternatives and generate solutions to an individual component of a mission. A composite CS is a collection of component CSs that, when linked, form and define a mission. Once a plan is formed, it is represented within the system using an execution matrix. An execution matrix is a time-phased list of tasks that must be performed to accomplish a mission.

Reece and Kelly (1996) describe a software architecture for individual human CGAs in a DSE/DVE. Their layered architecture has four levels; 1) physical, 2) feedback control, 3) action selection, and 4) problem solving. The physical level manages the physical interaction of the CGA with the DSE/DVE. All of the data and procedures needed to portray the individual actor are defined at this level, including the CGA's motion model. Their perception model for the human includes a visual and an aural component; the visual model represents primary and peripheral fields of view, whereas the hearing model is a simple propagation model that considers sound volume. Weapon aiming and firing are handled at the physical level and are treated as atomic actions. The physical level also contains a combatant fatigue model. The control level provides feedback control to the activities/actions in the physical level that require feedback, which in their initial implementation is limited to motion control. At the action selection level, knowledge is used to select actions to be performed. Decisions at the action selection level initiate all physical and problem solving activities. The action selection level consists of a hierarchy of subtasks that permit the entity to decompose its tasks into primitive activities, where a task is domain-specific knowledge about action. The action level also contains knowledge about what subtasks can accomplish a task in a given situation, how to select subtasks, and an inference engine. The problem solving level performs planning.

Butler (1998) discusses the Joint Simulation System (JSIMS) software architecture and their approach to composability. JSIMS is based upon a reference architecture consisting of three tiers; the virtual tier, the model tier, and the physical tier. The virtual tier contains the description of the synthetic environment, performs rendering, and handles networking tasks. The model tier organizes the model. The physical tier describes the physical resources, such as computers, networks, facilities, and software, that support the operation of the simulation environment. The two key concepts in the JSIMS approach to composability are the means of composability and the levels of composability. The means of composability addresses the technologies used to achieve composability, which are assembly, extension, and parameterization. Assembly refers to the process of putting together a system from pre-existing parts. Extension refers to extending the functionality of an existing system, and is accomplished by overriding functions and by selective feature activation. Parameterization allows JSIMS to compose systems by altering system parameters. JSIMS has three levels of composability, which are the component, package, and simulation levels. JSIMS uses a core software subsystem to support the architecture and to support composition; basically all composition efforts are based upon attaching changed parts to the stable core. The layers of the architecture are the basic unit of decomposition, with the top layer devoted to providing components for mission space modeling, system lifecycle management, real-world system interfaces, and applications adopted from any of the modeling and simulation resource repositories. The system core software forms the framework for the system and provides the base classes that can be re-used. To aid in achieving composability, JSIMS provides a Common Object Model that describes how models fit together, the legal model combinations, interoperability conditions, and the limitations that apply to each model.

Because of the computational cost of modeling human behavior and the current inability of ModSAF to portray degradation in human performance due to environmental effects, LaVine *et al.* (1999a,b) proposed a client-server approach to addressing CGA behavioral representation. They advocate a client-server approach because it permits improvements to behavioral representation without requiring integration directly with existing software and it does not affect the performance of the CGA software. In their approach, they employed ModSAF as the baseline CGA software and transferred ownership of certain CGA attributes to a CGA server. The server computes the visual and operational performance changes that should result from world-state changes. When the computations are complete, the server returns the computed appearance and operational parameters to the entity that initially subscriber to the server and requested the computations.

In general, software architectures have gradually evolved from a state where the software architecture is closely tied to the reasoning system to a point where the architecture design and implementation are independent of the supported reasoning system(s). Most architectures rely on the programmer to be very familiar with the design and implementation in order to make changes to the implementation and they all seem to have high coupling and a low degree of information hiding. There does seem to be a move toward systems with greater encapsulation of functionality/objects and a more component-based approach. The most popular current architectural approach is the use of software modules as the foundation for the architecture in ModSAF. Software agents continue to gain in popularity as an architectural solution. Software layering, sometimes used in conjunction with object-orientation or components, has been used by a few systems to help control the complexity of the software. Nevertheless, the question of which architectural approach(es) best suit the CGA domain remains an open problem.

6 Major CGF Systems and Decision Mechanisms

In this section we examine some of the significant reasoning system architectures that have been used in CGAs. A reasoning system architecture must describe the type of reasoning methodology employed, how information is provided to the decision mechanisms, knowledge storage, planning, and how the actions to be performed as a result of decisions are circulated to the rest of the application. In this section, we will examine the most important aspects of several prominent reasoning system architectures and the decision mechanisms employed by each. Some of the major systems were discussed above in the architectures subsection, those discussions will not be repeated here.

TacAir-Soar (Tambe *et al.* 1995a,b, Jones 1999) provides a broad suite of capabilities for aircraft CGAs. TacAir-SOAR is a symbol-processing, rule-based system constructed upon the SOAR architecture. SOAR uses production rules to store long-term knowledge, rules are also used to propose, select, and apply operators. All tasks in SOAR (Newell *et al.* 1991, Rosenbloom *et al.* 1988, 1993, Laird *et al.* 1987) are treated as attempts to achieve goals in a problem space(s). Each problem space consists of a set of states and a set of operators. States represent situations. Operators act as goals, which are decomposed by SOAR's rules into primitive operations that execute as soon as they are discovered. The operators perform the basic acts in the system. Operator selection, application, and termination are determined by the system's knowledge. Whenever operator-selection knowledge is insufficient to permit selection of the next action, an impasse occurs. The impasse leads to the creation of a subgoal to determine the operator that should be selected next. If an operator is too complex, an impasse also occurs, which leads to the creation of a subgoal to resolve the complex operator. SOAR's dynamic goal decomposition scheme leads to the dynamic creation of a goal hierarchy. The goal for the TacAir-SOAR project is to produce humanlike behaviors in pilot CGAs that operate within military DVEs. The TacAir-SOAR system only provides the decision-making and human behavior modeling capabilities. All other capabilities required for the system to participate in a DSE/DVE are provided by ModSAF, including the network interface, dynamics models, sensor models and dead reckoning.

The challenges of managing interacting goals to achieve appropriate behaviors using SOAR have also been addressed by Jones *et al.* (1994). In their view, the complexity of the CGA operational environment inevitably forces a CGA to attend to a large number of goals at the same time. In the course of their work, they identified five issues that must be addressed by a CGA software application. These issues are representation of: 1) agent goals, 2) agent goals, 3) reasoning about the effectiveness of different actions for achieving combinations of goals, 4) learning, and 5) knowledge base updating. As a result of these considerations, a major question they had to address was how to map agent goals into the system architecture. The solution they adopted was to use a mixed representation for agent goals, some goals are represented explicitly and other are represented implicitly. The architectural mechanisms in SOAR act to support the explicit goals, the implicit goals are achieved using behavior modifiers. They examined two approaches to mapping goals, in one all agent goals are mapped into the architectural goal hierarchy. In the other approach, no goals are mapped and all agent reasoning takes place in service of a single, broad architectural goal; all other goals are descriptions of the agent's current situation.

One of the earliest discussions of CGA reasoning system architectures is the hybrid system described by Chaib-draa, Paquet, and Lamontagne (Chaib-draa and Paquet 1993). Their architecture, called RPD, allows an individual agent to combine reactive, planning, and deliberative (RPD) decision-making. The RPD decision-making architecture has five principal components; 1) perception-action, 2) recognition and identification, 3) decision-making, 4) planning, and 5) knowledge base. A procedural reasoning system determines the procedures to run. A rule based reasoning component is used to recognize situations and classify them in terms of the CGA's goals. A case-based reasoning component is used to adapt old situations to meet new demands. Decision-making using cognitive maps is employed to reflect decision maker beliefs. Case-based reasoning is used for re-planning. The architecture has three layers, one layer each for reactive reasoning, planning, and deliberative reasoning. The reactive level is used to tie perception directly to action and to tie perception to identification to action. The planning level is used to mediate the connection between perception and action and supports handling familiar situations.

The architecture developed by Becket and Badler (1993), uses a four level hierarchy of asynchronous control levels to model behavior. Symbolic reasoning is used at the top of the hierarchy and behavioral control is used at the bottom. Symbolic reasoning is used at the top of the hierarchy to perform abstract reasoning and planning and operates as an incremental planner. Behavioral control is used at the bottom of the hierarchy to provide motion control and to perform precise tasks. The hierarchy enables the system to address the individual shortcomings of the two approaches. For example, symbolic reasoning is brittle and inefficient for use in imprecise tasks. On the other hand, behavioral control lacks the capability to provide long-range planning and can not handle abstract goals. Below the symbolic reasoning level in the hierarchy lies the object-specific reasoning level. The object-specific reasoner is used to resolve issues of intention that arise when applying an action to a situation. The third level executes reasoning about behavioral abstraction and couples the higher levels and their planning functions to the lower levels wherein actions are performed. The fourth, and lowest level, executes perceptual, control and motor behaviors. The levels of the hierarchy operate asynchronously but are only partially de-coupled, each level is required to monitor the output of the levels below it.

Carter *et al.* (1993) present an architecture and a testbed for agents that can employ multiple learning strategies based on genetic algorithms. In their use of genetic algorithms, the gene that they use consists of three segments; 1) memory, 2) reasoning, and 3) instinct. The architecture consists of six primary modules, 1) the genetic algorithm (GA) module, 2) the environment module, 3) the control module, 4) the case-based reasoning module, 5) the rule-based reasoning module, and 6) the instincts module. The control module selects actors from the GA pool and during execution determines which of the three learning strategies will be used to make the next decision for the CGA. The case-based reasoning module maintains a history of CGA operations within the testbed and compares the case library with the current situation for the three closest matches with the history. These closest matches are then proposed as solutions for the current situation. The rule-based reasoning module uses a set of rules to determine the solution for the current decision. The instincts module contains encoding of reactive responses to basic world situations.

McIntyre and Middleton (1995) argue that CGAs must provide correct behavioral cues to the human and other CGA participants in a DSE/DVE. To provide these cues, they assemble scenarios using decision trees to represent the options that a CGA can execute in response to the actions of other actors in the DSE/DVE. Their work was performed within the context of the Integrated Unit Simulation System, IUSS. In IUSS, a mission is represented as a task network, where the tasks follow the form defined in the Battlefield Operating

Systems/Tasks in the Army Training and Evaluation Program (BOS-T/ARTEP). This form of task representation permits the system to dynamically adjust task performance throughout the execution of a scenario, which can be exploited to improve CGA behaviors. In their work, they focus on two types of fidelity as related to behaviors, resolution in kind and resolution in degree. Resolution in kind refers to the types of detail to be included. Resolution in degree refers to the accuracy of measurements. In IUSS, all the decisions and behaviors of the simulated soldiers reduce to a set of yes/no decisions concerning environmental state and appropriate response. These decision processes are arranged in a hierarchy, or decision tree, that serves to provide successively finer resolution of decision/classification choices, down to the level of individual procedures or options. In the tree, the factors that describe the environment are cues to the soldier's behaviors and the behaviors are specified by the mission task structure. In IUSS, the decision trees hold scenario mission descriptions that provide behavioral detail down to the level of individual taskings, taskings are supplemented/augmented by first level (leaf) decision trees that define individual behaviors within a task.

As noted by Calder and Chamberlain (Calder 1998), as simulation exercises grow in complexity and scope there is a corresponding need to represent more diversified operations and to improve the ability of command agents to perform mission planning and adapt to changing situations in the DSE/DVE. In their approach, each CE is encapsulated within a single component-scale object that is composed of a number of objects that perform commander functions. The embedded objects perform communications, process events, plan progress monitoring, plan context maintenance, perform replanning, and perform terrain analysis functions. Using these embedded objects, a CE can perform a variety of commander functions. They stress that development of command agents requires extensive access to subject matter experts. Calder and Chamberlain report that a key capability for their future CE framework would be provision of composability at all levels, especially composability of behaviors.

Howard and Lee (1998) discuss the architectural features of the Canonical Commander Model (CCM) they developed. The CCM extends ModSAF and inserts CGA commanders into military simulation environments. The architectural features of the CCM system include a modular design, parameterized software modules, an embedded expert system, and an execution-time coordination table. Their architecture has three levels, perception, reasoning, and action. At the perception level, sensor input and messages are used in conjunction with prior commands to perform analysis. The reasoning level performs planning and action selection based upon perceptions and prior planning using a fuzzy logic server. The action level uses the action selection and planning outputs to generate commands and operations orders for subordinate units. Planning is performed within a single planning unit that uses a hierarchical task network representation that can plan to achieve multiple goals with multiple constraints against an ongoing adversary that can force re-planning. Re-planning can take the form of either adopting standard operating procedures to address a situation or in re-planning of a portion of the mission. Knowledge is represented within the system using fuzzy logic tables, with each table corresponding to a single decision.

The expense and difficulty encountered in developing and extending knowledge bases led to the concept of composable behaviors for CGAs, as discussed by von der Lippe and Courtemanche (1998). Their project, called Composable Behavioral Technologies (CBT), allows users to compose custom behaviors for CGAs from a repository for behaviors. Their system consists of five components; 1) Behavior Repository, 2) Behavior Engine, 3) Semi-Automated Forces (SAF) Specific Services, 4) Execution Engine; and 5) Execution Tool. These components exist and execute independently of the CGA so that they can provide greater re-usability.

The only architectural location for CGA dependencies is in the SAF Specific Services component. The Behavior Repository provides persistent storage for behaviors. The Behavior Editor allows the user to construct a Behavior Representation Grammar (BRG) graphically using a logic diagram approach. The Execution Tool allows a user to create a mission and schedules the execution of behaviors. The SAF Specific Services component provides synchronous and asynchronous communication between the CBT and the CGA system that it is supporting. The BRG contains the instructions for executing composite behaviors and implements a control-flow behavior model.

The quest for realism in the behavioral modeling of command entities is the focus of the work reported by Gillis (1998). Gillis presents the Command, Control and Communications Simulation (C3SIM) system and a human behavior model hosted within this constructive simulation. Gillis argues that any simulation representation of unit behaviors must consider combat psychology, which includes the stress of battle, fatigue, time pressure, danger, individual differences, and level of training. Unit behavior must also represent superior performance brought about by individual differences and superior training. To support achieving the capabilities Gillis sets as his project's objectives, the behavior model used by Gillis is based on a cognitive reservoir theory that maintains the number of effective performance units for the entity. The model uses an effectiveness ratio computed using the product of interactions between battlefield stressors, circadian rhythm, performance use, confidence building events, training, and experience. The effectiveness ratio is then divided by the number of potential courses of action, which yields a probability of a correct decision. A timing variable is also used in conjunction with the effectiveness ratio to determine the time lag to complete/perform situation awareness activities and for course of action selection. The model also incorporates physical stressors and task stressors. The stressors are moderated by training, personality, unit cohesion, leadership, and social supports.

Hill *et al.* (1998) (Jones *et al.* 1996) describe the application of the SOAR architecture to rotary wing aircraft (RWA) Army helicopter company simulation. RWA SOAR uses ModSAF for communication and helicopter activity. The RWA agents control the helicopter, manipulate simulated lasers and other weapons, sense the terrain, perform terrain avoidance, and communicate with peers and superiors using the Command and Control Communications System Interface Language (CCSIL). The architecture for the agents is fundamentally the same as the SOAR architecture, where tasks and goals are represented as operators. Because the humans in a helicopter company must cooperate to accomplish their tasks, they implemented a teamwork capability for RWA-SOAR that relies upon team synchronization mechanisms and plan monitoring/repair capabilities to perform team tasks. Planning and plan repair is accomplished using hierarchical task-decomposition planning, with plan representation accomplished using a hierarchical task network (Gratch 1996).

Busetta *et al.* (1999), present the JACK Intelligent Agent framework. The JACK framework is a software framework that uses light-weight agents and strong typing and is built using the Java language. The JACK architecture is based upon the Belief-Desire-Intention (BDI) architecture for human behavior modeling. JACK agents are not bound to a specific communications language, instead they use an object-oriented middleware, such as CORBA, and message-passing structures, such as the Distributed Interactive Simulation (DIS) protocols, to communicate. The authors extended the JAVA™ compiler to convert the extensions into Java classes and statements. Finally, the authors also defined a set of JAVA™ classes to manage concurrency between agent tasks, support multiple agents, enable specification of default behaviors, and permit specification of a communication infrastructure. The BDI architecture used by JACK employs a model of human behavior that considers humans to have beliefs about the world, desires to satisfy, and intentions to act. Therefore, the activity of a rational agent consists in performing the actions it intends to execute until it is forced to re-evaluate its intentions by events in the world. BDI systems differ from purely deductive systems and other traditional reasoning systems in that they model intentionality, which allows it to effectively model behaviors centered on standard operating procedures. In JACK, the mental state is defined by determining the external events that drive the agent, the goals that the agent can set for itself, the beliefs that influence the adoption of plans, and the procedures required to accomplish tasks.

The Synthetic Forces Behavioral Architecture described by Page and Widdowson (1999), has standard interfaces between components so that behavioral models can be re-used. The objective of this work is to define a means for supporting a distributed behavioral model and to provide a capability for communication of goals and world views between CGAs. The Behavioral Architecture defines a mechanism inter-application communication and defines a standard model representation format for the messages used to perform the communication. Within the architecture, which is assumed to be a distributed system, application locations are not fixed on the network and instead applications are given the tools they need to locate the other applications that they need. In general, in their approach every communication between two models requires two translations, one from the first model's representation to the standard format and another from the standard format to the second model's representation.

As the preceding summaries indicate, researchers have investigated a wide number of reasoning system architectures. In general, reasoning system architectures have been developed to limit the decision space and to provide timely information flows for the reasoning system, supporting a single decision-making approach, and enabling composability of behaviors. The more popular/widespread approaches for reasoning system architectures are those that are based on rules and task frames or that use production systems coupled with a hierarchy of operators. Decision trees and case-based reasoning have also been investigated, usually in conjunction with task frames. Few of these architectures support learning by the reasoning system. In the effort to achieve composition of behaviors and to control the complexity of decision systems, some have explored the use of primitive behaviors that are then scaled or combined to achieve complex behaviors. Other approaches to composability have included separating the behavior mechanisms from the remainder of the CGA's decision mechanisms and then using a grammar to structure the composition. However, to date few researchers have addressed the issue of how atomic/elementary behaviors should be identified. The most prevalent approach is one where behaviors are partitioned using an object-orientation.

7 Human Behavior Modeling, Skills Modeling, Decision Modeling, and Artificial Intelligence Technologies

Human behavior modeling addresses the problem of computing realistic behavior and reactions for a CGA by developing behavior models that can be used to compute a reasonable analog of the output of the human decision-making process at a human scale rate of time. Human behavior modeling requires the acquisition of knowledge about the domain models and information that humans use in the decision-making process, such as doctrine, tactics, and knowledge models. In this section we discuss human behavior models that have been developed.

A human behavior model must have the following characteristics: 1) correct output modeling at a human-scale rate of time, 2) unpredictability, and 3) certifiability. A central component of achieving these capabilities is doctrinal fidelity, which addresses the need to represent the doctrine and tactics of the modeled actor, whether it is a single person, an aggregate, or a command and control element. Acquiring information about the doctrine is merely the first step in this process, the doctrine must also be represented in a computable manner and must be certified as to its correctness. The first component, correct outputs modeling, forces the CGA to react at a human-scale rate of time and requires that the decisions appear to be made by a human. Unpredictability addresses the need for CGAs to behave so that human opponents can not detect patterns in the behavior and then use these patterns to defeat the CGA. Predictable patterns in CGA behaviors lead to negative training because the human operators will learn to defeat the automated opponent by exploiting these patterns of behavior. The third component of human behavior modeling is that the behavior be certifiable. To be certified, a CGA software system must be able to be measured against and compared to the exhibited behaviors of a human in a comparable situation.

The Action/Cognition Behavior Model (ACBM) described by Landweer (1993), performs three main functions; scenario control, force simulation, and scenario development. What Landweer calls generic functions are used to represent human functions; a generic function allows a CGA to perform a simple activity, complex behaviors are composed from these simple functions. To increase the realism of the behaviors of ACBM CGAs, the actors do not have direct access to the ground truth (actual state) of the simulation world. Instead, each actor has a representation of the DSE/DVE that is based upon its defined perceptual capabilities. The cognitive functions available to an ACBM CGA are 1) notice, 2) digest, 3) react, 4) ponder, and 5) review. The notice function allows the initial recognition of information. The digest function integrates the information into the actor's representation of the DSE/DVE. The react function performs evaluation of the immediate, short-term status of the actor and its resources. The ponder function uses perceptions from long-term memory to evaluate long-term changes in state and resources. The review function discards perceptions that are out of date. To store information and associate the information with a time frame, iconic, short-term, medium-term, and long-term memory are used. Within the ACBM, thinker objects associated with each cognitive function are used to move information from one stage of memory to another.

LaVine *et al.* (1993) address the issue of modeling the performance degradation over time that is caused by environmental stressors. The methodology that they developed to assemble a model has three major parts; 1) a task classification taxonomy, 2) a questionnaire to determine the relationship between performance and physiological response based on stressor and time since exposure, and 3) a task network model. The task taxonomy consists of five skills, which are attention, perception, psychomotor, physical, and cognitive skills. In their model, attention is the ability to attend to a stimulus for extended periods. Perception is the ability to detect and categorize stimuli. The psychomotor skill aspect models the ability to maintain select characteristics of an environment within a desired range of conditions. The physical skill component models the ability to perform muscular work. Cognitive skill is the ability to apply concepts and rules to the environment to generate plans and actions. The ability to perform each skill is rated on a seven-point scale by a subject matter expert using a questionnaire. The data from the questionnaires are used to develop functions that describe degradations in performance of specific tasks over time.

Weaver and Muller (1994) address the subject of modeling military decision-making. Their approach is composed of four parts, the intrinsic characteristics of the decision, the perceived environment, a neural network (for decision-making), and three output variables that scale the decision. The two external inputs into the decision are the intrinsic classification characteristics of the decision and the environment for the decision. These data are then combined using a model of human decision-making to characterize the decision, in their system they use a neural network. They chose neural networks for decision-making because they claim that the neural networks' non-linear characteristics match the characteristics of human decision-making and that neural networks can operate well with incomplete and contradictory data. The neural network operates primarily as the memory for the system. The system has three output variables, the neural network can provide either a specific outcome value or a probability distribution for each of the three variables. The first of the variables is the timeliness of the decision, the second is the quality of the decision, and the third output is the quantitative aspect of the decision.

Parsons (1994) uses fuzzy logic control technology to simulate command decision-making in the Maneuver-Warfare Analytical and Research System (MWARS). In his system, Parsons concentrated on the operational level of warfare and not the strategic or tactical levels. In Parson's approach, fuzzy logic is used to perform analysis on intelligence data and on situation reports. A fuzzy logic controller is used to correlate and to store incoming intelligence data and to convert these inputs into a value called combat power. Combat power indicates what the CGA commander believes to be true about the distribution and effectiveness of enemy forces. To correlate incoming data, the fuzzy logic controller uses time, location, velocity, sensor quality, and combat power values for the enemy forces to compute fuzzy values for report quality and level of duplication, with the lowest quality and highest level of duplication reports being the ones discarded from the intelligence reports database. Using the database, the CGA commander computes route assessment values and area assessment values to assess troop movement routes and enemy strength. Another fuzzy logic controller is used to compute updated assessments from the situation reports. Based on the situation reports, the fuzzy controller computes fuzzy values for mission status and psychological status using values for the current intelligence assessment and the unit status values. The output of the commander's decision-making operation is a set of fuzzy assessment of alternatives. Non-fuzzy rules are used to defuzzify the assessments and convert the assessments into mission orders, location assignments, and air/fire support allocations.

Fogel *et al.* (1996) discuss the use of evolutionary programming for human behavior modeling in CGAs. They employed a Valuated State Space matrix to score potential actions at each time step and to represent the enemy's mission. They use the matrix in conjunction with evolutionary programming to generate alternative courses of action. The mission is represented as a hierarchy of measurable parameters and a normalizing function that operates across the levels of the hierarchy. The Valuated State Space and the normalizing function together indicate what to measure, how accurate the measurement must be, and how to combine the measurements into a single assessment of a situation. The approach also identifies open problems for the CGA and assigns values to prospective solutions. To solve problems, they use evolutionary programming. The authors report indications that the CGA does learn increasingly better behaviors and generates adaptive plans by using evolutionary programming. However, their work was conducted within simple environments with only one actor.

Rajput and Karr (1996), investigated achieving cooperative behavior between entities using a Decentralized Control Architecture (DCA). In their approach, entities cooperate directly without direct supervision by a commander. The commander CGA makes the behavioral decisions for its unit and conveys decisions to subordinates, then each subordinate executes the decision independently. By using this approach, the authors claim that unit behaviors emerge as a result of direct cooperation. In addition, the CGA implementations are modular, which eases the problems associated with verifying behaviors. Their approach to achieving cooperative behavior is based upon finite state machines coupled with the use of events within simulated radio messages and observation to enable cooperation. To minimize synchronization problems that can occur when internal and external events are interleaved for execution, they maintain separate queues for each type of event, with execution priority given to internal events. They report that they can achieve complex behaviors using their approach by assembling a hierarchy of finite state machines that each has their own set of responsibilities and that cooperate and communicate with each other to achieve the objectives and tasks assigned to the commander. As a result, in their system there are capabilities for inter-actor and intra-actor cooperation to achieve assigned tasks.

Adamson and Joshi (1996) discuss the use of genetic algorithms to achieve realistic behaviors in CGAs. The objective of their work was to enable a CGA commander to provide general instructions at a level of detail similar to that in the real world. They report that genetic algorithms can succeed in developing realistic behavior in pursuit of primary goals such as survival and tactical objectives.

Zachary *et al.* (1998) describe an approach to the behavioral simulation of a team of Naval watchstanders to provide real-time evaluation of trainee activities. They use the COGNET representational system coupled with the GINA simulation environment as the foundation for their system. COGNET is a computational cognitive framework that has four parts, a theoretical part, a description language, knowledge elicitation methods, and behavioral/cognitive simulation tools. The COGNET theory proposes that cognitive processes are the result of the operation of a computational mechanism on a set of symbols that represent a portion of the world. In COGNET, human information processing is modeled using a cognitive architecture and a set of symbols operated upon by the architecture. The cognitive architecture separates human information processing into three components; perception, cognition, and motor action. The perceptual process receives information from the environment, turns it into internal symbolic information, and places the symbolic information into the information store. The cognitive process manipulates the symbolic representation using internal knowledge. The internal knowledge indicates how to "chunk" the environment, how to achieve task goals within different contexts, and how to recognize those situations when a given task goal is relevant. The mental model in COGNET is implemented as a series of blackboard panels, in their implementation they used twelve panels with more than ninety levels and 100 perceptual demons. The COGNET/GINA system dynamically generates these human training performance standards by computing behavioral predictions, thereby permitting greater variability in the scenarios used for training.

Jones *et al.* (1998) describe a model of pilot fatigue that was incorporated into the TacAir SOAR system. Their model is based upon research findings that indicate that fatigue impacts response time and accuracy in tasks and that fatigue arises from a sleep loss component, a circadian rhythm component, and a secondary rhythm component. Their model has two functions, one function describes attentional lapses and the other function describes slowing of response time. Fatigue effects were modeled by introducing artificial delays into SOAR processing.

Booker *et al.* (1998, 1999) present a decision-making framework for CGAs that is based upon the concept of motivational control. The concept of motivational control is based upon the observation that internal influences are powerful guides to how some action selection mechanisms operate. Their work is based upon the d-r-k model, in their implementation of the model the strength of a motivation is indicated by the deviation of some internal variables from user-defined preferences and the actor is motivated to reduce deficits by taking action. In the model, given a set of deficits, d, availability parameters, r, and accessibility parameters, k, the optimal solution to reducing the deficits is to formulate the problem as a control problem and compute the d-r-k products for each possible combination of parameters. The optimal solution is the d-r-k product with the largest value. Within the model, stimulus cues can be used to attenuate availability and accessibility values and so simulate preferences and handicaps. In moving the d-r-k model into the CGA arena, they focused on characterizing deficits as doctrinal factors that influence tactical operations. They treat the repertoire of tactical behaviors as resources and threats as stimuli.

The Land Formation Model (LFM) described by Grant *et al.* (1999), models decision-making behavior, physical behavior, communication, and interaction between individuals, between organizations and between the constituent parts of organizations. The model was based upon CABLE, which is a distributed agent architecture that uses multiple, co-operating software agents to represent human interaction. In CABLE, agents communicate via a service-based interface (a service defines a set of methods that can be called by other agents). The LFM architecture provides functional unit agents (FUNs), a coordinator to manage the passage of time, and a presenter to manage the user interface. Data logging and simulation state restoration services are also incorporated into the architecture. Organizations are portrayed using: 1) an Organization Manager model, 2) an Intelligent Behavior model, 3) a Physical Behavior model, 4) a communications model, 5) a logistics model, 6) a movement model, 7) an environment model, 8) a route planner, and 9) a master event list. To interact with the DSE/DVE, an actor creates one or more Actions for physical behavior or for organization behavior. Actions are sent to the agent assigned the role(s) called for by the action, upon receipt the action is processed by the Intelligent Behavior modeler for the recipient agent. This approach separates reasoning about the need for an action from its execution and the recipient of a directive can respond to the Action directive in the same way as to those directives that it generates internally.

Heinze *et al.* (1999), investigated improving the fidelity of individual agents by using a model of human cognition that separates the skill-based components from the knowledge-based components of human behavior. To represent the knowledge-based component they use the belief-desire-intention (BDI) model and for the skill-based component they use machine learning. In the BDI architecture, agent operation is defined by four processes: 1) processing incoming data to develop situation awareness, 2) assessing the situation, 3) selecting a tactic based upon the assessment, and 4) employment of standard operating procedures to realize the tactic. The BDI module is responsible for realizing cognition related to plans, beliefs, goals, intentions, and other cognitive processing. They supplemented the BDI model by adding skill-based perception and skill-based action to the model. Skill-based perception allows the agent to recognize situations based upon previous experience. Skill-based action gives the agent the ability to perform complex sequences of behaviors with little cognitive effort and is responsible for sub-cognitive processes such as experience, memory, recognition, and learning. Because skill and knowledge based activities may have to be executed simultaneously, their architecture directs all incoming information to both the BDI module and the skill module to permit each module to respond to the world state. To enable the agent to react at a skill-level in response to the environment, the BDI module also receives data from the skill module. The data moving from the skill to the BDI module contains situational descriptions that indicate that the skill module has recognized some aspect of the environment. This data is then factored into the reasoning processes of the agent. To enable the skill module to efficiently perform its recognition task, the BDI module provides the skill module with information concerning its mental state.

Because of the complexities involved in modeling human performance and behavior, Laughery and Dahn (1999) and Belyavin (1999) developed the Integrated Performance Modeling Environment (IPME). The IPME builds upon a suite of human performance modeling tools and provides an environment for their integration.

The features of IPME that assist in modeling human performance include the ability to portray environmental variables that shape human performance, the simulation of operator traits and states, the ability to define performance shaping factors, the prediction of operator performance workload measurement, and a task scheduler to schedule operator tasks within a task network. The task network that they use is an extension of task analysis that enables the task network to be used as a predictive model. The performance of the tasks for a CGA are interrelated using a set of shared variables that the user can define, along with the tasks and the task network, using the IPME graphical user interface. IPME can be used to model stressors that arise directly from the environment (such as heat, noise, and vibration) and those that arise form the context of the task (such as fear and sleep deprivation).

Hudlicka and Billingsley (1999) present a methodology for modeling individual differences in human behavior, which they call the Methodology for Analysis and Modeling of Individual Differences (MAMID). The MAMID approach consists of: 1) identification of the cognitive structures and processes that mediate performance, 2) design of a parameterized model for these moderators, 3) identification of the cognitive, affective, and personality factors that mediate performance, and 4) encoding of the parameters and knowledge bases for the behavior model. They argue that a methodology is needed because existing models and simulations generally only portray normative behaviors and so do not portray the full range of variability of normal human behaviors. The MAMID cognitive architecture has an attention process that controls cue detection, a situation assessment process that mediates information integration, a decision-making and action selection process, an action execution process, and an action monitoring process. Within this architecture, two techniques are used to reflect individual behavior. The techniques are the manipulation of model content to reflect desired characteristics and the manipulation of model processing parameters to reflect a particular style of behavior.

The perceived importance of training as a determinant of military performance led to the research reported by Archer *et al.* (1999). They believe that models of humans in combat should reflect the training levels of each human, but the difficulty and complexity of modeling training led them to develop a simpler model. They identified eight skill categories (taxons) that form their skill taxonomy; 1) planning, 2) maneuvering, 3) gunnery/weapons employment, 4) communications, 5) emplacement/displacement, 6) command and control, 7) doctrine and tactics, and 8) navigation. They then identified variables that affected performance in these skills. The variables were chosen based upon several criteria: 1) documented capability for affecting training, 2) ability to address important aspects of training, 3) simplicity, 4) generalizability to all skill categories, and 5) ability to affect the elapsed time and accuracy of performance. To model the affect of training on performance, they consider the peak performance achievable for a task, the contribution toward peak performance made by the type and quality of each type of training, the minimum amount of each type of training required to attain peak performance, the task difficulty, and the elapsed time since training occurred. Because learning is not a linear process, their model reflects the fact that greater performance improvement occurs during the initial training on a task than occurs in later training. To compute the effect of training on performance, they determine each skill relevant to a task and then for each skill compute each skill component's value (percentage of peak performance) based upon the number of hours of training received in each type of training venue (classroom, field, simulator, etc.) for the skill. This computation yields the percentage of peak performance achieved for a skill. To determine overall task performance, each skill value is multiplied by the percentage that the skill contributes to task performance. The resulting weighted skills contributions are summed to indicate the percentage of peak performance achieved on a task. The paper contains a number of algorithms and equations that present their approach in detail.

Hill (1999) presents a model of perceptual attention for human behavior. The need for this type of model arises from the need to improve the modeling of cognition and observable behavior. Perceptual attention modeling requires the consideration of visual behaviors such as search, fixation, tracking, and grouping and that these visual behaviors must be supported by both cognitive and motor behaviors. As the foundation for the model, the author discusses attention modeling because it supports the control of perception and limits the amount of information that must be processed. Focus of attention helps humans to cope with environments where there is too much information for the visual system to process because attention focuses processing power on a specific region of a scene to the exclusion of the remainder of the scene. The author's model of attention treats attention like a zoom lens. In the zoom lens metaphor, there is a reciprocal relationship between magnification/level of detail and the size of the viewing field. Low levels of magnification result in greater field of view and lower level of detail. Conversely, increasing magnification increases the amount of detail but shrinks the field of view.

Powell (1998) investigated performing planning in the face of incomplete information and poorly-defined goals and objectives. Powell notes that humans generally operate successfully within this type of environment, but current CGA systems do not. He reports that the current methodology is a linear progression from information acquisition, goal definition, plan development, plan evaluation, and finally plan execution. The approach that he advocates interleaves these activities and includes activities for sub-problem decomposition, plan critiquing, and plan repair to permit the CGA to partially plan based upon incomplete information and fuzzy goals. Powell recommends limiting the planning software to producing and evaluating only two alternative plans since the experts he observed only produced, at most, two plans.

In conclusion, there has been a wide variety of human behavior models that have been proposed. In general, researchers have tried modeling human behavior by accounting for some combination of attention, intent, situation assessment, decision-making, skills, training, fatigue, environmental stressors, education, experience, and motivation. Very few systems have attempted to model more than a few of these factors and most models are generally assembled using an ad hoc approach instead of employing a strong methodology. Early efforts at human behavior modeling used finite state machines, recently these efforts have been coupled with agents to try to improve the fidelity of the modeled behaviors. Concurrent control and cognition in conjunction with arbitration as well as case-based reasoning coupled with machine learning have also been investigated. Intelligent agents have also been used to model behavior by having one agent responsible for each type of behavior and then arbitrating or scaling their individual outputs. Early implementations generally relied upon a blackboard approach to model cognition.

Recently, neural networks and fuzzy logic have both been evaluated for their ability to capture and portray a wide range of diverse human behaviors. Recent work in human behavior modeling for CGAs has addressed situation awareness, teamwork, skill level, training, learning, and psychological factors to a limited degree. Models of the effects of fatigue have been developed, but most of these efforts have been ad hoc and have not produced a methodology or structure for the model. The cognitive models that have been reported use a variety of techniques such as motivational control, beliefs-desires-intentions, and traits and states. The level of training has been modeled by determining the skills relevant to a task and then assigning the CGA an appropriate level of skill given the level of training that is to portrayed.

Having completed our review of research and progress in the CGA field, we will close the chapter by revisiting the main points in the chapter and by putting forward some suggestions for further work. The next section contains a short overview of the highlights of the chapter, our recommendations for further work, and some concluding remarks.

8 Conclusions and Suggestions for Further Research

This concludes our introduction to the field of computer-generated forces and the major technologies that are used to assemble CGFs. At the beginning of thc chapter we introduced the key concepts and decision-making systems and so provided a backdrop for the remainder of the chapter. Concepts discussed included the use of doctrine, major CGF components, behavior modeling, CGF fidelity, and CGF capabilities. Section two presented a discussion of the basic requirements for CGF systems and the implications of these requirements for CGF performance and operation. The discussion of the CGF requirements pointed out the need for careful selection of decision-making machinery for CGF and for the need for a broad set of capabilities for CGF systems. In section three we reviewed the state of the art for CGFs by examining significant previous research and CGF projects and discussing the significant CGF software architectures that have been proposed. Briefly, the discussion in Section three indicated a broad field of research with significant emphasis given to a few key capabilities, such as human behavior modeling and reasoning systems, that are central to improving CGF performance. The discussion of CGF software architectures also indicated the broad ranging research that has been conducted and the number of alternatives that have been investigated in the effort to improve the quality and reduce the cost of CGFs. The architectures have gradually grown in complexity and capabilities over the last few years, with steady adoption of cutting edge technologies being the hallmark of CGF software architectures.

Section four is the heart of the chapter in that it contains the discussion of the major CGF systems that have been proposed and/or developed. We briefly summarized how the CGF systems operate, the capabilities that they possess, and the decision mechanisms they use. As with software architectures, a wide array of approaches have been investigated in the effort to improve the capabilities of CGFs, with the major approaches being the use of rule-based systems, case-based systems, and finite-state machines for reasoning. Other approaches for decision-making have been investigated but have not seen widespread adoption and their capabilities and utility for CGFs remains uncertain though promising. In section five we presented a discussion of human behavior modeling, skills modeling, and decision modeling as they relate to CGFs. The discussion in this section indicates the need for proper human behavior modeling in CGF and the challenges that must be addressed when trying to model human behavior. As in the other main components of CGF technology, a significant number of approaches have been investigated and no clear-cut superior system has been reported to date. Most systems have modeled only a small portion of the spectrum of human behavior and, to date, no compelling evidence that indicates the priority to be associated with each aspect of human behavior has been reported in the CGF literature.

Before discussing recommendations for future work, we will first briefly summarize the current state of CGA technologies. The current most widespread CGA application is ModSAF, but it does not scale in terms of computing power and lacks the ability to correctly portray complex human behaviors. The approach of using task frames coupled with a finite state machines in ModSAF results in a system that is not robust in its response to unexpected situations. Generally, significant efforts to improve ModSAF behaviors do not yield commensurate improvements in performance because a ModSAF CGA's response is strictly limited to what has been explicitly programmed into it. In ModSAF there is no capability for interpolation or extrapolation of behaviors or for learning, as a result the performance of a ModSAF entity is costly to extend. ModSAF also requires human assistance to monitor the performance of the entities, which is a barrier to achieving realistic behaviors.

Autonomy and fidelity remain important objectives for all CGA systems, and human behavior modeling is an important technology for achieving both of these objectives. However, human behavior models and CGAs in general are not ready to provide autonomy and fidelity in a timely manner for a number of reasons. Chief among these reasons is that the current approach to developing human behavior models does not produce robust, scaleable models. The decision mechanisms and knowledge bases are also narrowly focused and difficult to reuse and are not housed within architectures that support the development or use of hybrid reasoning, comprehensive human behavior models, or composite knowledge bases. Because the size and complexity of the knowledge bases will have to greatly increase to meet demands for improved performance, composability coupled with a learning capability for CGAs appears to be necessary.

The desire for composability, flexibility, rapid construction, verification and validation of systems, coupled with the need for high fidelity CGAs, points to the need for a generalized software and knowledge base architecture. This architecture should support a variety of decision-making systems, enable assembly and composition of knowledge bases, and support experimentation, evaluation and development of human behavior models, decision support systems, information distribution systems, simulations, and CGA systems. The software architecture would be the software skeleton for all of the software components and system capabilities.

As a starting point for developing the architecture, we believe that the architecture should be object-oriented and software component and framework based. A common CGA architecture should support a variety of decision-making systems, enable assembly and composition of knowledge bases, and support experimentation, evaluation and development of human behavior models, decision support systems, information distribution systems, simulations, and CGA systems. The architecture must support a number of needed capabilities. First, and foremost, the CGAs must be able to operate autonomously and independently adapt to new situations and to changes in mission. These new CGAs must possess information fidelity, therefore they must accurately model information flows and information quality within their software so that the decision mechanisms operate upon data whose volume, quality, timeliness, and comprehensiveness matches the characteristics of the data available in the real world. The CGAs must be able to exhibit the natural variability in performance and behavior that arises from the nature of the mission, skill, level of training, and experience on the part of the actor and due to the characteristics of its environment. CGAs must be able to engage in cooperative planning in a manner that reflects skill, experience, and position in the command hierarchy. Finally, CGAs must be able to automatically detect critical or existence-threatening situation(s), automatically respond to the situation(s), and disseminate a warning about the situation(s) to the rest of its organization.

Clearly, an important objective for CGA research is to develop systems that are adaptable to a variety of circumstances and robust in their response to situations as well as exhibit realistic unpredictability and variability in behaviors. These objectives point to a need for representations for intent, deception, and adaptability in CGAs. Redressing current shortfalls will require advances in a number of technologies and capabilities. First is the assembly of knowledge bases that are more comprehensive and that contain "first principle" information in a format suitable for advanced reasoning techniques (such as fuzzy logic) so that the CGA has the information that it needs to respond appropriately and effectively. Extensive knowledge bases must be coupled with adaptable decision-making and behaviors that overcome the brittleness and rigidity of current systems. A promising approach to achieving better reasoning systems is the development of hybrid reasoning systems. A hybrid reasoning system combines different types of reasoning systems in an architecture that supports composition. Additionally, behavior moderators such as morale, fatigue, stress, fear, information operations, and intent must be modeled and included within all types of CGAs. There is also a need for composable and tailorable behaviors; that is, there is a need for behaviors that can be assembled from atomic elements to achieve the objectives of a specific simulation environment. Improved human behavior models are not needed just for individuals, there is a corresponding need for improvement in group behavior modeling. Group behavior improvement must be based upon an accurate modeling of group psychology and group intent as well as the group's mission, objectives, group goals and group tasking. The commander, or leader, of the group must also be modeled. Because low-level military operations are increasing in importance and frequency, human behavior models must be able to perform in low-level military operational environments and effectively perform in operations other than war (OOTW) environments (Fields and Spradlin 2000, Johns and Silverman 2001, Rigg *et al.* 2000, Taylor *et al.* 2001, Wallace *et al.* 2000, Weaver *et al.* 2001).

The first step in improving human behavior models is the development of an ontology and research methodology for the categorization and acquisition of the knowledge needed to develop human behavior models and to combine multiple behavior models into a single system (as discussed in (Banks *et al.* 1998a,b, Banks and Stytz 1999a,b,c,d, 2000)). Ontology development requires a precise definition for all of the terms and phrases used to define a knowledge base and human behavior model. In addition to the development of the ontology and methodology, there is a parallel need for software tools and intelligent agents to support variable resolution modeling, interoperability, and behavioral representation. The desired human behavior models should incorporate behavior moderators, including intent, skills, fatigue, emotion, experience, training, and shared goals. Because there is a need for CGAs at all levels of a command hierarchy, human behavior models that all operate and reason at the strategic level are needed. To achieve this capability, modeling of the political decision making process and strategic reasoning must be performed and the results integrated into upper command level human behavior models.

To produce accurate and robust human behavior models in a timely manner as well as enable model verification a human behavior model development environment is needed. The environment would not be a CGA or human behavior modeling solution, instead it would be used to develop better human behavior models. The development environment should support knowledge acquisition as well as system creation, reasoning system selection, evaluation, experimentation, and development activities. The environment must support the development and use of advanced and hybrid decision-making systems, it should support multiple reasoning techniques, and enable the reasoning techniques used by a CGA to be specialized for certain aspects of its domain. The common modeling and development environment would support all decision-making mechanisms, support experimentation and evaluation of subsystems and systems, support human behavior model development, and enable human behavior model interoperability.

The human behavior model development environment should have intelligent agents to assist the developer in the assembly and evaluation of human behavior models and the systems that use them. To support and complement the capabilities provided by the common development environment, a suite of performance and evaluation metrics as well as evaluation scenarios are needed. We believe surveys such as the one presented here and the selection of systems to use in simulation environments would be more valuable if there were standards against which to measure the performance and capabilities of systems and individual CGAs. This task will be complicated and require a broad-based standardization effort, and may require the development of parallel capabilities for measurement and evaluation for different types of CGAs based upon their reasoning mechanisms, behavior moderators, accuracy of sensor and dynamics models, terrain model, and domain of activity. Some components of CGA performance, such as average time to compute a decision, knowledge base size, active memory requirement, and average agreement with subject matter experts, should be able to be quickly standardized and gathered. Other measures that are specific to a domain, type of behavior moderators, or form of reasoning may take more effort to standardize but would be an invaluable aid in evaluating and selecting CGAs and systems for use in a simulation application.

The research and development conducted in information distribution, decision support systems, CGAs, simulation systems, and human behavior modeling has demonstrated an ability to support a wide variety of needs for analysis and training. However, the current capabilities for systems and human behavior modeling are inadequate to satisfy these needs. Therefore, research into better CGAs, information distribution, decision support systems, simulations, and human behavior models for CGAs and their information distribution, decision support systems, simulations, and constructive models is needed. CGAs must operate realistically without human control or supervision, and humans should only be involved in their operation in the most dire of circumstances. However, the development of CGA autonomy is not possible given the current state of technology and will require a significant, long-term research commitment in many areas.

The selection of the proper decision-making machinery and human behavior models for CGAs for a given application area remains an issue for the CGA community, and there is no clear-cut solution to this problem in sight. However, in light of the difficulties involved in the selection process, some efforts have recently been undertaken to develop metrics and evaluate CGA components with the objective of being able to characterize their optimal performance and application regimes (Gluck and Pew 2001a,b, Gugel *et al.* 2001a,b).

The cost of CGF development and of research remains an issue, and one that we believe significantly impedes progress. However, this issue may gradually resolve itself due too the continued improvements in computing power. In addition, an efficient, cost-effective computing platform for research may be gaming computer platforms, as discussed in by a number of authors (Laird 2000, 2002, Laird and vanLent 2001, Jacobsen and Hwang 2002, Kaminka *et al.* 2002). The low cost, widespread availability, relatively simple interfaces to the gaming software, and capability for visually inspecting the results and operation of the CGA rapidly would make for a rapid development cycle and the capability for conducting experiments that are readily evaluated by subject matter experts.

In conclusion, in the field of computer-generated actors, we believe the technologies of that software architecture, decision-making system, and human behavior modeling are the keys to improving CGAs and fostering their use within more application areas and in fields where the ability of replicate human behaviors must be of the highest quality. We believe that a long-term commitment to developing improved human behavior models, decision-making systems, and software architectures as well as knowledge bases and the systems that use them is needed in order to significantly improve the quality of CGAs. The first steps toward this necessary and worthwhile improvement have been taken, it remains to be seen if the researchers can achieve the degree of success needed to maintain and even broaden support for research and development of CGAs.

References

Adamson, J. (1996), "The CAEN wargame for OOTW applications," *Proceedings of the 6th Conference on Computer Generated Forces and Behavior Representation*, Orlando, FL, 23-25 July, pp. 49-56.

Adamson, J. and Joshi, K.G. (1996), "Genetic algorithms and force simulation," *Proceedings of the 6th Conference on Computer Generated Forces and Behavioral Representation*, Orlando, FL, 23-25 July, pp. 237-242.

Adkins, M.K. (1996), "Polling vs. event-driven Computer Generated Forces (CGF) architectures," *Proceedings of the 6th Conference on Computer Generated Forces and Behavioral Representation*, Orlando, FL, 23-25 July, pp. 313-318.

Allen, P. and Kalphat-Lopez, M. (1998), "Implementation of Patriot system model in ModSAF," *Proceedings of the 7th Conference on Computer Generated Forces and Behavior Representation*, Orlando, FL, 12-14 May, pp. 189-197.

Allsop, D.J., Harrison, A., and Sheppard, C. (2001), "The representation and storage of military knowledge to support the automated production of CGF command agents," *Proceedings of the 10th Conference on Computer Generated Forces*," Norfolk, VA, 15-17 May, pp. 159-170.

Archer, R., Walters, B., Yow, A., Carolan, T., Laughery, R., and Gillis, P. (1999), "Training as a performance shaping factor in computer generated forces," *Proceedings of the 8th Conference on Computer Generated Forces and Behavioral Representations*, Orlando, FL, 11-13 May, pp. 505-513.

Aronson, J. (1994), "The SimCore tactics representation and specification language," *Proceedings of the 4th Conference on Com-*

puter Generated Forces and Behavioral Representation, Orlando, FL, 4-6 May, pp. 187-193.

Balzer, R. (1996), "Generating computer generated forces," *Proceedings of the 6th Conference on Computer Generated Forces and Behavioral Representation,* Orlando, FL, 23-25 July, pp. 181-188.

Banks, S.B., Stytz, M.R., Hutson, L.J., and Silver, S.M. (1998a), "A computable combat psychology model for computer generated forces," *The 1998 Fall Simulation Interoperability Workshop*, Orlando, FL., 13-18 Sep., pp. 35-45.

Banks, S.B., Hutson, L.J., Stytz, M.R., and Santos, E. Jr. (1998b), "Incorporation of multiple skill levels into a domain-independent computer generated force architecture," *7th Conference on Computer Generated Forces and Behavioral Representation*, Orlando, FL, 12 - 14 May, pp. 497-508.

Banks, S.B. and Stytz, M.R. (1999a), "Considerations for the next generation of Air Force computer-generated actors," *Proceedings of the 4th International SIMTECT Conference*, Melbourne, Australia, 29 Mar – 1 Apr, pp. 175-180.

Banks, S.B. and Stytz, M.R. (1999b), "An approach to enhance human behavior modeling for computer-generated actors," *Proceedings of the 4th International SIMTECT Conference*, Melbourne, Australia, 29 Mar – 1 Apr, pp. 199-204.

Banks, S.B. and Stytz, M.R. (1999c), "User modeling for distributed virtual environment intelligent agents," *SPIE 13th Annual International Symposium on Aerospace/Defense Sensing and Controls: AeroSense '99, Modeling and Simulating Sensory Response for Real and Virtual Environments,* vol. 3694, Orlando, FL, 4-9 April, pp. 118-127.

Banks, S.B. and Stytz, M.R. (1999d), "Inclusion of user modeling techniques for human behavior representation in a domain-independent computer generated force architecture," *8th Conference on Computer Generated Forces and Behavioral Representation*, Orlando, FL, 12 - 14 May, 457-468.

Banks, S.B. and Stytz, M.R. (2000), "Advancing the state of human behavior representation for modeling and simulation: technologies and techniques," *Proceedings of the 9th Conference on Computer Generated Forces and Behavioral Representations*, Orlando, FL, 16-18 May, pp. 87-100.

Banks, S.B. and Stytz, M.R. (2001), "The next generation of human behavior representation for modeling and simulation: the HBR common development environment and framework," *Proceedings of the 10th Conference on Computer Generated Forces*," Norfolk, VA, 15-17 May, pp. 177-188.

Baxter, J.W. and Horn, G.S. (1999), "A model for co-ordination and co-operation between CGF agents," *Proceedings of the 8th Conference on Computer Generated Forces and Behavioral Representations*, Orlando, FL, 11-13 May, pp. 101-111.

Becket, W. and Badler, N.I. (1993), "Integrated behavioral agent architecture," *Proceedings of the 3rd Conference on Computer Generated Forces and Behavioral Representation,* Orlando, FL, 17-19 March, pp. 57-68.

Belyavin, A. (1999), "Modeling the effect of stress on human behavior," *Proceedings of the 8th Conference on Computer Generated Forces and Behavioral Representations*, Orlando, FL, 11-13 May, pp. 481-487.

Berkowitz, J.P. (1996), "Considerations for the use of entity-based simulations for tactical decision making training," *Proceedings of the 6th Conference on Computer Generated Forces and Behavior Representation*, Orlando, FL, 23-25 July, pp. 131-139.

Bergenthal, J., Lin, D., Macannuco, D., and Patel, R. (2000), "A simulation framework for multiple forces and sides," *Proceedings of the 9th Conference on Computer Generated Forces and Behavioral Representations*, Orlando, FL, 16-18 May, pp. 321-328.

Beugnard, A., Jezequel, J.M., Plouzeau, N., and Watkins, D. (1999), "Making components contract aware," *IEEE Computer*, vol. 32, no. 7, July, pp. 38-45.

Bimson, K., Marsden, C., McKenzie, F., and Paz, N. (1994), "Knowledge-based tactical decision making in the CCTT SAF prototype," *Proceedings of the 4th Conference on Computer Generated Forces and Behavioral Representation,* Orlando, FL, 4-6 May, pp. 293-303.

Bohrer, K., Johnson, V., Nilsson, A., and Rubin, B. (1998), "Business process components for distributed object applications," *Communications of the ACM*, vol. 41, no. 6, June, pp. 43-48.

Bokma, A. and Slade, A. (1993), "Developing large-scale agent-based systems: an example from air traffic control," *Proceedings of the 3rd Conference on Computer Generated Forces and Behavioral Representation,* Orlando, FL, 17-19 March, pp. 21-32.

Booker, L.B., Burke, C.D., and Whittaker, G.M. (1998), "Toward motivational control of tactical behaviors," *Proceedings of the 7th Conference on Computer Generated Forces and Behavior Representation*, Orlando, FL, 12-14 May, pp. 521-531.

Booker, L.B., Burke, C.D., and Hughes, J.D. (1999), "Motivational control of tactical behaviors: interim results," *Proceedings of the 8th Conference on Computer Generated Forces and Behavioral Representations*, Orlando, FL, 11-13 May, pp. 17-24.

Bouche, J.-P., Floch, J.-P., and Michel, M. (1999), "Using command agents for military planning," *The 1999 Spring Simulation Interoperability Workshop*, Orlando, FL., Sept. 12-17, pp. 503-510.

Bowden, F., Davies, M., and Dunn, J.M. (1995), "Representing role-based agents using coloured Petri nets," *Proceedings of the 5th Conference on Computer Generated Forces and Behavioral Representation,* Orlando, FL, 9-11 May, pp. 513-520.

Bowden, F., Gabrisch, C., and Davies, M. (1997), "Simulations of air defence C3I using the distributed interactive C3I effectiveness DICE) simulation," *Proceedings of the 2nd International SIMTECT Conference*, Canberra, Australia, 17-20 Mar, pp. 71-76.

Bowden, F. and Davies, M. (1998), "Design methodology and implementation technologies for C3I artificial agents," *Proceedings of the 3rd International SIMTECT Conference*, Adelaide, Australia, 2-6 Mar, pp. 111-116.

Brasse, M., Huiskamp, W., and Stroosma, O. (1999), "A component architecture for federate development," *The Fall Simulation Interoperability Workshop*, Orlando, FL, 12-17 Sep., pp. 168-174.

Braudaway, W. (1993), "A blackboard approach to computer generated forces," *Proceedings of the 3rd Conference on Computer Generated Forces and Behavioral Representation,* Orlando, FL, 17-19 March, pp. 11-20.

Brown, A.W. and Wallnau, K.C. (1998), "The current state of CBSE," *IEEE Software*, vol. 15, no. 5, September/October, pp. 37-46.

Briggs, R. and White, B. (1997), "Knowledge acquisition process for Close Combat Tactical Trainer (CCTT)," *Proceedings of the*

2nd International SIMTECT Conference, Canberra, Australia, 17-20 Mar, pp. 341-351.

Burke, C.D. (1998), "A "paper map" metaphor for terrain reasoning," *Proceedings of the 7th Conference on Computer Generated Forces and Behavior Representation*, Orlando, FL, 12-14 May, pp. 109-117.

Busetta, P., Ronnquist, R., Hodgson, A., and Lucas, A. (1999), "Light-weight intelligent software agents in simulation," *Proceedings of the 4th International SIMTECT Conference*, Melbourne, Australia, 29 Mar – 1 Apr, pp. 169-174.

Butler, B. (1998), "Simulation composability for JSIMS," *Proceedings of the 7th Conference on Computer Generated Forces and Behavior Representation*, Orlando, FL, 12-14 May, pp. 393-406.

Calder, R.B., Smith, J.E., Courtemanche, A.J., Mar, J.M.F., and Ceranowicz, A.Z. (1993), "ModSAF behavior simulation and control," *Proceedings of the 3rd Conference on Computer-Generated Forces and Behavioral Representation*, Orlando, FL, 347-356.

Calder, R.B., Carreiro, R.L., Panagos, J.N., Vrablik, G.R., Wise, B.P., Chamberlain, F.L., and Glasson, D.P. (1996), "Architecture of a command forces command entity," *Proceedings of the 6th Conference on Computer Generated Forces and Behavior Representation*, Orlando, FL, 23-25 July, pp. 19-30.

Calder, R.B. (1998), "Command agents in simulation and training," *Proceedings of the 7th Conference on Computer Generated Forces and Behavior Representation*, Orlando, FL, 12-14 May, pp.557-568.

Calder, R.B. and Drummey, J. (1999), "Definition of a military intelligent agent architecture," *Proceedings of the 8th Conference on Computer Generated Forces and Behavioral Representations*, Orlando, FL, 11-13 May, pp. 551-562.

Calder, R.B., Comanor, K., and Richards, S. (1999a), "Course of action analysis using constraint satisfaction," *Proceedings of the 8th Conference on Computer Generated Forces and Behavioral Representations*, Orlando, FL, 11-13 May, pp. 263-274.

Calder, R.B., Kleiner, M.S., and Carey, S.A. (1999b), "From domain knowledge to behavior representation," *The 1999 Spring Simulation Interoperability Workshop*, Orlando, FL., Sept. 12-17, pp. 363-372.

Carey, S.A. and Kleiner, M.S. (1998), "Multi-echelon mission-to-task decomposition," *Proceedings of the 7th Conference on Computer Generated Forces and Behavior Representation*, Orlando, FL, 12-14 May, pp. 67-73.

Carr, F.H. and Roberts, J.D. (1998), "MITRE Applique Interface: closing the gap between battlefield C4I systems and simulations," *Proceedings of the 7th Conference on Computer Generated Forces and Behavior Representation*, Orlando, FL, 12-14 May, pp.485-496.

Carter, M.E., Dean, C., Pollatz, C., Brink, A., Canody, D., Barry, P., Wallace, T., Dockery, J., and Littman, D. (1993), "Architecture for intelligent virtual agents that employ multiple learning strategies: or smart tanks avoid black holes," *Proceedings of the 3rd Conference on Computer Generated Forces and Behavioral Representation,* Orlando, FL, 17-19 March, pp. 263-270.

Carver, D.E., Parsons, G.M., and Naff, W.T. (1996), "Computer generation of joint Theater Missile Defense (TMD) Assets," *Proceedings of the 6th Conference on Computer Generated Forces and Behavioral Representation,* Orlando, FL, 23-25 July, pp. 283-289.

Ceranowicz, A. (1994), "ModSAF capabilities," *Proceedings of the 4th Conference on Computer Generated Forces and Behavioral Representation,* Orlando, FL, 4-6 May, pp. 3-8.

Ceranowicz, A., Coffin, D., Smith, J., Gonzalez, R., and Ladd, C. (1994), "Operator control of behavior in ModSAF," *Proceedings of the 4th Conference on Computer Generated Forces and Behavioral Representation,* Orlando, FL, 4-6 May, pp. 9-16.

Ceranowicz, A. (1998), "Evolutionary CGF development," *Proceedings of the 7th Conference on Computer Generated Forces and Behavior Representation*, Orlando, FL, 12-14 May, pp. 421-431.

Ceranowicz, A., Nielsen, P.E., and Koss, F.V. (2000), "Behavioral representation in ModSAF," *Proceedings of the 9th Conference on Computer Generated Forces and Behavioral Representations*, Orlando, FL, 16-18 May, pp. 501-512.

Chaib-draa, B. and Paquet, E. (1993), "Integrating reaction, planning, and deliberation in architecture for multiagent environment," *Proceedings of the 3rd Conference on Computer Generated Forces and Behavioral Representation,* Orlando, FL, 17-19 March, pp. 45-55.

Chandra, A. (1997), "A computational architecture to model human emotions," *Proceedings of Intelligent Information Systems (IIS'97)*, Grand Bahama Island, Bahamas, 8-10 December, pp. 86-89.

Chandrasekaran, B. and Josephson, J.R. (1999), "Cognitive modeling for simulation goals: a research strategy for computer-generated forces," *Proceedings of the 8th Conference on Computer Generated Forces and Behavioral Representations*, Orlando, FL, 11-13 May, pp. 117-126.

Cheek, M.L. and Eamon, J.C. (1999), "Testbed for evaluating automated C2 decision models," *Proceedings of the 8th Conference on Computer Generated Forces and Behavioral Representations*, Orlando, FL, 11-13 May, pp. 285-292.

Cicalese, C.D.T. and Rotenstreich, S. (1999), "Behavioral specification of distributed software component interfaces," *IEEE Computer*, vol. 32, no. 7, July, pp. 46-53.

Cisneros, J.E., Karr, C.R., McCauley-Bell, P.R., and Rajput, S. (1996), "Threat analysis using fuzzy set theory," *Proceedings of the 6th Conference on Computer Generated Forces and Behavioral Representation,* Orlando, FL, 23-25 July, pp. 455-462.

Cooper, J.W. (1998), "Using design patterns," *Communications of the ACM*, vol. 41, no. 6, June, pp. 65-68.

Coplien, J.O. (1999), "Reevaluating the architectural metaphor: toward piecemeal growth," *IEEE Software*, vol. 16, no. 5, September-October, pp. 40-44.

Coradeschi, S., Karlsson, L., and Torne, A. (1996), "Intelligent agents for aircraft combat simulation," *Proceedings of the 6th Conference on Computer Generated Forces and Behavioral Representation,* Orlando, FL, 23-25 July, pp. 93-99.

Coulter, K.J. and Laird, J.E. (1996), "A briefing-based graphical interface for exercise specification," *Proceedings of the 6th Conference on Computer Generated Forces and Behavior Representation*, Orlando, FL, 23-25 July, pp. 203-207.

Courtemanche, A.J. and Monday, P. (1994), "The incorporation of validated combat models into ModSAF," *Proceedings of the 4th Conference on Computer Generated Forces and Behavioral Representation,* Orlando, FL, 4-6 May, pp. 129-140.

Courtemanche, A.J. and Ceranowicz, A. (1995), "ModSAF development status," *Proceedings of the 5th Conference on Computer Generated Forces and Behavioral Representation,* Orlando, FL, 9-11 May, pp. 3-13.

Courtemanche, A.J., Hamilton, S.E., and Monday, P. (1995), "Representation of missiles in ModSAF," *Proceedings of the 5th Conference on Computer Generated Forces and Behavioral Representation,* Orlando, FL, 9-11 May, pp. 267-274.

Courtemanche, A.J. (1999), "Design patterns for computer generated forces," *Proceedings of the 8th Conference on Computer Generated Forces and Behavioral Representations*, Orlando, FL, 11-13 May, pp. 25-36.

Courtemanche, A.J. and Burch, R.B. (2000), "Using and developing object frameworks to achieve a composable CGF architecture," *Proceedings of the 9th Conference on Computer Generated Forces and Behavioral Representations*, Orlando, FL, 16-18 May, pp. 49-62.

Cox, K.F. (1998), "A framework-based approach to HLA federate development," *The Fall Simulation Interoperability Workshop*, Orlando, FL, 13-18., pp. 1059-1068.

Cox, M. and Flaglien, B.A. (1999), "Agent-based automation of CGF operator tasks in a combined simulation environment," *Proceedings of the 8th Conference on Computer Generated Forces and Behavioral Representations*, Orlando, FL, 11-13 May, pp. 203-209.

Craft, M.A. and Karr, C.R. (1996), "Testing future weapons systems using CGF systems," *Proceedings of the 6th Conference on Computer Generated Forces and Behavioral Representation,* Orlando, FL, 23-25 July, pp. 141-150.

Cross, N. and Ronnquist, R. (1999), "A Java agent environment for simulation and modeling," *Proceedings of the 4th International SIMTECT Conference*, Melbourne, Australia, 29 Mar – 1 Apr, pp. 163-167.

Cunningham, C.T. (1994), "Development of intelligent simulations at LLNL," *Proceedings of the 4th Conference on Computer Generated Forces and Behavioral Representation,* Orlando, FL, 4-6 May, pp. 345-352.

Cusack, M.A. and Hoare, P. (1999), "Component based development in HLA using Java," *The Spring Simulation Interoperability Workshop*, Orlando, FL, 14-19 March, pp. 233-242.

Davies, M. and Gabrisch, C. (1995), "The Distributed Interactive C3I Effectiveness (DICE) simulation project: an overview," *Proceedings of the 5th Conference on Computer Generated Forces and Behavioral Representation,* Orlando, FL, 9-11 May, pp. 15-27.

Deutsch, S. (1993), "Notes taken on the quest for modeling skilled human behavior," *Proceedings of the 3rd Conference on Computer Generated Forces and Behavioral Representation,* Orlando, FL, 17-19 March, pp. 359-365.

Devro, L. and Gonzalez, A.J. (2001), "Collaborative teamwork representation in context-based reasoning," *Proceedings of the 10th Conference on Computer Generated Forces*," Norfolk, VA, 15-17 May, pp. 289-300.

Digre, T. (1998), "Business object component architecture," *IEEE Software*, vol. 15, no. 5, Sep./Oct., pp. 60-69.

Dompke, U.K.J., Baeyer, A.V., Heinbokel, T., Heineken, E., and Foyse, J. (1999a), "Human factors in staff level decision making," *The 1999 Spring Simulation Interoperability Workshop*, Orlando, FL., Sept. 12-17, pp. 429-433.

Dompke, U.K.J., Baeyer, A.V., and Heineken, E. (1999b), "Problems of validating human behavior models," *The 1999 Fall Simulation Interoperability Workshop*, Orlando, FL., Sept. 12-17, pp. 151-153.

Donohoe, P. (1999), *Software Architecture*, Kluwer Academic Publications, Boston, Mass.

Dubas, B.J., Waag, G.L., Wertman, C.C., and Starr, P.N. (1999), "OPFOR Perception of the Battlespace (OPB): challenges for intelligence community support to military training simulations," *The 1999 Fall Simulation Interoperability Workshop*, Orlando, FL., 12-17 Sept., pp. 810-818.

Edwards, M. and Stytz, M.R. (1996), "The Fuzzy Wingman: an intelligent companion for DIS-compatible flight simulators", *The SPIE/SCS Joint 1996 SMC Simulation Multiconference: 1996 Military, Government, and Aerospace Simulation Conference*, vol. 28, no. 3, New Orleans, Louisiana, 77 - 82.

Eggleston, R.G., Young, M.J., and McCreight, K.L. (2001), "Modeling human work through distributed cognition," *Proceedings of the 10th Conference on Computer Generated Forces*," Norfolk, VA, 15-17 May, pp. 99-104.

Ehrhart, L.S. and Bigbee, A.J. (1998), "Supporting decision-makers in OOTW course of action analysis," *Proceedings of the 7th Conference on Computer Generated Forces and Behavior Representation*, Orlando, FL, 12-14 May, pp. 435-444.

Feinerman, L.E., Prochnow, D.L., and King, R.A. (2001), "Icarus: an HLA federation of HBR models," *Proceedings of the 10th Conference on Computer Generated Forces*," Norfolk, VA, 15-17 May, pp. 105-112.

Fields, M.A. and Spradlin, G. (2000), "Modeling civilian crowds in a battlefield simulation," *Proceedings of the 9th Conference on*

Computer Generated Forces and Behavioral Representations, Orlando, FL, 16-18 May, pp. 451-458.

Fogel, L.J. V, Porto, W., and Owen, M. (1996), "An intelligently interactive non-rule-based computer generated force," *Proceedings of the 6th Conference on Computer Generated Forces and Behavioral Representation,* Orlando, FL, 23-25 July, pp. 265-271.

Foss, W. and Robinson, B. (1999), "Review of the Army Ground Command Entity for the OneSAf Testbed," *Proceedings of the 8th Conference on Computer Generated Forces and Behavioral Representations*, Orlando, FL, 11-13 May, pp. 363-370.

Franceschini, R.W. and Cox, A. (1998), "Review of US Army Ground Command Forces for OneSAF," *Proceedings of the 7th Conference on Computer Generated Forces and Behavior Representation*, Orlando, FL, 12-14 May, pp. 333-337.

Franceschini, D.J., Nocera, D.L., Stober, D.R., Napravnik, L.J., Allen, P.M., and McCulley, G. (1998), "Recent developments in ModSAF," *Proceedings of the 7th Conference on Computer Generated Forces and Behavior Representation*, Orlando, FL, 12-14 May, pp. 37-44.

Franceschini, D., Zimmerman, J., and McCulley, G. (1999), "CGF system composability through dynamically loadable modules," *Proceedings of the 8th Conference on Computer Generated Forces and Behavioral Representations*, Orlando, FL, 11-13 May, pp. 341-347.

Funge, J. (1999), "Cognitive modeling for computer generated forces," *Proceedings of the 8th Conference on Computer Generated Forces and Behavioral Representations*, Orlando, FL, 11-13 May, pp. 491-497.

Gagne, D. (1995), "Realistic doctrinal behaviors in computer generated forces through plurality," *Proceedings of the 5th Conference on Computer Generated Forces and Behavioral Representation,* Orlando, FL, 9-11 May, pp. 521-527.

Gagnon, F., Howells, P.B., and Simon, E. (1999), "On the use of frameworks for real-time simulation applications," *The 1999 Fall Simulation Interoperability Workshop*, Orlando, FL., Sept. 12-17, pp. 483-486.

Galanis, G., Jennings, A., and Beckett, P. (1997), "Towards a general model of perception for simulation," *Proceedings of the 2nd International SIMTECT Conference*, Canberra, Australia, 17-20 Mar, pp. 21-26.

Gamma, E., Helm, R., and Vlissides. (1995), "*Design Patterns: Elements of Reusable Object Oriented Software*.. Addison-Wesely Reading, Mass.

Garcia-Crespo, A., Domingo, P., Paniagua, F., and Rodriguez-Galan, R. (1997), "The Norman attention model implementation using fuzzy logic," *Proceedings of Intelligent Information Systems (IIS'97)*, Grand Bahama Island, Bahamas, 8-10 December, pp.241-245.

Ge, Z., James, J., and Nerode, A. (1995), "A multiple agent hybrid control architecture for automated forces: design and software implementation," *Proceedings of the 5th Conference on Computer Generated Forces and Behavioral Representation,* Orlando, FL, 9-11 May, pp. 45-52.

Gelenbe, E. (1999), "Modeling CGF with learning stochastic finite-state machines," *Proceedings of the 8th Conference on Computer Generated Forces and Behavioral Representations*, Orlando, FL, 11-13 May, pp. 113-115.

George, G.R., Conquest, M., Brooks, B., and Bell, H. (1998), "Integration of AirSF to the modular control equipment: an effective distributed mission training approach for C2 operators," *Proceedings of the 7th Conference on Computer Generated Forces and Behavior Representation*, Orlando, FL, 12-14 May, pp. 377-386.

George, G.R., Breitbach, R.A., Brooks, R.B., Steffes, R., Bell, H.H., and Bruhl, C.C. (1999), "Synthetic forces in the Air Force Command and Control distributed mission training Environment, current and future," *Proceedings of the 8th Conference on Computer Generated Forces and Behavioral Representations*, Orlando, FL, 11-13 May, pp. 319-331.

Gillis, P.D. (1998), "Realism in computer generated forces command entity behaviors," *Proceedings of the 7th Conference on Computer Generated Forces and Behavior Representation*, Orlando, FL, 12-14 May, pp. 455-464.

Gillis, P.D. and Hursh, S.R. (1999), "Using behavior moderators to influence CGF command entity effectiveness and performance," *Proceedings of the 8th Conference on Computer Generated Forces and Behavioral Representations*, Orlando, FL, 11-13 May, pp. 237-251.

Glenn, F., Stokes, J., Wolf, J.J. and Scolaro, D. (1998), "A simulation model of the operator of an air defense weapons systems," *Proceedings of the 7th Conference on Computer Generated Forces and Behavior Representation*, Orlando, FL, 12-14 May, pp. 357-363.

Gluck, K.A. and Pew, R.W. (2001a), "Overview of the Agent-based Modeling and Behavior Representation (AMBR) Model Comparison Project," *Proceedings of the 10th Conference on Computer Generated Forces*," Norfolk, VA, 15-17 May, pp. 3-6.

Gluck, K.A. and Pew, R.W. (2001b), "Lessons learned and future directions for the AMBR Model Comparison Project," *Proceedings of the 10th Conference on Computer Generated Forces*," Norfolk, VA, 15-17 May, pp. 113-124.

Goldman, S. (1996), "Knowledge acquisition and delivery: constructing intelligent software command entities," *Proceedings of the 6th Conference on Computer Generated Forces and Behavior Representation*, Orlando, FL, 23-25 July, pp. 31-36.

Goldschen, A.J., Harper, L.D., and Anthony, E.R. (1998), "The Role of Speech in a Distributed Simulation: The STOW-97 CommandTalk System," *Proceedings of the 7th Conference on Computer Generated Forces and Behavior Representation*, Orlando, FL, 12-14 May, pp.621-630.

Gonzalez, A.J. and Ahlers, R. (1995), "Context-based representation of intelligent behavior in simulated opponents," *Proceedings of the 5th Conference on Computer Generated Forces and Behavioral Representation,* Orlando, FL, 9-11 May, pp. 53-62.

Gonzalez, A.J. and Murillo, M. (1999), "Validation of human behavior models," *The 1999 Spring Simulation Interoperability Workshop*, Orlando, FL., Sept. 12-17, pp. 50-57.

Goss, S., Heinze, C., and Pearce, A. (1999), "Recognizing user intentions in a virtual environment," *Proceedings of the 4th International SIMTECT Conference*, Melbourne, Australia, 29 Mar – 1 Apr, pp. 247-253.

Grama, C., Gonzalez, A., Pollak, E., Brasch, R., and Wartski, J. (1998), "Automated plan generation for mid-echelons in the military command decision process," *The 1998 Spring Simulation Interoperability Workshop*, Orlando, FL., 9-13 March., pp.636-645.

Grant, S., Fisher, M., and Pearce, P.V. (1999), "The LFM organisation model," *Proceedings of the 8th Conference on Computer Generated Forces and Behavioral Representations*, Orlando, FL, 11-13 May, pp. 155-166.

Gratch, J. (1996), "Task-decomposition planning for command decision making," *Proceedings of the 6th Conference on Computer Generated Forces and Behavior Representation*, Orlando, FL, 23-25 July, pp. 37-45.

Gratch, J., Marsella, S., Hill, R.W., and Stone, G. (1999), "Deriving priority intelligence requirements for synthetic command entities," *Proceedings of the 8th Conference on Computer Generated Forces and Behavioral Representations*, Orlando, FL, 11-13 May, pp. 301-307.

Gratch, J. (2000), "Modeling the interplay between emotion and decision-making," *Proceedings of the 9th Conference on Computer Generated Forces and Behavioral Representations*, Orlando, FL, 16-18 May, pp. 399-412.

Gugel, S.A., Pratt, D.R., and Smith, R.A. (2001a), "Using Q-learning to control individual combatants," *Proceedings of the 10th Conference on Computer Generated Forces*," Norfolk, VA, 15-17 May, pp. 131-140.

Gugel, S.A., Pratt, D.R., and Smith, R.A. (2001b), "The implementation of multiple behavioral approaches in the CGF test-bed," *Proceedings of the 10th Conference on Computer Generated Forces*," Norfolk, VA, 15-17 May, pp. 255-266.

Harmon, S.Y., Yang, S.C., and Tseng, D.Y. (1994), "Command and control simulation for computer generated forces," *Proceedings of the 4th Conference on Computer Generated Forces and Behavioral Representation,* Orlando, FL, 4-6 May, pp. 263-273.

Harmon, S.Y. (1998), "Evaluating and comparing intelligence," *Proceedings of the 7th Conference on Computer Generated Forces and Behavior Representation*, Orlando, FL, 12-14 May, pp.599-610.

Harmon, S.Y. and Youngblood, S.M. (1999), "Validation of human behavior representations," *The 1999 Spring Simulation Interoperability Workshop*, Orlando, FL., Sept. 12-17, pp. 277-284.

Hartzog, S.M. and Salisbury, M.R. (1996), "Command Forces (CFOR) program status report," *Proceedings of the 6th Conference on Computer Generated Forces and Behavior Representation*, Orlando, FL, 23-25 July, 1996, pp.11-17.

Heinze, C., Lloyd, I., Goss, S., and Pearce, A. (1999), "Collaborating cognitive and sub-cognitive processes for the simulation of human decision making," *Proceedings of the 4th International SIMTECT Conference*, Melbourne, Australia, 29 Mar – 1 Apr, pp. 239-246.

Hepplewhite, R.T. and Baxter, J.W. (1996a), "Broad agents for intelligent battlefield simulation," *Proceedings of the 1st International SIMTECT Conference*, Melbourne, Australia, 25-26 Mar, pp. 107-112.

Hepplewhite, R.T. and Baxter, J.W. (1996b), "Broad agents for intelligent simulation," *Proceedings of the 6th Conference on Computer Generated Forces and Behavior Representation*, Orlando, FL, 23-25 July, pp. 319-327.

Hepplewhite, R.T. and Baxter, J.W. (1998), "Planning and search techniques to produce terrain dependent behaviours," *Proceedings of the 7th Conference on Computer Generated Forces and Behavior Representation*, Orlando, FL, 12-14 May, pp. 99-109.

Herbsleb, J.D. and Grinter, R.E. (1999), "Architectures, coordination, and distance: Conway's Law and beyond," *IEEE Software*, vol. 16, no. 5, September-October, pp. 63-70.

Hicinbothom, J.H. and Lyons, D.M. (1999), "Synthetic forces for team training," *Proceedings of the 8th Conference on Computer Generated Forces and Behavioral Representations*, Orlando, FL, 11-13 May, pp. 333-338.

Hicinbothom, J.H. (2001), "Maintaining situation awareness in synthetic team members," *Proceedings of the 10th Conference on Computer Generated Forces*," Norfolk, VA, 15-17 May, pp. 231-244.

Hieb, M.R., Tecuci, G., Pullen, J.M., Ceranowicz, A., and Hille, D. (1995), "A methodology and tool for constructing adaptive command agents for computer generated forces," *Proceedings of the 5th Conference on Computer Generated Forces and Behavioral Representation,* Orlando, FL, 9-11 May, pp. 135-146.

Hieb, M.R., Tecuci, G., and Pullen, J.M. (1996), "Training a ModSAF command agent through demonstration," *Proceedings of the 6th Conference on Computer Generated Forces and Behavior Representation*, Orlando, FL, 23-25 July, pp. 243-254.

Hill, R.W., Chen, J., Gratch, J., Rosenbloom, P., and Tambe, M. (1998), "Soar-RWA: planning, teamwork, and intelligent behavior for synthetic rotary wing aircraft," *Proceedings of the 7th Conference on Computer Generated Forces and Behavior Representation*, Orlando, FL, 12-14 May, pp. 177-188.

Hill, R.W., Gratch, J., Rosenbloom, P.S., and Whitney, R. (2000), "Flexible group behavior: lessons learned building virtual commanders," *Proceedings of the 9th Conference on Computer Generated Forces and Behavioral Representations*, Orlando, FL, 16-18 May, pp. 203-214.

Hille, D., Hieb, M.R., and Tecuci, G. (1994), "CAPTAIN: building agents that plan and learn," *Proceedings of the 4th Conference on Computer Generated Forces and Behavioral Representation,* Orlando, FL, 4-6 May, pp. 411-422.

Hoagland, D.G., Martin, E.A., and Brett, B.E. (1998), "Improving the DoD's requirements process through inclusion of realistic operaor performance," *The 1998 Fall Simulation Interoperability Workshop*, Orlando, FL., Sept. 13-18, pp. 419-424.

Holmes, R. (1996), "Representative CGF behaviour in a DIS virtual environment through digital communication," *Proceedings of the 1st International SIMTECT Conference*, Melbourne, Australia, 25-26 Mar, pp. 187-192.

Howard, M. and Lee, C. (1998), "Architecture of a generic command entity," *Proceedings of the 7th Conference on Computer Generated Forces and Behavior Representation*, Orlando, FL, 12-14 May, pp.569-579.

Hudlicka, E. and Billingsley, J. (1999), "Representing behavior moderators in military human performance models," *Proceedings of the 8th Conference on Computer Generated Forces and Behavioral Representations*, Orlando, FL, 11-13 May, pp. 423-433.

Ioerger, T.R., Volz, R.A., and Yen, J. (2000), "Modeling cooperative, reactive behaviors on the battlefield with intelligent agents," *Proceedings of the 9th Conference on Computer Generated Forces and Behavioral Representations*, Orlando, FL, 16-18 May, pp. 13-24.

Jackson, P.M. (1996), "Design considerations for interoperability in mission training systems," *Proceedings of the 1st International SIMTECT Conference*, Melbourne, Australia, 25-26 Mar, pp. 193-198.

Jackson, L.A. and Pearman, G.M. (1999), "A low level distributed simulation architecture," *The 1999 Fall Simulation Interoperability Workshop*, Orlando, FL., 13-17 Sep., pp.863-867.

Jacobson, J. and Hwang, Z. (2001), "Unreal tournament for immersive interactive theater," *Communications of the ACM*, vol. 45, no. 1, January, pp. 39-42.

Johns, M. and Silverman, B.G. (2001), "How emotions and personality effect the utility of alternative decisions: a terrorist target selection case study," *Proceedings of the 10th Conference on Computer Generated Forces*," Norfolk, VA, 15-17 May, pp. 55-64.

Johnson, W.L. (1994), "Agents that explain their own actions," *Proceedings of the 4th Conference on Computer Generated Forces and Behavioral Representation,* Orlando, FL, 4-6 May, pp. 87-95.

Jones, R.M., Tambe, M., Laird, J.E., and Rosenbloom, P.S. (1993), "Intelligent automated agents for flight training simulators," *Proceedings of the 3rd Conference on Computer Generated Forces and Behavioral Representation,* Orlando, FL, 17-19 March, pp. 33-42.

Jones, R.M. and Laird, J.E. (1994), "Multiple information sources and multiple participants: managing situational awareness in an autonomous agent," *Proceedings of the 4th Conference on Computer Generated Forces and Behavioral Representation,* Orlando, FL, 4-6 May, pp. 511-517.

Jones, R.M., Laird, J.E., Tambe, M., and Rosenbloom, P.S. (1994), "Generating behavior in response to interacting goals," *Proceedings of the 4th Conference on Computer Generated Forces and Behavioral Representation,* Orlando, FL, 4-6 May, pp. 317-324.

Jones, R.M., Laird, J.E., and Nielsen, P.E. (1996), "Moving intelligent automated forces into theater-level scenarios," *Proceedings of the 6th Conference on Computer Generated Forces and Behavior Representation*, Orlando, FL, 23-25 July, pp. 113-117.

Jones, R.M., Laird, J.E., and Neville, K. (1998), "Modeling of pilot fatigue with a synthetic behavior model," *Proceedings of the 7th Conference on Computer Generated Forces and Behavior Representation*, Orlando, FL, 12-14 May, pp. 349-356.

Jones, R.M. (1999), "Graphical visualization of situational awareness and mental state for intelligent computer-generated forces," *Proceedings of the 8th Conference on Computer Generated Forces and Behavioral Representations*, Orlando, FL, 11-13 May, pp. 219-222.

Jones, R.M., Laird, J.M., Nielsen, P.E., Coulter, K.J., Kenny, P., and Voss, F.V. (1999), "Automated intelligent pilots for combat flight simulation," *AI Magazine*, vol. 20, no. 1, Spring, pp. 27-42.

Kaminka, G.A., Veloso, M.M., Schaffer, S., Sollitto, C., Adobbatti, R., Marshall, A.N., Scholer, A., and Tejada, S. (2001), "GameBots: a flexible test bed for multiagent team research," *Communications of the ACM*, vol. 45, no. 1, January, pp. 43-45.

Karr, C.R., Rajput, S., Cisneros, J.E., and Lee, H.L. (1995), "Automated mission planning in ModSAF," *Proceedings of the 5th Conference on Computer Generated Forces and Behavioral Representation,* Orlando, FL, 9-11 May, pp. 159-168.

Karr, C.R. and Holbrook, R. (1999), "Modeling command and control in Warsim 2000," *Proceedings of the 8th Conference on Computer Generated Forces and Behavioral Representations,* Orlando, FL, 11-13 May, pp. 309-317.

Katz, A. (1993), "Intelligent player – first principle foundations," *Proceedings of the 3rd Conference on Computer Generated Forces and Behavioral Representation,* Orlando, FL, 17-19 March, pp. 329-334.

Keepence, B. and Mannion, M. (1999), "Using patterns to model variability in product families," *IEEE Software*, vol. 16, no. 4, July/August, pp. 102-108.

Keirsey, D., Krozel, J., Payton, D., and Tseng, D. (1994), "Case-based computer generated forces," *Proceedings of the 4th Conference on Computer Generated Forces and Behavioral Representation,* Orlando, FL, 4-6 May, pp. 307-316.

Kenny, P.G., Durfee, E.H., and Kluge, K.C. (1996), "Mission planning and coordinated execution for unmanned vehicles," *Proceedings of the 6th Conference on Computer Generated Forces and Behavior Representation*, Orlando, FL, 23-25 July, pp. 329-336.

Kim, Y. and Hill, R.W. (2000), "How long can an agent look away from a target," *Proceedings of the 9th Conference on Computer Generated Forces and Behavioral Representations*, Orlando, FL, 16-18 May, pp. 35-40.

Kocabas, S., Oztemel, E., Uldudag, M., and Noc, N. (1995), "Automated agents that learn and explain their own actions: a progress report," *Proceedings of the 5th Conference on Computer Generated Forces and Behavioral Representation,* Orlando, FL, 9-11 May, pp. 63-68.

Kocabas, S. and Oztemel, E. (1998), "AISim: an intelligent agent for distributed interactive simulation," *Proceedings of the 7th Conference on Computer Generated Forces and Behavior Representation*, Orlando, FL, 12-14 May, pp. 257-261.

Koenig, S. and Liu, Y. (1999), "Simulating high-stake decisions," *Proceedings of the 8th Conference on Computer Generated Forces and Behavioral Representations*, Orlando, FL, 11-13 May, pp.499-504.

Kraus, M.K., Franceschini, D.J., Tolley, T.R., Napravnik, L.J., Mullaly, D.E., and Franceschini, R.W. (1996), "CCTT SAF and ModSAF behavior integration techniques," *Proceedings of the 6th Conference on Computer Generated Forces and Behavior Representation*, Orlando, FL, 23-25 July, pp. 159-170.

Kuijpers, N., van Gool, P., and Jense, H. (1998), "A component architecture for simulator development," *The 1998 Spring Simulation Interoperability Workshop*, Orlando, FL., 9-13 March., pp.1336-1344.

Kuokka, D.R. (1993), "A framework for integrating autonomous agents," *Proceedings of the 3rd Conference on Computer Generated Forces and Behavioral Representation,* Orlando, FL, 17-19 March, pp. 181-190.

Lacy, L. and Dugone, T.D. (2001), "Representing computer generated forces behaviors using extensible markup language (XML) techniques," *Proceedings of the 10th Conference on Computer Generated Forces*," Norfolk, VA, 15-17 May, pp. 171-176.

Laird, J.E., Newell, A., and Rosenbloom, P.S. (1987), "SOAR: An architecture for general intelligence," *Artificial Intelligence*, vol. 33, no. 1, pp. 1-64.

Laird, J.E., Jones, R.M., and Nielsen, P.E. (1994), "Coordinated behavior of computer generated forces in TacAir-SOAR," *Proceedings of the 4th Conference on Computer Generated Forces and Behavioral Representation,* Orlando, FL, 4-6 May, pp. 325-332.

Laird, J.E., Johnson, W.L., Jones, R.M., Koss, F., Lehman, J.F., Nielsen, P.E., Rosenbloom, P.S., Rubinoff, R., Schwamb, K., Tambe, M., vanDyke, J., vanLent, M., and Wray, R.E. III. (1995), "Simulated intelligent forces for air: the SOAR/IFOR project 1995," *Proceedings of the 5th Conference on Computer Generated Forces and Behavioral Representation,* Orlando, FL, 9-11 May, pp. 27-36.

Laird, J.E., Coulter, K.J., Jones, R.M., Kenny, P.G., Koss, F.V., and Nielsen, P.E. (1998a), "Integrating intelligent computer generated forces in distributed simulations: TacAir Soar in STOW-97," *The 1998 Spring Simulation Interoperability Workshop*, Orlando, FL., 9-13 March., pp.1195-1201.

Laird, J.E., Jones, R.M., and Nielsen, P. (1998b), "Lessons learned from TacAir-Soar in STOW-97," *Proceedings of the 7th Conference on Computer Generated Forces and Behavior Representation*, Orlando, FL, 12-14 May, pp. 25-35.

Laird, J.E. (2000), "An exploration into computer games and computer generated forces," *Proceedings of the 9th Conference on Computer Generated Forces and Behavioral Representations*, Orlando, FL, 16-18 May, pp. 241-250.

Laird, J.E. and vanLent, M. (2001), "Human-level AI's killer application: interactive computer games," *AI Magazine*, vol. 22, no. 2, Summer, pp. 15-26.

Laird, J.E. (2002), "Research in human-level AU using computer games," *Communications of the ACM*, vol. 45, no. 1, January, pp. 32-35.

Labbiento, A. (2000), "ITEMS: a CGF implementation for an attack helicopter crew trainer," *Proceedings of the 9th Conference on Computer Generated Forces and Behavioral Representations*, Orlando, FL, 16-18 May, pp. 283-288.

Landweer, P. (1993), "Action/cognition behavior model for computer generated forces," *Proceedings of the 3rd Conference on Computer Generated Forces and Behavioral Representation,* Orlando, FL, 17-19 March, pp. 335-345.

Lankester, H. (1995), "Multi-application command agents," *Proceedings of the 5th Conference on Computer Generated Forces and Behavioral Representation,* Orlando, FL, 9-11 May, pp. 169-177.

Lannon, G.P., Klein, H.A., and Timian, D.H. (2001), "Integrating cultural factors into threat conceptual models," *Proceedings of the 10th Conference on Computer Generated Forces,*" Norfolk, VA, 15-17 May, pp. 45-54.

Lattimore, P.J. and Reicken, M.E. (1993), "A virtual battlespace language," *Proceedings of the 3rd Conference on Computer Generated Forces and Behavioral Representation,* Orlando, FL, 17-19 March, pp. 87-93.

Laughery, R. and Dahn, D. (1999), "A tool for building interoperable simulation models – the integrated performance modeling environment," *Proceedings of the 8th Conference on Computer Generated Forces and Behavioral Representations*, Orlando, FL, 11-13 May, pp. 395-399.

LaVine, N.D., Laughery, R.K., Young, B., and Kehlet, R. (1993), "Semi-Automated Forces (SAFOR) crew performance degradation," *Proceedings of the 3rd Conference on Computer Generated Forces and Behavioral Representation,* Orlando, FL, 17-19 March, pp. 405-415.

LaVine, N.D., Kehlet, R., O'Connor, M.J., and Jones, D.J. (1999a), "Transferring ownership of ModSAF behavioral attributes," *The 1999 Spring Simulation Interoperability Workshop*, Orlando, FL., Sept. 12-17, pp. 612-621.

LaVine, N.D., Kehlet, R., and Peters, S.D. (1999b), "A client-server approach to CGF behavioral representation," *Proceedings of the 8th Conference on Computer Generated Forces and Behavioral Representations*, Orlando, FL, 11-13 May, pp. 411-421.

Liang, Y., Robichaud, F., Fugere, B.J., and Ackles, K.N. (2001), "Commander behavior and course of action selection in JWARS," *Proceedings of the 10th Conference on Computer Generated Forces*," Norfolk, VA, 15-17 May, pp. 387-400.

von der Lippe, S.R., McCormack, J.S., and Kalphat, M. (2000), "Embracing temporal relations and command and control in composable behavior technologies," *Proceedings of the 9th Conference on Computer Generated Forces and Behavioral Representations*, Orlando, FL, 16-18 May, pp. 183-192.

Lorenzo, M., Riggs, B., Rizik, P.D., Christiansen, A.D., Florek, D., and Petty, M.D. (2000), "Next generation CGF requirements to support sensor development," *Proceedings of the 9th Conference on Computer Generated Forces and Behavioral Representations*, Orlando, FL, 16-18 May, pp. 413-420.

Mall, H., Bimson, K., McCormack, J., and Ourston, D. (1995), "Command entity cognitive behaviors for SAF and CGF," *Proceedings of the 5th Conference on Computer Generated Forces and Behavioral Representation,* Orlando, FL, 9-11 May, pp. 203-209.

Marshall, H., Chandler, E.V., McEnany, B.R., and Thomas, J.G. (1996), "SAF and manned simulators correlation issues in CCTT," *Proceedings of the 6th Conference on Computer Generated Forces and Behavior Representation*, Orlando, FL, 23-25 July, pp. 355-365.

McCauley, B., Tackaberry, M., Elsaesser, C., Highes, J., Pemberton, B., and Smith, E. (1998), "Using multi-agent adversarial planning for command agents in the WARSIM testbed," *The 1998 Fall Simulation Interoperability Workshop*, Orlando, FL., 13-18 Sep., pp.1311-1315.

McCoramack, J.S. and von der Lippe, S. (1999), "Embracing temporal relations in composable behavior technologies," *The 1999 Fall Simulation Interoperability Workshop*, Orlando, FL., 12-17 Sep., pp.567-573.

McIlroy, D. and Heinze, C. (1996), "Air combat tactics implementation in the Smart Whole Air Mission Model," *Proceedings of the 1st International SIMTECT Conference*, Melbourne, Australia, 25-26 Mar, pp. 125-130.

McIlroy, D., Heinze, C., Appla, D., Busetta, P., Tidhar, G., and Rao, A. (1997), "Towards credible computer-generated forces," *Proceedings of the 2nd International SIMTECT Conference*, Canberra, Australia, 17-20 Mar, pp. 79-84

McIntyre, R.T. III and Middleton, V.E. (1995), "A behavioral approach to fidelity requirements for simulation of dismounted combatants," *Proceedings of the 5th Conference on Computer Generated Forces and Behavioral Representation,* Orlando, FL, 9-11 May, pp. 495-499.

McKenzie, F., Catanzaro, J., and Petty, M.D. (2001), "A personality-based command decision-maker: results and recommendations," *Proceedings of the 10th Conference on Computer Generated Forces*," Norfolk, VA, 15-17 May, pp. 351-362.

Meliza, L.L. and Vaden, E.A. (1995), "Measuring entity and group behaviors of semi-automated forces," *Proceedings of the 5th Conference on Computer Generated Forces and Behavioral Representation,* Orlando, FL, 9-11 May, pp. 181-192.

Meyer, B. and Mingins, C. (1999), "Component-based development: from Buzz to Spark," *IEEE Computer*, vol. 32, no. 7, July, pp. 35-37.

Might, R.J. and Cannon, J.B. (1999), "Linking user requirements to techniques for representing human behavior in models and simulations," *Proceedings of the 8th Conference on Computer Generated Forces and Behavioral Representations*, Orlando, FL, 11-13 May, pp. 181-192.

Murray, G., McIllroy, D., and Stewart, S. (1999), "Representing the human component in operational analysis simulation," *Proceedings of the 4th International SIMTECT Conference*, Melbourne, Australia, 29 Mar – 1 Apr, pp. 265-270.

Naikar, N., Sanderson, P., and Lintern, G. (1999), "Work domain analysis: identification of training needs for simulator acquisition," *Proceedings of the 4th International SIMTECT Conference*, Melbourne, Australia, 29 Mar – 1 Apr, pp. 65-68.

Newell, A., Yost, G.R., Laird, J.E., Rosenbloom, P.S., and Altmann, E. (1991), "Formulating the problem space computational model." In *CMU Computer Science: A 25th Anniversary Commemorative*, ed. Rashid, R.F. pp. 255-293, New York: ACM Press/Addison-Wesley.

Oakes, M.R. (1999), "Threat environments in modeling and simulation: quo vadis?," *The 1999 Fall Simulation Interoperability Workshop*, Orlando, FL., 13-17 Sep., pp.968-977.

Ogata, M., Higashide, A., and Yasuda, T. (1999), "Trial of human decision support simulation using GUI models in CORBA and HLA," *The Fall Simulation Interoperability Workshop*, Orlando, FL, 12-17 Sep., pp. 57-64.

Oztemel, E. and Kocabas, S. (1996), "Design principles for intelligent agents in distributed interactive simulation," *Proceedings of the 1st International SIMTECT Conference*, Melbourne, Australia, 25-26 Mar, pp. 103-106.

Pace, D.K. (1999), "Use of Subject Matter Experts (SMEs) in simulation evaluation," *The 1999 Fall Simulation Interoperability Workshop*, Orlando, FL., Sept. 12-17, pp. 126-134.

Page, I. and Widdowson, M. (1999), "Synthetic forces behavioral architecture," *Proceedings of the 8th Conference on Computer Generated Forces and Behavioral Representations*, Orlando, FL, 11-13 May, pp. 89-100.

Parsons, J.D. (1994), "Using fuzzy logic control technology to simulate human decision-making in warfare models," *Proceedings of the 4th Conference on Computer Generated Forces and Behavioral Representation,* Orlando, FL, 4-6 May, pp. 519-529.

Patrone, D.M. and Nardo, A. (2000), "A domain and simulation-independent architecture for creating simulated object behaviors," *Proceedings of the 9th Conference on Computer Generated Forces and Behavioral Representations*, Orlando, FL, 16-18 May, pp. 63-72.

Penney, R.W. (1996), "Drilling CGF agents in METT-T: an alternative approach to conventional AI," *Proceedings of the 6th Conference on Computer Generated Forces and Behavioral Representation,* Orlando, FL, 23-25 July, pp. 519-530.

Perry, N. (1997), "Coordinate frame limitations on the use of legacy simulators in a DIS environment," *Proceedings of the 2nd International SIMTECT Conference*, Canberra, Australia, 17-20 Mar, pp. 169-172.

Pittman, J.A., Smith, I., Cohen, P., Oviatt, S., and Yang, T.C. (1996), "Quickset: a multimodal interface for military simulation," *Proceedings of the 6th Conference on Computer Generated*

Forces and Behavior Representation, Orlando, FL, 23-25 July, pp. 217-224.

Pongracic, H., Appla, D., and McIlroy, D. (1997a), "Are intelligent computer generated forces now available for human-in-the-loop simulations?," *Proceedings of the 2nd International SIMTECT Conference*, Canberra, Australia, 17-20 Mar, pp. 85-90.

Pongracic, H., Marlow, D., and Triggs, T. (1997b), "Issues in cost-effectiveness and fidelity of simulation," *Proceedings of the 2nd International SIMTECT Conference*, Canberra, Australia, 17-20 Mar, pp. 221-226.

Powell, G.M. (1998), "A cognitive task analysis of army corps-level planning," *Proceedings of the 7th Conference on Computer Generated Forces and Behavior Representation*, Orlando, FL, 12-14 May, pp. 47-58.

Pratt, D.R., Bhargava, H.K., Culpepper, M., and Locke, J. (1994), "Collaborative autonomous agents in the NPSNET virtual world," *Proceedings of the 4th Conference on Computer Generated Forces and Behavioral Representation,* Orlando, FL, 4-6 May, pp. 177-186.

Pratt, D. (1996), "Next generation computer generated forces," *Proceedings of the 6th Conference on Computer Generated Forces and Behavior Representation*, Orlando, FL, 23-25 July, pp. 3-8.

Pratt, D.R. (1998), "Turning the corner," *Proceedings of the 7th Conference on Computer Generated Forces and Behavior Representation*, Orlando, FL, 12-14 May, pp. 509-519.

Pratt, D.R. (2001), "Use of evolutionary algorithms for runtime decision making," *Proceedings of the 10th Conference on Computer Generated Forces*," Norfolk, VA, 15-17 May, pp. 333-342.

Pratt, S.M., Barham, P.T., Fulford, D.A., and Koehling, J. (1999), "Starting points for representing humans within the High Level Architecture (HLA)," *The 1999 Fall Simulation Interoperability Workshop*, Orlando, FL., 12-17 Sept., pp. 523-529.

Pratt, D.R., Courtemanche, A.J., Campbell, C., and Moyers, J. (2000), "An empirical evaluation of programming languages for computer generated forces," *Proceedings of the 9th Conference on Computer Generated Forces and Behavioral Representations*, Orlando, FL, 16-18 May, pp. 151-164.

Psotka, J. (1998), "Technologies for representing cognition and emotion in CGF," *The 1998 Spring Simulation Interoperability Workshop*, Orlando, FL., 9-13 March., pp.1154-1158.

Puckett, J.M., Johnson, K., and Wise, B. (1999), "Using finite state machines for behavior representation with composable object models in NASM," *The 1999 Spring Simulation Interoperability Workshop*, Orlando, FL., Sept. 12-17, pp. 208-212.

Rajput, S., and Karr, C.R. (1996), "A new mechanism for cooperative behavior in ModSAF," *Proceedings of the 6th Conference on Computer Generated Forces and Behavior Representation*, Orlando, FL, 23-25 July, pp. 189-199.

Rajput, S., Karr, C.R., Cisneros, J.E., and Parsons, R.J. (1996), "Learning the selection of reactive behaviors," *Proceedings of the 6th Conference on Computer Generated Forces and Behavior Representation*, Orlando, FL, 23-25 July, pp. 255-264.

Reece, D.A. and Kelly, P. (1996), "An architecture for computer generated individual combatants," *Proceedings of the 6th Conference on Computer Generated Forces and Behavioral Representation,* Orlando, FL, 23-25 July, pp. 337-344.

Rigg, G., Morley, R., and Hepplewhite, R.T. (2000), "Themis: The objective assessment of CGF performance," *Proceedings of the*

9th Conference on Computer Generated Forces and Behavioral Representations, Orlando, FL, 16-18 May, pp. 133-138.

Roberts, D.W., Mitchell, C.M., Thurman, D.A., and Chappell, A.R. (1999), "Computational models of team decision-making," *The 1999 Fall Simulation Interoperability Workshop*, Orlando, FL., 12-17 Sept., pp. 993-1000.

Roberts, D.W. and Crawford, K.A. (2000), "The high level architecture as a means to integrate and reuse human behavior models," *Proceedings of the 9th Conference on Computer Generated Forces and Behavioral Representations*, Orlando, FL, 16-18 May, pp. 329-332.

Rosenbloom, P.S., Laird, J.E., and Newell, A. (1988), "Meta-levels in SOAR." In *Meta-Level Architectures and Reflections*, eds. Maes, P. and Nardi, D. pp. 227-240, Amsterdam, North Holland.

Rosenbloom, P.S., Lee, S., and Unruth, A. (1993), "Bias in planning and explanation-based learning." In *Machine Learning Methods for Planning*, ed. Minton, S. pp. 197-232, San Francisco, CA: Morgan Kaufman.

Rosenbloom, P.S., Johnson, W.L., Jones, R.M., Koss, F., Laird, J.E., Lehman, J.F., Rubinoff, R., Schwamb, K.B., and Tambe, M. (1994), "Intelligent automated agents for tactical air simulation: a progress report," *Proceedings of the 4th Conference on Computer Generated Forces and Behavioral Representation,* Orlando, FL, 4-6 May, pp. 69-78.

Roza, M., Voogd, J., Jense, H., and Gool, P.V. (1999), "Fidelity requirements specification: a process oriented view," *The 1999 Fall Simulation Interoperability Workshop*, Orlando, FL., Sept. 12-17, pp. 210-220.

Rubinoff, R. and Lehman, J.F. (1994), "Natural language processing in an IFOR pilot," *Proceedings of the 4th Conference on*

Computer Generated Forces and Behavioral Representation, Orlando, FL, 4-6 May, pp. 97-104.

Sale, N., Usher, T., Page, I., and Wonnacott, P. (1999), "Multiple representations in synthetic environments," *Proceedings of the 8th Conference on Computer Generated Forces and Behavioral Representations*, Orlando, FL, 11-13 May, pp. 349-361.

Salisbury, M.R., Booker, L.B., Seidel, D.W., and Dahman, J.S. (1995), "Implementation of Command Forces (CFOR) simulation," *Proceedings of the 5th Conference on Computer Generated Forces and Behavioral Representation,* Orlando, FL, 9-11 May, pp.423-430.

Sapaty, P.S. (1996), "Distributed modeling of cooperative behavior by mobile agents," *Proceedings of the 6th Conference on Computer Generated Forces and Behavioral Representation,* Orlando, FL, 23-25 July, pp. 599-613.

Schwamb, K.B., Koss, F.V., and Keirsey, D. (1994), "Working with ModSAF: interfaces for programs and users," *Proceedings of the 4th Conference on Computer Generated Forces and Behavioral Representation,* Orlando, FL, 4-6 May, pp. 395-400.

Severinghaus, R., Williams, K., and Clare, T. (1998), "Behavior representation using cognitive task analysis modeling – bridging the gap between decision processes and model development requirements," *The 1998 Fall Simulation Interoperability Workshop*, Orlando, FL., 13-18 Sep., pp.296-304.

Shaw, M. and Garlan, D. (1996), *Software Architecture: Perspectives on an Emerging Discipline*, Prentice Hall, Upper Saddle River, NJ.

Sherman, R.H. (1994), "Using computer generated forces to support cooperative mission planning," *Proceedings of the 4th Con-*

ference on Computer Generated Forces and Behavioral Representation, Orlando, FL, 4-6 May, pp. 37-49.

Siksik, D.N. (1993), "Intelligent computer generated forces through expert systems," *Proceedings of the 3rd Conference on Computer Generated Forces and Behavioral Representation,* Orlando, FL, 17-19 March, pp. 3-10.

Siksik, D.N. (2000), "STRIVE: an open and distributed architecture for CGF representation," *Proceedings of the 9th Conference on Computer Generated Forces and Behavioral Representations,* Orlando, FL, 16-18 May, pp. 41-48.

Smith, S.H. (1993), "ILLISH: Intermediate Level Language, Interpreted for Script Handling," *Proceedings of the 3rd Conference on Computer Generated Forces and Behavioral Representation,* Orlando, FL, 17-19 March, pp. 95-105.

Steuart, S. and Murray, G. (1996), "Air mission modeling - an application for artificial intelligence," *Proceedings of the 1st International SIMTECT Conference*, Melbourne, Australia, 25-26 Mar, pp. 113-118.

Stone, G.F. and Pettitt, B.L. (1998), "Future battlespace implications to computer-generated forces," *Proceedings of the 7th Conference on Computer Generated Forces and Behavior Representation*, Orlando, FL, 12-14 May, pp. 389-392.

Stytz, M.R., Adams, T., Garcia, B., Sheasby, S.M., and Zurita, B. (1996a), "Developments in rapid prototyping and software architecture for distributed virtual environments," *IEEE Software*, vol. 14, No. 5, Sep-Oct, pp. 83-92.

Stytz, M.R., Banks, S.B., and Santos, E. (1996b), "Requirements for intelligent aircraft entities in distributed environments," *18th Interservice/Industry Training Systems and Education Confer-*

ence, Orlando, Florida, 3 - 5 December, publication on CD-ROM.

Stytz, M.R. and Banks, S.B. (1998a), "Developing simulation object models for complex, multi-actor computer generated force applications: deficiencies and lessons learned," *7th Conference on Computer Generated Forces and Behavioral Representation*, Orlando, FL, 12 - 14 May, pp. 135-148.

Stytz, M.R. and Banks, S.B. (1998b), "Requirements for the next generation of air force models," *The 1998 Fall Simulation Interoperability Workshop*, Orlando, FL., 13-18 Sep., pp. 46-54.

Stytz, M.R., Banks, S.B. and Santos, E. Jr. (1998), "The distributed mission training integrated threat environment project: goals, architecture, and current status," *7th Conference on Computer Generated Forces and Behavioral Representation*, Orlando, FL, 12 - 14 May, pp. 155-168.

Stytz, M.R. and Banks, S.B. (1999a), "The distributed mission training integrated threat environment system architecture, rules, and design overview," *The Spring Simulation Interoperability Workshop*, Orlando, FL, 14-19 March, pp. 858-867.

Stytz, M.R. and Banks, S.B. (1999b), "Requirements and HLA considerations for computer-generated actors in distributed mission training," *The SCS 1999 Advanced Simulation Technology Conference: 1999 Military, Government, & Aerospace Simulation Conference*, San Diego, CA, 11-15 April, pp. 33-38.

Stytz, M.R. and Banks, S.B. (1999c), "Robust computer generated actor migration," *Proceedings of the 1999 Fall Simulation Interoperability Workshop*, Orlando, Florida, 15-20 September, pp. 695-703.

Stytz, M.R. and Banks, S.B. (2000), "The development of computer generated actor frameworks for distributed simulation,"

Proceedings of the 9th Conference on Computer Generated Forces and Behavioral Representations, Orlando, FL, 16-18 May, pp. 483-494.

Stytz, M.R. and Banks, S.B. (2001), "The dynamic adaptive threat environment architecture for computer-generated actors," *Proceedings of the 10th Conference on Computer Generated Forces*," Norfolk, VA, 15-17 May, pp. 189-204.

Surdu, J.R., Cox, B.A., Ragsdale, D.J., Yen, J., and Pooch, U.W. (1998), "Implementing military units in simulation as intelligent agents," *Proceedings of the Military, Government, and Aerospace Simulation Conference*, Boston, MA, 5-9 April, pp. 190-198.

Takvorian, Y. (1998), "Activities as a framework for modeling behaviours in a CGF environment," *Proceedings of the 3rd International SIMTECT Conference*, Adelaide, Australia, 2-6 Mar, pp. 117-122.

Tambe, M. and Rosenbloom, P.S. (1994), "Event tracking in complex multi-agent environments," *Proceedings of the 4th Conference on Computer Generated Forces and Behavioral Representation,* Orlando, FL, 4-6 May, pp. 473-484.

Tambe, M., Schwamb, K., and Rosenbloom, P.S. (1995a), "Building intelligent pilots for simulated rotary wing aircraft," *Proceedings of the 5th Conference on Computer Generated Forces and Behavioral Representation,* Orlando, FL, 9-11 May, pp. 39-44.

Tambe, M., Johnson, W.L., Jones, R.M., Koss, F., Laird, J.E., Rosenbloom, P.S., and Schwamb, K. (1995b), "Intelligent agents for interactive simulation environments," *AI Magazine*, vol. 16, no. 1, 15-40.

Tambe, M. (1996), "Flexible teamwork for intelligent simulated pilots," *Proceedings of the 6th Conference on Computer Generated Forces and Behavioral Representation,* Orlando, FL, 23-25 July, pp. 591-598.

Taylor, G., Koss, F., and Nielsen, P. (2001), "Special operations forces IFORs," *Proceedings of the 10th Conference on Computer Generated Forces,*" Norfolk, VA, 15-17 May, pp. 301-306.

Tenney, Y.J. and Spector, S.L. (2001), Comparison of HBR models with human-in-the-loop performance in a simplified air traffic control simulation with and without HLA protocols: task simulation, human data, and results," *Proceedings of the 10th Conference on Computer Generated Forces,*" Norfolk, VA, 15-17 May, pp. 15-26.

Tolk, A. (1999), "Adjustable rulesets versus neural nets for order generation in closed combat simulation models," *Proceedings of the 8th Conference on Computer Generated Forces and Behavioral Representations*, Orlando, FL, 11-13 May, pp. 167-173.

Towne, D.M. (1999), "Automated production of instructionally appropriate scenarios," *Proceedings of the 8th Conference on Computer Generated Forces and Behavioral Representations*, Orlando, FL, 11-13 May, pp. 195-201.

Tuttle, C., Lacy, L., Levine, B., O'Brien, L., and Sheehan, J. (1999), "Knowledge acquisition tool – supporting the development of a knowledge base," *Proceedings of the 8th Conference on Computer Generated Forces and Behavioral Representations*, Orlando, FL, 11-13 May, pp. 175-180.

Tyler, J.G., Brisbois, G.R., Canova, B.S., Christensen, P.H., Lee, M.D., Pack, D.L., and Peterson, E.L. (1998), "Enhancing individual combatant behavior models through knowledge representation," *The 1998 Fall Simulation Interoperability Workshop*, Orlando, FL., Sept. 12-17, pp. 758-767.

Tyler, J.G. (2000), "Agent-based reasoning models for command decision makers," *Proceedings of the 9th Conference on Computer Generated Forces and Behavioral Representations*, Orlando, FL, 16-18 May, pp. 215-230.

VonderLippe, S.R. and Courtemanche, A.J. (1998), "Imerim results in the development of user composable behaviors," *Proceedings of the 7th Conference on Computer Generated Forces and Behavior Representation*, Orlando, FL, 12-14 May, pp. 545-553.

VonderLippe, S.R. and Courtemanche, A.J. (1999), "Advances in the execution engine for composable behaviors," *Proceedings of the 8th Conference on Computer Generated Forces and Behavioral Representations*, Orlando, FL, 11-13 May, pp. 37-45.

Vrablik, R., Carey, S.A., Chamberlain, F., Longfritz, M.J., and Kleiner, M. (1998), "Advanced synthetic command forces architecture," *The 1998 Spring Simulation Interoperability Workshop*, Orlando, FL., 9-13 March., pp.176-184.

Vrablik, R. and Longfritz, M.J. (1999), "An engine for performing behavior decision-making," *The 1999 Spring Simulation Interoperability Workshop*, Sept. 12-17, Orlando, FL., pp. 354-362.

Wallace, P., Walkden, K., and Lintern, G. (1999), "An analysis of F/A-18A pilot training tasks," *Proceedings of the 4th International SIMTECT Conference*, Melbourne, Australia, 29 Mar – 1 Apr, pp. 15-19.

Wallace, S.A., Laird, J.E., and Coulter, K.J. (2000), "Examining the resource requirements of artificial intelligence architectures," *Proceedings of the 9th Conference on Computer Generated Forces and Behavioral Representations*, Orlando, FL, 16-18 May, pp. 73-86.

Weaver, W.B. (1993), "Behavioral representation completeness in CGF implementations," *Proceedings of the 3rd Conference on*

Computer Generated Forces and Behavioral Representation, Orlando, FL, 17-19 March, pp. 375-386.

Weaver, W.B. and Mullen, W.J. (1994), "Simulating generic military decision making with an empirically-trained neural network," *Proceedings of the 4th Conference on Computer Generated Forces and Behavioral Representation,* Orlando, FL, 4-6 May, pp. 531-540.

Weaver, R., Silverman, B.G., Shin, H., and Dubois, R. (2001), "Modeling and simulating terrorist decision-making: a performance moderator function approach to generating virtual opponents," *Proceedings of the 10th Conference on Computer Generated Forces,*" Norfolk, VA, 15-17 May, pp. 39-44.

Webber, B. and Badler, N. (1993), "Virtual interactive collaborators for simulation and training," *Proceedings of the 3rd Conference on Computer Generated Forces and Behavioral Representation,* Orlando, FL, 17-19 March, pp. 199-204.

Williams, K., Severinghaus, R., and Clare, T. (1998a), "A modified GOMS cognitive task analysis technique for creating computational models of adversary behavior," *Proceedings of the 7th Conference on Computer Generated Forces and Behavior Representation*, Orlando, FL, 12-14 May, pp. 75-83.

Williams, K., Severinghaus, R., and Clare, T. (1998b), "Behavior representation using cognitive task analysis modeling – bridging the gap between decision processes and model development requirements," *The 1998 Fall Simulation Interoperability Workshop*, Orlando, FL, 13-18 Sep., pp. 296-304.

Woodward, S. (1999), "Evolutionary project management," *IEEE Software*, vol. 32, no. 10, October, pp. 49-59.

Wu, W. and Yoshikawa, H. (1997), "A pilot study on Human Cognitive Reliability (HCR) by human model simulation," *Proceed-*

ings of Intelligent Information Systems (IIS'97), Grand Bahama Island, Bahamas, 8-10 December, pp. 95-99.

Yufik, Y.M., Beadling, R.G., and Hu, T.H.G. (1997), "Operator modeling in complex large-scale systems," *Proceedings of the Military, Government, and Aerospace Simulation Conference*, Atlanta, GA, 6-10 April, pp. 231-237.

Zachary, W., Ryder, J., Hicinbothom, J., Santarelli, T., Scolaro, J., Szczepkowski, M., and Cannon-Bowers, J. (1998), "Simulating behavior of tactical operators and teams using COGNET/ GINA," *Proceedings of the 7th Conference on Computer Generated Forces and Behavior Representation*, Orlando, FL, 12-14 May, pp. 365-376.

Zhang, W. and Hill, R.W. (2000), "Situation awareness and assessment – an integrated approach and applications," *Proceedings of the 9th Conference on Computer Generated Forces and Behavioral Representations*, Orlando, FL, 16-18 May, pp. 365-376.

Zimm, A.D. (1999), "Modeling maneuver warfare: incorporating human factors and decisionmaking in combat modeling," *Proceedings of the 8th Conference on Computer Generated Forces and Behavioral Representations*, Orlando, FL, 11-13 May, pp. 225-236.

Chapter 4

Using Automaton Models to Rapidly Illuminate Tactical Options

M.K. Lauren and R.T. Stephen

The continuing evolution and complexity of the battlefield present enormous problems for analysts. However, progress in the field of Complexity Science suggests that many aspects of complicated phenomena are best understood in terms of the multiple interactions of relatively simple rules. In this chapter, we describe an approach to battlefield modeling based on the use of simple cellular automaton agent entities. The agent entities in the model have autonomy to act independently of their initial instructions in response to situations that arise during the simulation. This makes the events in the simulation highly non-linear, and promotes a wide variation of outcomes. It also provides the possibility of modeling groups of combatants with differing personalities, and further, allows the modeling of scenarios that are not concerned solely with weapons fire. At the same time, the simplicity of the model makes it much easier to experiment with variations on scenarios than is possible with traditional, computationally expensive methods. This allows the parameter space to be thoroughly mapped and high payoff areas to be identified quickly. The simple approach that these kinds of methods represent is potentially an avenue by which analysts and military officers can explore innovative ideas without tying up manpower and hardware for extended periods.

1 Introduction

The Cold War brought a false sense of security for the defense modeling communities of the major Western powers. They knew who their potential enemies were, and how they would conduct war.

By contrast, the current climate is far less predictable and the roles of the major militaries have become increasingly blurred. Certainly, the events of September 11, 2001, drew everyone's attention to the fact that war need not be waged by conventional means only. It also brought the term "asymmetric warfare" into the limelight.

Broadly speaking, asymmetric warfare refers to aggressive action that does not involve one conventional military force pitched against another.

It is a type of warfare described as asymmetric because an attack under such a doctrine is usually directed at a target with no suitable defenses. An example would be to try to damage a military force by destroying its unarmed supporting infrastructure, rather than attacking it directly.

The US Marine Corps has been one of the first organizations to recognize the need to address this changing military landscape. It has produced a concept it calls the Three Block War. That is, a Marine may have to engage in intense combat, be involved in separating groups of protagonists, and engaged in humanitarian assistance roles all in the same day (Brandstein 1999).

New Zealand's Defence Technology Agency became involved in the efforts to address this problem as a result of its dealings with the Marine Corps (Lauren 2001).

New Zealand's armed forces, in keeping with international best practice, are adapting analysis methods to justify force structure and acquisition. However, all of the New Zealand Army's deployments of recent times have been for peacekeeping and humanitarian operations. For these, analysis methods designed for the Cold War are virtually useless.

Consequently, New Zealand and the Marine Corps have a mutual interest in trying to discover analysis methods that allow us to deal with this new environment.

Cold War-style modeling tools appear to be weak in dealing with (at least) four areas:

1. **Non-linearities:** Under the right conditions, small changes in the course of events can have a dramatic impact on the outcome.
2. **Intangibles:** How do such things as personality and morale affect the outcome?
3. **Innovative foes:** How do you find potential weakness in your force structure that could be exploited by an innovative foe?
4. **And, non-traditional actions:** How do you deal with restrictive rules of engagement and the presence of large numbers of non-combatants?

1.1 Models

Most of our investigations have been based around the use of Agent-Based Models (ABMs). ABMs have the characteristic of containing entities that are controlled by decision-making algorithms. That is, entities within an ABM combat model make their own decisions, based on how they perceive their environment, as opposed to the modeler explicitly determining their behavior in advance.

Our chosen method to address the four problem areas outlined above is to employ Cellular Automaton (CA) models. CA models are considered the simplest class of ABMs.

While the more sophisticated ABMs employ complex decision-making algorithms, CA models typically rely on simple "local" rules. With CA models, the details of the individual entities are not stressed as heavily as the relationships between them. Because these relationships are usually non-linear, the output of the model can often seem to be more than simply the sum of its components.

Conventional combat models are very much the sum of their components. The outcomes are usually strongly dependent on the number and killing power of the weapons in the model. By contrast, because agents can cooperate, the outcome of a situation is often not as strongly affected by tangible elements, but rather dependent on the personalities of the agents and whether they are able to cooperate effectively.

Additionally, due to their non-linear nature, the outcomes of CA simulations can be much more complex and unpredictable than conventional models.

Examples of the early forms of CA models are the famous Ising model of magnetic spin alignment in physics, and Conway's Game of Life.

These models typically consist of a homogeneous lattice of cells that may be occupied by discrete entities, with a maximum of one entity per cell (Ilachinski 2001). Typically these entities evolve dynamically according to fixed rules that are local, in the sense that the rules only apply to the entity and the cells immediately adjacent to them.

For example, in the Game of Life, entities are born or die depending on how many other entities occupy the neighboring cells.

The investigation of complex phenomena using CA models forms part of an emerging discipline often referred to as "Complexity Science" (Waldrop 1992). It turns out that many apparently complex behaviors in the natural world can be explained as the combination of a few simple rules. Examples are the flocking behavior of birds and the defensive swirls that schools of fish adapt. The patterns these creatures form can appear amazingly complex and give the illusion that some central intelligence is producing them. But mathematical simulations such as CA models have shown that it is the propensity of the individuals to obey these simple rules that produces these complex patterns.

Thus relatively simple models can often provide much more convincing simulations of real phenomena than highly detailed models of the individual components that fail to capture the interaction between the components.

In essence, this is what we are striving for from our CA combat model: producing complex and realistic-appearing behavior using simple rule sets.

For combat modeling, strictly local behavioral rules and homogeneous lattices are too restrictive to allow reasonable representation of warfare.

A significant step towards a useful CA combat model came with the development of the widely distributed ISAAC (Irreducible Semi-Autonomous Adaptive Combat) model by Andy Ilachinski from the Center for Naval Analyses (Ilachinski 2000).

This model allows interactions over much longer ranges than typical CA models, and introduces realistic battle features such as communications and objectives. These differences between ISAAC and traditional CA models raise the question of whether ISAAC is still a true CA model. Our view is that whether it does qualify as a CA model is largely a matter of semantics. It still retains the qualities that are of importance to us, i.e., it uses simple rules and representations of the individual components, allowing us to focus on the patterns produced by the whole.

Since our first investigations with ISAAC, the model has seen many improvements, evolving into the EINSTein (Enhanced ISAAC Neural Simulation Tool) model, and has spawned a series of similar models.

Despite these developments, we felt we needed to develop our own CA model in order that we could properly address issues of concern to us. The parameters within our model, called MANA (Map Aware Non-uniform Automata) (Lauren and Stephen 2002), can be divided into three basic types:

- The first, the personality weightings, determine an automaton's propensity to move towards or away from certain things, such as friendly or enemy units, waypoints, terrain types, or "remembered" enemy locations.
- The second type, called constraints, act as conditional modifiers to these weightings. For example, the Advance parameter prevents an automaton from moving towards its objective without a minimum number of friendly units accompanying it.
- A final set of parameters describes the basic capabilities of automata, such as weapons range, sensor range, movement rate, single-shot kill probability, and stealth.

So far this parameter set bears a superficial similarity to those used in ISAAC. However, as well as differences in the algorithm used to calculate movement, MANA has several features that are unique to it.

- MANA allows a range of triggers to cause a change in an entity's personality. For example, a contact with an enemy, a shot being fired, reaching a goal point, and becoming injured may all change an automaton's personality.
- MANA also places greater emphasis on global interactions, by providing each side with a "memory map", which is used to mark the locations of enemies. The entities thus react to both enemies they see and enemies they "remember".

This last feature is an extremely simple yet effective method by which one can provide the entities with a "memory". This allows the model to produce sophisticated behavior rivaling that possible with much more complicated ABMs.

On the memory map, enemy locations are recorded and entities are drawn to or repelled from them just as they are for enemies they can see directly. The location information is removed after a set period of time, representing that it has dated and is no longer of importance. The contact information also distinguishes between enemies of differing threat levels. Hence entities may be attracted to one kind of enemy, but repelled by another.

Another feature is a settable level of Brownian randomness. MANA entities use a penalty function to rank possible moves, based on their personality rules. If several moves have a similarly low penalty, a move is chosen at random from the best moves. A "movement precision" parameter sets how wide the margin should be for accepting similarly good moves. By introducing this randomness we are notionally representing small differences in the personalities of individual automata, which might otherwise have exactly the same parameter set. In other words, an automaton put into a certain situation will not always act the same way, even if its parameters do not change.

Although, like ISAAC, MANA has a fixed set of parameters and rules, it is still able to produce an amazingly wide range of realistic looking behavior. Unfortunately, it is difficult to back this claim up without seeing the model. Nevertheless, we hope that the following sections of this chapter will provide some evidence as to the utility of the model.

1.2 Philosophy

An important part of our philosophy in choosing CA style models is that we are embracing a relatively uncomplicated approach to agent modeling. That is not to say that MANA is not sophisticated, but rather, that it does not use sophisticated decision-making algorithms.

The whole point of our approach is that the model be easy to set up and run. This sort of minimalist approach to modeling is often referred to as a distillation modeling. The term was coined duc to the fact that such models are designed to capture the "essence" of a particular situation, ignoring superfluous details.

One approach to using such models is to run them a large number of times, say, on a supercomputer (for example, the Marine Corps uses the Maui High Performance Computer Center for this purpose). This allows a thorough investigation of the parameter space, much more so than is possible with conventional computationally expensive models.

Our approach with MANA is slightly different from the dedicated distillation models. MANA is intended to be able to be used with varying levels of detail. While MANA can easily be used as a distillation model, it can also represent the characteristics of the entities in more detail if desired. For example, the latest version of MANA allows details such as terrain line-of-sight modeling and range-dependent weapon performance to be modeled.

On the other hand, MANA is relatively inexpensive computationally, and is able to be used as an effective analytical tool even with a desktop PC. MANA keeps the computation level down by using stochastic representations of the details of a real situation, rather than try to model them explicitly. That is to say, rather than modeling the elevation of a gun, the trajectory of its bullet, and the imperfect human doing the aiming, we simply represent all these things in terms of a single hit probability.

As with distillation models, MANA allows rapid and systematic exploration of the parameter space in order to build broad parameters landscapes from which it is possible to identify the most critical elements in a scenario. Though it is possible to do so with a desktop PC, it is generally preferable to use the computing power of a cluster of PCs to do this sort of extensive parameter exploration.

MANA's ease of use makes it suitable for exploring so-called co-evolving situations. That is, for a given scenario, an enemy that initially performs poorly may be able to change its tactics to greatly improve its performance, or alternatively, nullify the effectiveness of some new piece of equipment. In return, the friendly force may change its tactics to try to improve performance, and so on. Because MANA is relatively simple to set up, it allows rapid iteration of such evolution.

This contrasts with the approach of very detail-centric models. The detail inherent in such models not only increases the computational expense, but also dramatically prolongs the set-up time. This makes it impractical to rapidly explore the parameter space of the model, and hinders efforts to explore co-evolving situations. In effect, the extra detail ties the model to a so-called "point estimate". That is, only a very small area of the parameter space is examined, and the modeler tries to convince his/herself that that point in the parameter space represents a known force's performance with high fidelity.

Traditional point-estimate approaches narrow the modeler's view of the parameter space to the extent that it is impossible to determine the existence of nearby high-payoff areas in the parameter landscape.

The main advantage of CA models over more complicated ABMs is that they use fixed sets of parameters and rules. Combined with an easy-to-use Graphical User Interface (GUI), this makes set up possible for even inexperienced users of the model. Model behavior can be altered significantly simply by altering a couple of values in the edit boxes of the GUI.

Some other groups who use ABMs take a different approach. They seek to build a model capable of adding behavioral rules at will. However, the generality strived for by these groups tends to lead to modeling platforms that resemble high-level programming languages. This is due to the great difficulty associated with adding new rules that may be quite different from the existing ones.

Our view is that in order to make these changes, it may be just as easy to alter the code directly as it is to alter extremely complicated behavioral rules which resemble code.

Apart from anything else, the more detailed a model, the more obscure its workings, a problem that is compounded if the user is not the model designer. Certainly, we have seen examples of ABMs that require teams of people just to maintain the knowledge base necessary to run them.

1.3 Adding Personality

Behavior of troops in any kind of scenario plays an important role, but is often overlooked by analysts because human nature is a mathematical intangible.

As with many other intangibles in analytical combat models, it is often thrown away in favor of dwelling on infinitesimal details, almost always of little real importance to a given scenario.

Generally speaking, there seems to be a school of analysts who believe that just because they have an equation to describe some aspect of a scenario, then that aspect must be more "real" than the aspects of the scenario that cannot be so easily described.

We would argue that it is precisely the aspects that are most difficult to describe which are of the greatest importance. So much in war is intangible, and these intangibles are spoken about so often it is impossible to ignore them.

The greatest complaint that analysts have in introducing parameters which represent "intangibles" inevitably comes back to the question of validation. How do you know which intangibles you want to introduce into the model, how do you represent them within the model, and how do you know what value to use for these parameters?

However, this is where the rapid exploration of the parameter space proves critical. If you can explore enough variations of behavior, you can select just those cases that appear realistic. In other words, the best way to use the model is not to have preconceived ideas about what the parameter values should be, but rather to look at a range of parameters and choose the ones that produce the most interesting behavior. Our experience has certainly been that experimentation by trial and error with the parameters will lead to realistic behavior far faster than trying to determine what likely parameter values are in advance. In this way, validation does not appear to be such an important issue.

It is important to bear in mind that what we are actually modeling is the outcome *assuming* that the combatants behaved in the way we modeled them, which is not necessarily what would happen in real life. It is easy enough to change the parameters to generate different behavior and see how this affects the outcome. This is essentially the idea of our approach: generate many different behaviors and see which ones are the most interesting.

It should also, however, be remembered that models like MANA are not intended to be able to describe every aspect of a military operation. Furthermore, there is no inbuilt "intelligence" which determines the plan the MANA entities are working to. Consequently, careful thought must be given when setting up a scenario.

There must be a clear idea of which aspects of warfare the scenario is addressing, and what the entities are trying to do. Though such an approach may seem pre-potted, the non-linear nature of the model ensures that, regardless of the modeler's preconception, a startlingly large number of outcomes are possible.

Since it is only necessary to have simple behavioral rules to achieve this, it seems almost pointless to make the rules more complicated than necessary. Turning the question around, if even simple rules produce complicated and unpredictable behavior, then what hope is there of relating observed behavior back to specific aspects of a complicated rule set?

MANA's limitations must also be noted. It is currently not designed to examine careful formation fighting, for example.

Nor do the entities within MANA always behave in a sensible manner. They often exhibit actions that might be described as "mistakes". However, we see this as a strength of MANA.

Rather than having heavily pre-potted behavior representing a "perfect way" of doing things, it may be highly useful to be able to ask which kinds of formations appear to produce the best results, and what are the consequences of making certain kinds of mistakes. Placing too many constraints upon the model eliminates these possibilities.

This leads us to the possibility of learning information from extreme outcomes (i.e. where the force being modeled has performed exceptionally well or poorly). To explore this kind of question, we have found it useful to record the random seed number for each run, and go back and visually examine the best and worst cases to see what the highest pay-off tactics are, or what kind of things might lead to disaster. These outliers have the potential to provide the greatest insights into the strengths and vulnerabilities of a force or set of tactics.

2 Applications

In this section we discuss a specific application of the MANA model to representing training levels and command and control within a small unit. This is done using a generic scenario designed to represent a "peacekeeping" environment.

In the scenario, a peacekeeping battalion acting as part of a larger coalition force has been tasked to protect a province which has been designated a safe area. Anti-coalition forces are active in the neighboring provinces, and have been attempting to conduct covert operations within the battalion's area of responsibility.

These opposition units typically consist of semi-trained belligerent groups armed with automatic weapons. Many of its members were previously associated with the region's former military-controlled government, who are generally hostile to the presence of the coalition forces. These guerillas operate on a small enough scale for the government's present leadership to deny any association with them.

2.1 Scenario

The battalion has discovered that small groups of belligerents are attempting to infiltrate its safe area for the purpose of causing civil instability, and generally discrediting the efforts of the peacekeeping force.

In this scenario, a squad (Blue force) is on patrol in a 2km × 2km area near the boarder to the province. The squad is heading east towards an observation post. A belligerent group (Red force) has entered the area from the north, and is attempting to access an infiltration route to the south.

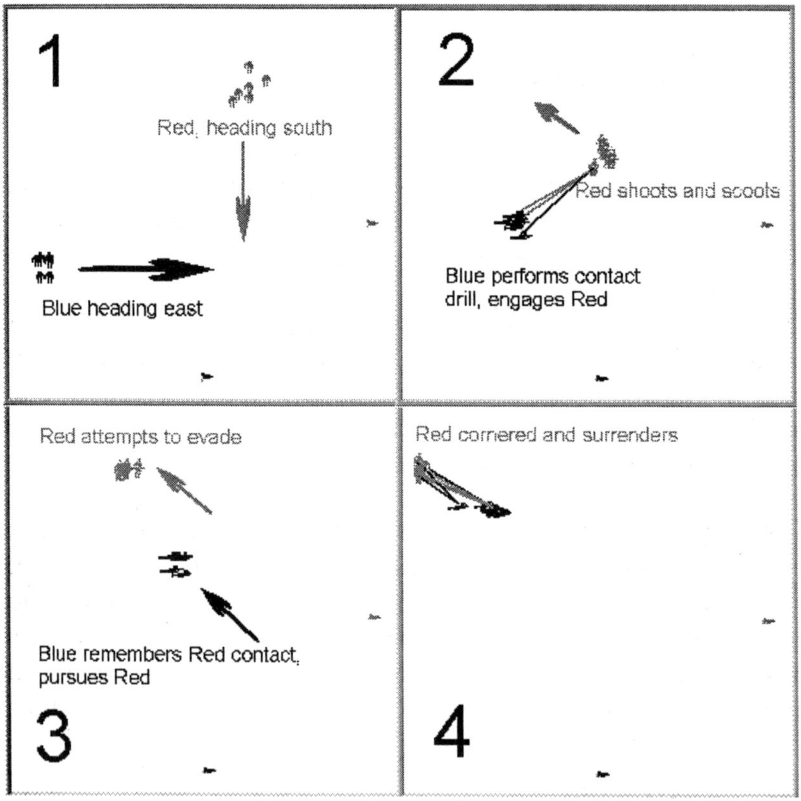

Figure 1. Typical evolution of a single run.

Recent encounters with belligerents in this area have resulted in exchanges of gunfire, and the battalion is currently operating on a shoot-on-sight basis. The local terrain is covered by uniformly distributed vegetation on undulating ground.

The squad consists of trained, but not necessarily combat practiced/ experienced, troops. They use standard dash-dive-crawl-shoot procedures on contact with hostiles, and make full use of cover. They are able to move on foot with a degree of stealth.

The belligerent group is well-armed but less well trained. They are consequently less adept at contact procedures, are not as stealthy, and less-accurate shooters.

The belligerents will use shoot-and-run tactics to evade the squad. The squad will attempt to pursue the belligerents to deter them from reaching their goal, and will do so until they loose their "trail".

Figure 1 shows how a single run might typically play out. Note that due to the non-linear nature of the MANA model, any given run may play out in a completely different way. While for the example shown the belligerent squad is trapped in the corner (notionally representing some impassible barrier), the Blue squad might just as easily lose contact with the belligerents before a decisive engagement can be achieved.

2.2 Control v Squad Mix

The purpose of this study was to compare the relative merits of troop quality and squad leadership (a.k.a. control).

Experienced troops were deemed to be better shooters, and stealthier. These two quantities were represented by the parameters that corresponded to kill probability and detection probability.

To represent leadership, we varied the tactics used by the squad. Two variations were examined: a rigid (i.e., soldiers stick to their own team) two-squad leapfrog method; and a more flexible leapfrog tactic (i.e. soldiers did not belong to one team or another) that allowed splitting into more than two teams.

Additionally, in the rigid action case, if one member of the team makes contact, the whole team goes to ground. In the fluid case, only the squad members in contact go to ground, the rest maneuver into contact.

The difference in the tactics was achieved by use of the MANA trigger states. The first tactic was achieved by splitting the squad into two teams, and giving each team this trigger rule: If one member of the team is shot at then all the members of the team react (i.e., take cover).

The second tactic was achieved by having only the individual who was shot at react, allowing the other members of the squad to maneuver into better firing positions.

To examine which of troop quality and leadership played the biggest role for the squad, we varied the number of "experienced" troops within the squad for both sets of tactics. The difference between the performance of the two-team leapfrog tactics and the fluid leapfrog tactics is examined in Figure 2 in terms of Red mission success (note that since this figure shows Red success, a high value is bad). It is notable that the "rigid action" case performed much worse than the fluid case for most troop mixes.

The same can be seen from the Blue casualty level analysis in Figure 3. Note that for this figure, the mixes with 0-2 experienced troops appear to outperform the mixes with 3-6 experienced troops. This is simply because the Blue squad was not engaging Red as often (an undesirable outcome), hence fewer casualties were likely but the mission failed more often.

It can be seen from Figure 3 that the rigid case has four to 10 times the casualties of the fluid case (depending on the mix). Meanwhile, the Blue squad prevents Red from achieving its mission much less than half of the time for all mixes of the rigid case, whereas Red is stopped nearly half the time for the fluid case. Also note that if the squad tactics are fluid, casualties do not change much with different mixes.

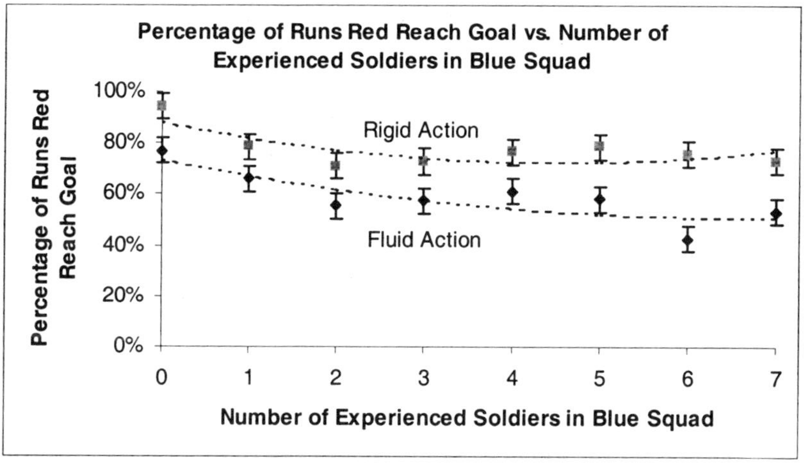

Figure 2. Relative performance of the rigid and fluid cases in terms of preventing Red from achieving its mission.

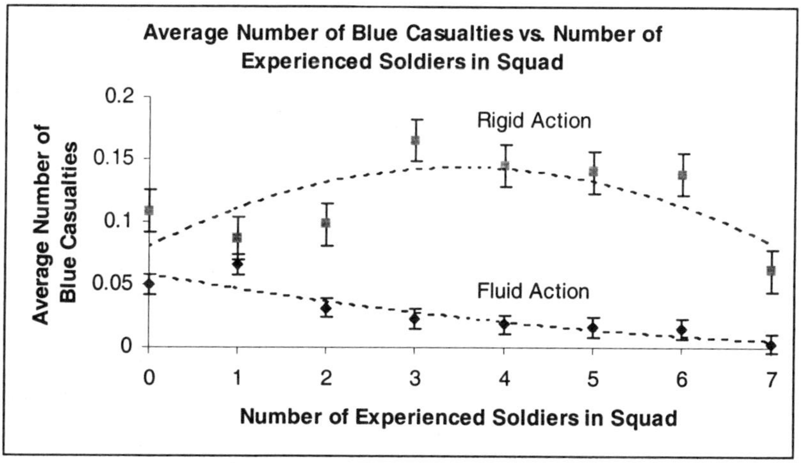

Figure 3. Relative performance of the rigid and fluid cases in terms of Blue casualties.

This highlights the improvement that can be obtained by a capable leader keeping good control of his squad. It seems entirely reasonable that an inexperienced section leader or poorly trained section would execute the contact tactics in a much less fluid way than a well-drilled and confident section or commander.

Thus, we have shown how an intangible, personality-driven effect such as leadership can produce a larger gain than tangible aspects such as shooting ability.

3 Parameter Landscapes

In this section we briefly describe how the model can be used to rapidly map parameter spaces.

One of the main advantages of agent-based models is that the autonomy of the agents means that if certain parameters change from run-to-run, the agents can often adapt to those changes in sensible ways.

Consequently we might systematically vary certain parameters without the need to redesign the scenario. This contrasts with traditional combat models, which usually rely on highly choreographed movement.

Using such automated and systematic variation of parameters it is possible to build parameter "landscapes" which graphically reflect the performance of one force. Example landscapes are shown in Figure 4.

Landscapes enable the analyst to quickly compare the merits of different parameters, to identify the parameter values that are likely to bring success, and to identify which parameters bring non-linear pay-offs. For example, the landscape in the figure was generated using a scenario in which a small reconnaissance unit had to iden-

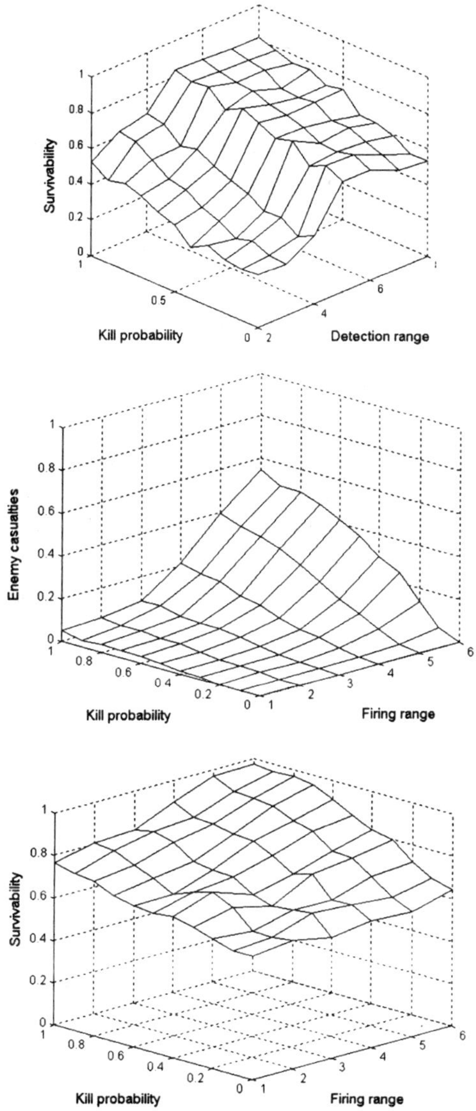

Figure 4. An example parameter landscape generated using a reconnaissance/counter-reconnaissance scenario. The landscape compares the relative importance of particular parameters, as well as allows rapid identification of the values required to ensure success.

tify a target protected by a large counter-reconnaissance force (Lauren and Baigent 2001).

In that study, we systematically varied the parameters associated with different kinds of equipment. This allows the determination of which equipment attribute is of most value in this scenario.

For example, in the landscape at the top of the figure, one can easily see that sensor equipment (providing a detection range advantage) potentially has a much greater impact on the recon force's survivability than improving its weaponry (represented by kill probability).

It can be quickly seen which values for detection range bring this payoff. In this case, the reconnaissance force's detection range needs to be 5 or better. It can also been seen that the effect of this parameter is non-linear, since survivability jumps nearly discontinuously once the detection range value increases above 4.

4 Technical Aspects of the MANA Approach

In this section, we discuss how one might build an automaton combat model using MANA as an example. To this end, we present a framework and a set of parameters and rules that we have found useful for generating interesting behavior for our analysis.

As discussed, in MANA each agent's artificial intelligence has been restricted as much as possible. The emphasis is on modeling the interactions of many simple agents rather than generating sophisticated behaviors for individual agents.

MANA was developed in the object-oriented programming language "Delphi". Agent-based models are well suited to object-oriented development methods, as there is a tight correspondence of the entities to be modeled and the software objects. An object-oriented design also makes changing the model easier.

The model consists of a few key classes (of objects). Note that the word "object" is used in the following text for clarity, but to be exact, it is referring to the class of objects.

Within our model, the sensor and weapon capabilities are simplified and integrated into the agent object. However, there is no reason why MANA could not be extended to have weapon objects and sensor objects that can be "owned" by agents, allowing more varied and complex models of these capabilities.

4.1 Objects

4.1.1 Battlefield Object

The Battlefield object contains an overview of the rectangular grid that represents the battlefield where the agents interact. It has a map of the terrain. It keeps track of the squads, and all the agents. It is responsible for moving all the agents, for running combat between agents, and recording progress data for model output (measures of effectiveness). It prevents agents from moving off the grid, into impassable terrain, or two agents occupying the same location.

At each time step the battlefield object decides an order for agents to move and fire, then calls each agent object to move itself. Tables 1 and 2 give examples of the type of data and functions used for the battlefield object.

4.1.2 Squad and Agent Objects

The squad object maintains a group of agents. It also has a set of properties that can be shared by the agents that belong to that squad, such as:

- a shared set of personalities that the squad agents can take on;
- a situational awareness map ("memory" map) of enemy contacts;
- waypoints that become the agent's goals, one after the other.

The agent object handles the data for individual agents, which includes the locations of the agent, what state it is in, and also detects the environment, fires and moves.

It contains the personality parameters that determine the weightings with which it moves toward or away from other objects. It has the ability to modify its personality if certain local conditions are met, but not its entire personality set, which is the job of the parent squad.

Table 1. Example of battlefield object parameters.

Battlefield Data	Description
Terrain	An object which maintains the terrain properties (going, cover, concealment, elevation) at each location on the battlefield
Width, Height	The dimensions of the battlefield grid
AllSquads	Storage for all the squad objects
AllAgents	Pointers to all the agent objects
Scenario	Records the scenario data and map filenames, scenario settings (run length, no. of runs, etc.), current time step
Casualties	Record of casualty data for each allegiance

Table 2. Example of battlefield object functions.

Battlefield Functions	Description
AddAgent	Adds a new agent to the AllAgents list
DelAgent	Removes a specified agent from the AllAgents list
DecideOrder	Randomises the order in which agents will be asked to move and fire
MoveAgents	Updates the situational awareness of all squads, then calls DecideOrder and asks each agent in turn to move and fire
AddSquad	Adds a new squad to the AllSquads list
RecordCas	Increments the count of casualties of the killed agent's allegiance
Display Events	Display agent movement, firing and interactions at the end of each step
File functions	Load and save scenario files

4.2 Parameter Set

Many of the parameters used in MANA are analogous to the parameters used in ISAAC. The personality weighting parameters for MANA are summarized in Table 3.

Table 3. Description of weighting factors.

Personality Element	Description	Controls propensity to move toward/away from
w1	Alive Friends	Agents of same Allegiance
w2	Alive Enemies	Agents of enemy allegiance
w3	Injured Friends	Injured agents of same allegiance
w4	Situation Awareness Friends	Agents of same allegiance, out of sensor range but on situational awareness map
w5	Next Waypoint	The next waypoint agent's squad has been assigned
w6	Enemy's Final Goal	The enemy's final goal (waypoint)
w7	Easy Terrain	Areas of Easy Going e.g. yellow road.
w8	Enemy Threat 1	Enemies in SA map which are of Threat Level 1
w9	Enemy Threat 2	Enemies in SA map which are of Threat Level 2
w10	Enemy Threat 3	Enemies in SA map which are of Threat Level 3
w11	Neutrals	Agents of neutral allegiance
w12	Cover	Areas which provide cover e.g. dense bush
w13	Concealment	Areas which provide concealment e.g. dense bush

These weightings play a significant role in determining each agent's movement. They combine with the constraint parameters r11, r12 and r13 (Table 4) to effectively determine the personality of the agent. The rest of the parameters are given in Table 4.

One of the most useful features of MANA is the state trigger, which allows the personalities of the automata to change under certain circumstances.

There are currently more than 20 triggers, and these are summarized in Table 5. For each trigger state, the squad object can maintain a set of personality, ranges/constraints, stealth, SSKPs, and threat level parameters. The trigger states can be individually enabled or ignored, a duration (in steps) can be set, as can the fallback state to which the squad members change (back to) after "duration" time steps have expired. When a squad event occurs, the squad is told to remember the event. The squad function "remember" simply tells its entire set of children agent objects to remember the agent event.

Table 4. Parameter definitions.

Parameter	Description
R1	Sensor Range
R2	Firing Range
R3	Max Targets Per Step
R4	Movement Speed
R5	Number of Hits to Kill
R6	Min Distance to Friends
R7	Min Distance to Enemy
R8	Min Distance to Next Waypoint
R9	Min Distance to Enemy Waypoint
R10	Min Distance to Easy Going
R11	Cluster Constraint
R12	Combat Constraint
R13	Advance Constraint
R14	Refuel (interaction) Range
R15	Prob. of refueling Enemy agent
R16	Prob. of refueling friend agent
R17	Prob. of refueling neutral agent
R18	Shot radius for area fire
R19	Min Distance to Neutrals

The personality, ranges and other properties of the agents are not changed until the agent "ego" function is called just before the agent next moves. The ego function checks the agent's memory for trigger events. If the current (non-default) state has expired, then the agent is told to remember the fallback state. If the non-default state is new (since the last time step), then the agent's properties, ranges, etc. are all changed to the new settings, read from the agent's parent squad properties.

Some spare states are allocated for use as intermediate fallback states. A future development might be to set agents to be attracted to other agents in certain states.

Table 5. State definitions.

State	Name in MANA	Description
1	Default State	Start State and Default Fallback State
2	Reach Waypoint	Agent State when Waypoint Reached
3	Taken Shot	Agent State When Hit
4	Shot At	Agent State When Shot At
5	Enemy Contact	Agent State When Enemy Contact Made
6	Squad Taken Shot	Whole Squad State When Squad Agent Hit
7	Squad Shot At	Whole Squad State When Squad Agent Shot At
8	Squad En Contact	Whole Squad State When Enemy Contact Made
9	Injured	Agent State When Injured
10	Squad Injured	Whole Squad State when a Squad Agent is Injured
11	Squad Dead	Whole Squad State when a Squad Agent is Killed
12	Attack	Agent State when a commander triggers an attack
13	Retreat	Agent State when an attack is over
14	Fuel Out	Agent State when fuel falls below zero
15	Refuel	Agent State triggered by another agent's refuelling action
16	Reach Final Wapoint	Agent State upon reaching the last waypoint
17	Refuel by En	Agent State triggered by an enemy agent's refuelling action
18	Refuel by Fr	Agent State triggered by a friend's refuelling action
19	Refuel by Neu	Agent State triggered by a neutral's refuelling action
20	Run Start	Agent State triggered at time-step 1 (useful for delays)
21-26	—	Reserved for future use
27	Squad En Contact 1	Whole Squad State When Enemy Contact of Threat 1 made
28	Squad En Contact 2	Whole Squad State When Enemy Contact of Threat 2 made
29	Squad En Contact 3	Whole Squad State When Enemy Contact of Threat 3 made
30-32	Spare	Spare states for use as intermediate fallback states

4.3 Movement

The most important action of an agent is to move. Agents in MANA move in an "instinctual" manner. The movement algorithm selects the grid square within its movement range that most satisfies its desire to move towards some entities and away from others. The current location is also considered, so the agent can stay put.

The movement algorithm consists of the following steps:

- Consider all moves within the movement range of the agent, including staying put.
- Eliminate moves into locations containing other agents or impassable terrain.

- Considering all the entities visible within range: decide on the most appealing of the permissible moves, using the personality weights.
- Impose behavior modifiers that change the basic behavior (e.g. minimum distance to others, cluster constraints, and so on). If a number of moves are nearly equal, then choose a move at random from the attractive moves.

A "movement precision" parameter sets how wide the margin should be for accepting similarly good moves. Setting the movement precision to a low value thus means that most often only the best move will be chosen and the movement will appear very deterministic. If the movement precision is large, the agents tend to wander about in a Brownian motion, as moves are selected at random.

The movement calculation is designed to keep the personality parameters balanced relative to one another and the tendency to move toward or away from an entity is constant with distance. For example, the weighting to move towards the next waypoint is the same whether the entity is three cells or 150 cells from it. These were pragmatic choices in order to allow the modeler to create a variety of behaviors by varying the personality parameters over a range of values. Also, to create realistic-looking behaviors, it is necessary to balance a number of personality parameters. It is therefore helpful for each parameter to affect agent behavior in a "fairly linear" way.

The penalty for moving to any grid location is the sum of 13 penalty calculations, corresponding to the 13 personality parameters listed in Table 3.

The algorithm used to calculate the penalty for a collection of entities within sensor range is the same for all 13 components. The general algorithm for calculating the penalty component associated with a candidate move is shown below:

```
Loop through all entities within range of current location
   OldDist = distance from current location to entity
   NewDist = distance from candidate location to entity
   If NewDist < Minimum then Direction = -1 else Direction = 1
   Sum = Sum + Direction * (NewDist+(100-OldDist))/100
   NumEntities = NumEntities + 1
Repeat loop
Penalty = Sum / NumEntities
```

The important term in the algorithm is:

```
(NewDist+(100-OldDist))/100
```

This treats all entities as if they are 100 units away. If NewDist < OldDist and a move closer is desirable, the penalty term will end up slightly less than 1.0. If NewDist < OldDist and a move away from the entity is desirable, a penalty term of slightly greater than 1.0 will result, i.e., greater penalty.

The penalty for moving towards or away from other agents is normalized by the number of agents.

For example, if I am an agent attracted to friendly agents, I seek to minimize my average distance to friendly agents within my sensor range. However, a number of constraints can modify this attraction. The minimum distance to friends/enemies range sets a radius to other agents, inside which the penalty is negated. The check against minimum distance is made for every agent within sensor range.

If an agent is within the minimum distance, its penalty is negated before adding to the sum for the penalty.

An example penalty calculation for enemy agents is illustrated in Figure 5 and Tables 6 and 7. Staying put at location $Blue_0$ produces a penalty component of 1.00000. The penalty component for Blue moving to $Blue_1$ is 1.00609 but for location $Blue_2$ it is less, at 1.00195. If the Blue agent's weighting towards the enemy is positive (attracted to the enemy) then the location with the least penalty is most attractive. Of the three locations shown, location $Blue_0$ (staying put) has the least penalty.

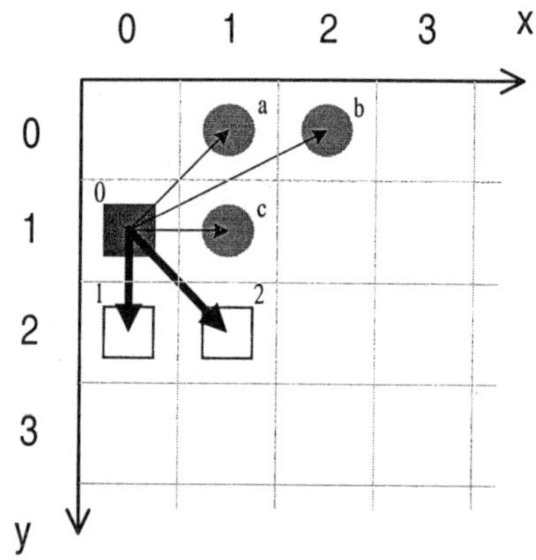

Figure 5. Sample grid for our penalty calculation.

Table 6. Penalty calculation for Blue candidate, location (0,2).

Enemy Agent	Location	OldDist (Dist to $Blue_0$)	NewDist (Dist to $Blue_1$)	(New + (100-Old)) / 100	Sum
Red_a	1,0	1.41421	2.23607	1.00821	1.00821
Red_b	2,0	2.23607	2.82843	1.00592	2.01414
Red_c	1,1	1.00000	1.41421	1.00414	3.01828
				Sum / 3 entities	1.00609

Table 7. Penalty calculation for Blue candidate, location (1,2).

Enemy Agent	Location	OldDist (Dist to $Blue_0$)	NewDist (Dist to $Blue_2$)	(New + (100-Old)) / 100	Sum
Red_a	1,0	1.41421	2.00000	1.00586	1.00586
Red_b	2,0	2.23607	2.23607	1.00000	2.00586
Red_c	1,1	1.00000	1.00000	1.00000	3.00586
				Sum / 3 entities	1.00195

A future improvement to the penalty calculation would be to make an agent better able to avoid nearby enemies. To maintain the relativity between penalty components, this could operate by counting a close entity as though it were many entities at the same position, and a far away entity as a fewer number of entities (or as a single entity at the extreme of the sensor range).

Further research needs to be done to determine what properties the movement algorithm needs to more realistically represent the actions of real combatants.

4.4 The Constraint Parameters

4.4.1 Cluster Constraint

The cluster constraint is designed to prevent agents from clustering in groups larger than a specified size. It operates on the attraction towards friends. Once the friend's penalty component of a move is calculated it is only counted if the number of friends within sensor range is less than the cluster constraint. Otherwise, it is negated, thus repelling friends if in a group larger than the cluster constraint size.

If the cluster constraint is zero or if the personality component towards friends is negative (repelled from friends), then the cluster constraint is ignored.

4.4.2 Combat Constraint

The combat constraint permits advancing toward the enemy only if a certain numerical advantage is met. Hence, the combat constraint modifies the penalty component for moving toward enemies. The numerical advantage is calculated as the number of friends minus the number of enemies in sensor range.

If the difference is greater than the combat constraint, then the move toward enemy penalty is used, otherwise it is negated, i.e. an attraction toward enemy becomes repulsion.

If the combat constraint is zero, or if the personality weight towards the enemy is negative (repelled from the enemy), then the combat constraint is ignored.

4.4.3 Advance Constraint

The advance constraint affects the penalty component for moving toward the next waypoint. If the number of friends within an agent's sensor range is fewer than the advance constraint, then the penalty toward the next goal is negated i.e. repelled from next goal, rather than attracted.

5 Conclusions

We have described how one can apply a cellular automaton model to combat scenarios. Our experience has been that CA models combined with parameter landscape techniques at worst provide useful insights, and at best have the potential to replace most other kinds of conventional analytical combat models.

Although cellular automaton models are simple compared with some agent models, we have found that this simplicity makes them extremely useful for quickly exploring concepts. It also allows the model to be easily understood and used by the service personal working with the analysts.

In saying this, our model as presented still requires much development to be of use for general applications. Certainly, the parameter set we have listed here is too limited to model many situations in enough detail.

Nonetheless, we feel that many of the shortcomings of the limited rule set presented can be overcome relatively easily. In the latest version, we have found that the addition of a few new rules, as well as adding some limited "global" intelligence (such as automated route planning or objective setting), dramatically improves the versatility of the model.

Though most of the studies we have used the model for are in the category of concept analysis, the model has great potential as a tactical option evaluation tool. This is our current thrust, and we have been able to make great progress in this regard with very simple extensions of the model.

Given that simple rules can produce an enormous range of realistic-looking combat behavior in a model, it is interesting to ask to what extent any real military action is simply the outcome of such simple rules.

We have plans to use the enhanced version of the model for operational decision making for small units, for analysis of force structure options (i.e., what equipment mix a certain unit should have), and for training and illustrative purposes.

References

Brandstein, A.G. (1999), "Operational synthesis: applying science to military science," *Phalanx*, vol. 32, no. 4, pp. 1, 30-31.

Ilachinski, A. (2000), "Irreducible semi-autonomous adaptive combat (ISAAC): an artificial-life approach to land warfare," *Military Operations Research*, vol. 5, pp. 29-46.

Ilachinski, A. (2001), *Cellular Automata: a Discrete Universe*, World Scientific Publishing Company.

Lauren, M.K. (2001), "Albert, serendipity and a Maori demigod – a cameo of progress in operational synthesis," *Phalanx*, vol. 34, no. 3, pp. 16-19.

Lauren, M.K. and Baigent, D.L. (2001). "Exploring the value of sensors to a recce unit using agent-based models," *Journal of Battlefield Technology*, vol. 4, no. 1, pp. 44-48.

Lauren, M.K. and Stephen, R.T. (2002), "Map-aware non-uniform automata (MANA) – a New Zealand approach to scenario modelling," *Journal of Battlefield Technology*, vol. 5, no. 1, pp. 27-31.

Waldrop, M. (1992), *Complexity: the Emerging Science at the Edge of Order and Chaos*, Simon and Schuster, New York.

Chapter 5

Multi-User Wireless Link for Real-Time Video Transfer for Defense Applications

W.M. Daniels, S. Forbes, and L.C. Jain

The Real Time Multi-User Wireless Video Link is being developed with the aim of enhancing the capabilities of soldiers by improving their ability to detect, acquire, locate and engage targets by day or night and in all visibility conditions. The primary objective of this work, undertaken as part of the 125 Soldier Combat System Project, is to improve a soldier's capability by removing the need for cables between a thermal weapon sight mounted on a weapon and a visual display mounted on a soldier's helmet. It improves their ability by enabling them to more easily travel through difficult terrain and enables them to navigate obstacles more easily. The future aims of this project are that it will enable video to be securely transferred to a remote base, enabling the base station to accurately monitor the situation. It could also be used to review the success of missions and for training purposes. This system also has many commercial applications for example monitoring in hazardous areas, surveillance, high quality video transmission/storage, law enforcement, fire-fighting and tele-medicine. In this chapter the requirements of this system and the development of prototype systems are discussed.

1 Introduction

The Soldier modernization concept is the Army's revolutionary program investigated under the 125 Soldier Combat System project. The aim is to integrate commercial off the shelf technologies into a complete soldier system for capability enhancement. One of the major requirements is to increase the lethality of the system by providing an improved capability to detect, acquire, locate and engage targets by day or night and in all visibility conditions.

The future combat helmet for the dismounted soldier is expected to include multiple capabilities such as visual display, auditory communication, a global positioning system, a visual link to weapon sight, and a shield from laser and fragmentation protection. The integration of these components is expected to give the individual soldier greater situational awareness and enhance fighting capabilities.

The aim of the Real Time Multi-User Wireless Video Link project is to improve the soldier's mobility by introducing a wireless link between a thermal weapon sight mounted on a weapon and a visual display mounted on the soldier's helmet. This system enhances the users' abilities by enabling them to see from covered positions, for example around buildings and over obstacles, minimizing the physical threat to the user. It helps the soldier take maximum advantage of the benefits that the Thermal Weapon Sight Provides. These include enabling the user to see under extreme conditions, through obscurants, like smoke, dust, or fog, or at night or in adverse weather conditions when visibility conditions are not optimal for the naked eye.

Video image transfer will be via a dedicated wireless video link. A wireless link improves the mobility, reliability and robustness of the system and makes it viable for use in the field. The absence of cables enables the user to move through dense terrain without being caught on foliage and avoids the need for the soldier to spend time disconnecting and then reconnecting the link. In Figure 1, we can see the soldiers observing around the corner of a building. It can be seen that this provides more flexibility for the soldier than just using the traditional cheek-to-stalk method.

Figure 1. Soldiers observing around the corner of a building.

In this chapter the development stages of this project will be discussed including technological issues encountered.

The Background section describes the first concept demonstrator produced at the Weapon Systems Division, DSTO, Salisbury. Subsequent sections describe the work undertaken on the project in collaboration between the Knowledge Based Intelligent Systems Engineering Centre at the University of South Australia and the Weapon System Division at the DSTO.

Deployment Scenario describes how the system is to be deployed in the field. General Structure explains the general structure of a digital wireless video system. Present Prototype details short-term development system, used to verify the modified concept and as a baseline for evaluating future work. The Future Prototype section outlines the approach that will be taken to develop the system for use in the field. Conclusions discuss the conclusions reached at this stage.

2 Background

A vast amount of work is being carried out throughout the world with the view to enhancing the future soldiers ability. Project LAND 125 – Soldier Combat System (also known as Project WUNDURRA) is Australia's soldier modernization program. Many programs have conducted trials using thermal sights and day-time cameras using cables for the transfer of the video signal. The DSTO concept demonstrator uses helical spiral cables as a backup device in the event of wireless transmission failure.

The use of cables reduces the mobility of the soldier. There is the potential for cables to get caught on obstacles; they may require disconnection and then reconnection executing maneuvers. This can be dangerous and time consuming, exposing the soldier to unnecessary risk. Springing cables snapping could cause potential injuries. The introduction of a wireless video link overcomes these problems.

High-speed wireless communication will be essential to the future soldier. Once the largest hurdle of providing enough bandwidth for real-time video transfer between the thermal weapon sight and the helmet mounted display has been overcome, the system can be expanded to enable short-range and long-range real time video transfer. For example it will enable forward observers to accurately relay more information in real-time to members of their section, section commanders and command centers (see section on deployment). By combining real-time video with other technologies like augmented reality, GPS and GIS the information becomes even more valuable.

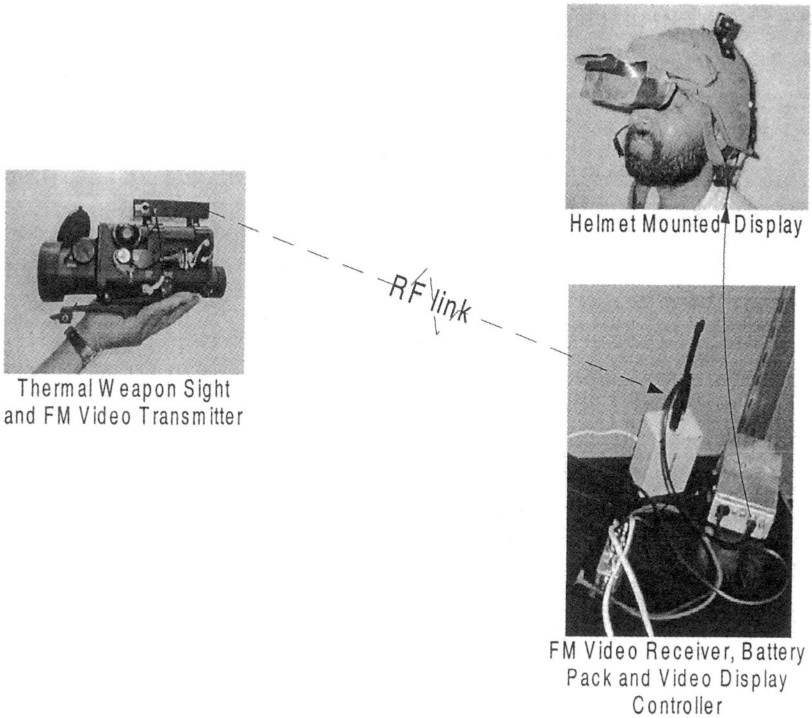

Figure 2. DSTO Concept Demonstrator.

3 DSTO Concept Demonstrator

The system developed by the Weapon Systems Division, DSTO, Salisbury SA, enables wireless RF transmission of video from a thermal imaging sight mounted on a weapon to a HMD. Analog transmission has been used in this system, assigning a separate frequency to each unit. This proves impractical for large numbers of users. The basic scheme of this system is show in Figure 2.

The Thermal Weapon Sight (TWS) produces an NTSC format video signal (Texas Instruments 1997):

- 352×240 pixels
- black and white (i.e., 256 colors = 8 bits/pixel)
- 29.97 fps (frames per second)
- $352 \times 240 \times 8 \times 29.97 = 20254924.8$ bps ≈ 19.3 Mbps
    - 1 Mega bit is defined as 2^{20} bits

The transmitter mounted on top of the TWS sends the video across the RF link using frequency modulation (FM) with a 165 mm antenna. The signal power is 80 mW.

This signal is usually transmitted over 3-5 m but can go up to 60m as observed in field trials. The signal is demodulated at the receiver mounted on the soldier's webbing. The receiver has a 150 mm "whip" antenna with BNC connector base. The signal is then displayed on the NTSC LCD screen mounted on the soldier's helmet.

Five of these systems were tested in the field using the frequencies shown in Table 1.

Table 1. Frequencies used in the first concept demonstrator. (Daniels *et al.* 1999).

Channel	Frequency
1	434 MHz
30	567 MHz
42	640 MHz
46	662 MHz
48	674 MHz

These successful trials identified the following limitations of the first concept demonstrators:

- Interference was observed when the transmitter and receiver units were in close proximity (centimeters apart);
- The five frequencies were chosen using trial-and-error methods to minimize the interference. It is not feasible to use this method to assign frequencies for our desired figure of one hundred users (i.e., a company);

- Interference was observed when the transmitter was in close proximity (i.e., less than 10cm) to other transmitters and the intra-section radio (operating at approximately 450 MHz).

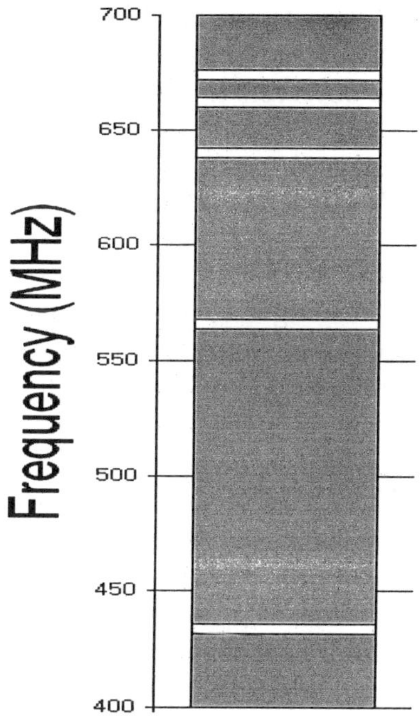

Figure 3. Graphical representation of the frequencies used. The white bands are the frequency bands used by the 4.2 MHz NTSC signal (Camera Interface Guide 1999).

4 Functional Requirements

The field trials held in Townsville training facility using the DSTO concept demonstrator showed that the concept was worth further development to overcome the limitations identified (see Section 3).

The basic functionality required for the system is to transfer a video stream from a camera (either a day-time camera or a thermal weapon sight) to a helmet mounted display in real-time. The system needs to be able to accommodate a large number of simultaneous users in close proximity to each other.

The long term objectives are for communication between section commanders, intra-section and long distances to base command, for example from forward observers to base command.

5 System Requirements

- **Video Input** : NTSC (black and white), 29.97 fps (analogue) While the system is designed with a thermal weapon sight the primary consideration, the input video signal may change specifications, for example to a PAL signal or digital video standard in the future.

- **Video display**: The DSTO concept demonstrator uses a Sony Glasstron with dual 1 inch LCD screens, which uses an NTSC (color) input. It is desirable to have VGA input, to avoid the need for conversion from VGA to NTSC, which consumes more power and adds further processing time. Future improvements in technology may provide significantly improved helmet mounted displays, for example retinal projection, involving projecting images directly onto the user's retina.

- **Video signal**: After qualitative testing it was found that the minimum acceptable frame rate was 15 frames per second (fps) and 352 × 240 pixels was the minimum resolution in grayscale (256 colors). It was found that the standard deviation between frames is the most significant factor, higher than frame rate. A large standard deviation between frames results in jittery video and this may make the video unwatchable. Higher resolution results higher quality but at the cost of greater bandwidth and greater processing time. Increasing the frame rate may not result in greater bandwidth usage depending on the type of video compression used but does require more processing due to the larger amount of data being processed.

- **Communication**: The system requires enough bandwidth to meet the video signal requirements. To meet the real-time nature we need the delay between the signal and display on the helmet mounted display to be minimized.
  This time is given by:
  $t = t_{input} + t_{tranmission} + t_{output}$
  = (video digitization/acquisition time
  + compression time + encoding time)
  + transmission time + (decoding time
  + decompression time + processing/display time).
  To minimize t we need to minimize t_{input}, $t_{tranmission}$, and t_{output}.
  theoretical t_{input}: to achieve our frame rate of 15 fps
  frame rate = 29.97 fps $\therefore$ 1 frame $\approx 33^1/_3$ ms
  frame rate = 15 fps $\therefore$ 1 frame $\approx 66^2/_3$ ms
  $t_{input\text{-}max} = 66^2/_3$ ms per frame

To ensure that the system remains operating at our nominal 15 fps we need to ensure that our input processing time takes less than $66^2/_3$ ms per frame. If this figure is exceeded buffers will overflow and require input frames to be dropped.

$\mathbf{t_{transmission}}$:
$f = 2.4\ \text{GHz} = 2.4 \times 10^9\ \text{Hz}$
$f = 400\ \text{MHz} = 400 \times 10^6\ \text{Hz}$

Speed of electromagnetic wave traveling through vacuum is speed of light = $3 \times 10^8\ \text{ms}^{-1}$ so if distance between 0.5-50m transmission time between $1^2/_3$ ns – $0.1^2/_3$ μs
$t_{transmission} \ll t_{input}$
$\therefore$ $t_{transmission}$ is negligible compared to t_{input}
$\therefore$ $t_{transmission} \approx 0$

t_{output}: to prevent the output buffers from overflowing output processing time needs to be equal to or less than t_{input}, i.e., symmetrical coding and decoding times, so the total time = $2 \times 66^2/_3$ ms = $133^1/_3$ ms

The limitation of these calculations is that we assume each individual frame is compressed and then transmitted. This is true for MJPEG compression which uses jpeg compression on each frame but not other techniques like MPEG-1 which is interlaced and uses two different types of frames, I and P frames.

However it does provide a rough guide to the working times.

- The system is required to support not less than a nominal figure of one hundred simultaneous users.
- Infra-red communication cannot be used as it would give away position because it is visible with thermal sight. The other problem is the multiple infrared beams would interfere with each other greatly due to the wavelength used.
- Reliability
  The two most critical areas of reliability is the power source used and the system architecture:
  - Existing batteries
    12V 5000mAH receiver battery for HMD plus internal HMD battery approx 2 hours gives a total of approximately 7 hours

  - Other batteries to consider:
    - TWS battery
    - Transmitter battery 9V

  This time seems to provide a reasonable working time.
  - Distributed system
    The system needs to work in a distributed varying outdoor environment. The system should work in ad-hoc mode, to avoid total system failure if the base station is damaged or malfunctions
- Maintainability and Lifecycle

The system is highly modular. The main benefits of this are:

- When components fail, just that component needs to be replaced or repaired
- New technology can be incorporated without major alterations to the system
- Scalable components can be added or removed depending on the mission
- The equipment can be phased in gradually, thus avoiding the need for a large financial outlay for completely new equipment.

- Availability
  High duty cycle – The system needs to operate continuously, the soldier cannot afford for the system to go into suspend mode to save power because it is not known when the system will be required to locate targets. Even a few milliseconds added reaction time could prove fatal.
- Soft systems – human factors
  - The biggest limitation of the system is the power usage. The common way to power such systems is to use batteries. Batteries comprise a large proportion of the total mass of the system
  - Looking at the helmet mounted display for extended periods is not desirable for the soldier – an improved display is necessary for long term usage.

6 Deployment Scenario

The primary objective is to transfer a real-time video signal from the thermal weapon sight (TWS) (or other video device) to the helmet mounted display (HMD) for each individual soldier. However the final aim is to enable video transfer between the soldiers in a given company and also to a remote base station.

The advantages of this such a system is that it will enable other soldiers and base command to quickly and accurately monitor what the soldier sees. Therefore this Video based weapon system could be used for evaluating field training in hostile situations.

The development approach is aimed to develop a the system for a company. Figure 4 shows the company structure where in such a system can be deployed using multi-access wireless technology.

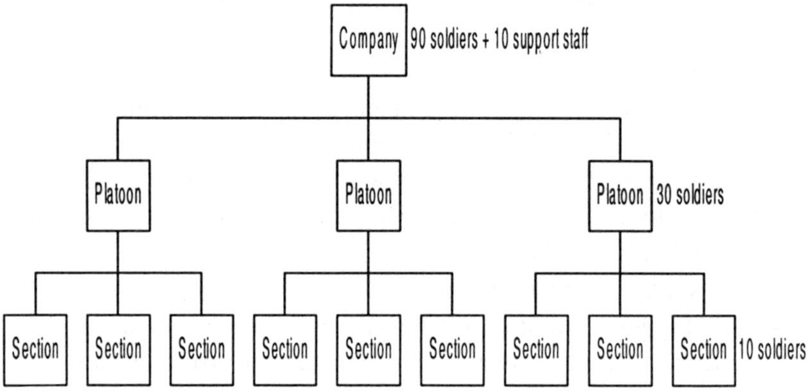

Figure 4. Company structure (Daniels *et al.* 1999).

It is possible exploit this hierarchical structure in the development of a video wireless system suitable for 100 users. Each group of 10 soldiers can forming a section can use the interference-free video link.

7 General Structure of the System

The system has two main components, the Communication system and the Video capture/compression/display system.

From the first concept demonstrator one can notice that some kind of frequency management scheme is required to allocate frequencies to each unit and to maximize the bandwidth utilization. Digital communication techniques can be used to address some of the interference issues. Reducing the video data rate by compression also reduces the bandwidth required to support our desired number of users.

The general structure of a system that overcomes the limitations identified in the first concept demonstrator is shown in Figure 5.

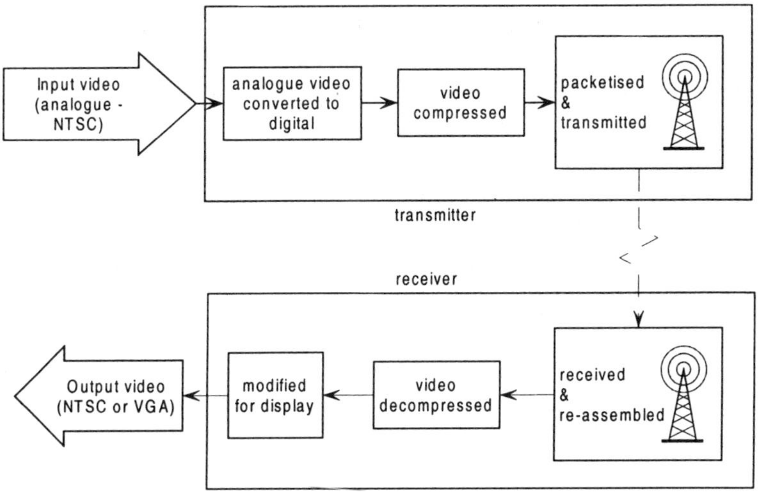

Figure 5. General system structure (Daniels *et al.* 1999).

The analogue NTSC input video signal is digitized and compressed. This reduces the data rate required for communications. This is then converted into packets and transmitted on a RF link using a frequency management scheme.

The packets are received and re-assembled, the video signal is decompressed and modified for display on an NTSC or VGA display.

Spread-spectrum techniques are suited to this kind of application. Spread-spectrum systems modulate the transmitted signal across a large bandwidth making the signal more resistant to interference, which usually occurs only in relatively small frequency bands (Katz 1996). They also are very secure, making interception difficult (Wi-LAN 1999). They are power rather than bandwidth limited, this makes it possible to have the transmitted signal weaker than background noise, provided the receiver has a high enough gain (Katz 1996). This makes interception almost impossible.

Some spread-spectrum systems include Direct Sequence-Spread Spectrum modulated Carrier Sense Multiple Access with Collision Avoidance (DS-CSMA/CA), Direct Sequence Code Division Multiple Access(DS-CDMA) and Spread Spectrum Frequency Hopping (FHSS) (Daniels *et al.* 1999).

These can be used (either directly or with modifications) to enable users to share the same bandwidth.

Time Division Multiple Access (TDMA) is another possible technique for bandwidth sharing.

Video compression enables the lower data rate and therefore reduced bandwidth. Some possible compression techniques include (Daniels *et al.* 1999):
- Motion Picture Experts Group (MPEG-1, MPEG-2, MPEG-4)
- Wavelet
- Random Neural Network (RNN) (Cramer *et al.* 1998).

Some of the other factors that need to be considered are:
- Quality of the video signal – needs to be high quality and transferred in real time
- Mass of the system – needs to be low to carry it possibly for long periods
- Robustness – needs to be able to withstand adverse operating conditions
- Power usage – low battery power (related to mass), and safety issues
- Prefer asynchronous data transfer to avoid the need to keep a global clock and synchronize the transmitters and receivers.

8 Communications System

8.1 Multiple-Access Techniques

To enable the system to support multiple simultaneous users in the limited frequency band available, without interference between units, some form of bandwidth sharing is required. This has been demonstrated in previous work in this area undertaken by the Weapon Systems Division of the Defence Science and Technology Organisation (DSTO) at Salisbury, South Australia (Wi-LAN 1999). The DSTO developed an analogue FM wireless link, with each transmitter/receiver pair using a fixed frequency. It was found that it is not feasible to assign individual frequencies to each unit due to the large guard bands required around the band being utilized to prevent interference between units and other radio devices. Some methods of achieving this are as follows:

- CSMA/CA (Carrier Sense Multiple Access with Collision Avoidance)
- TDMA (Time Division Multiple Access)
- FHSS (Frequency Hopping Spread Spectrum)
- DS/CDMA (Direct Sequence/Code Division Multiple Access)

8.2 Spread Spectrum Systems

These systems, with modifications, are all spread spectrum systems. In spread spectrum systems the transmitter signal is spread across a frequency band by modulating it with a unique spreading code word. The signal is then recovered by de-spreading at the receiver. Spread spectrum systems are more resistant to multi-path effects, and more tolerant of interference (Cramer *et al.* 1998). Spread spectrum systems also provide security features, for example detection avoidance and anti-jamming (Imatex 1999). These systems are power rather than bandwidth limited (Cramer *et al.* 1998), making them particularly suitable for mobile systems where power usage is critical. For high values of the processing gain.

(>1000), data can be transmitted at power levels below the ambient noise, making them very secure.

CSMA/CA using Direct Sequence Spread Spectrum Modulation (DSSS), TDMA with FHSS, FHSS and DS/CDMA are all spread spectrum systems.

8.3 Carrier Sense Multiple Access/Collision Avoidance (CSMA/CA)

This method of multiple access involves the transmitter listening to the communication channel. If no other transmitter is using the channel then the transmitter transmits. If the channel is being used, then the transmitter waits a period of time before trying again.

The key advantage of this technique is the high data transfer rates. Currently available PCMCIA cards have rates of up to 11 Mbps (Katz 1996). Multiple channels are available (13 channels (Katz 1996) in the PCMCIA card being considered for the prototype system), so by having ten users per channel (this number may depend on the operational conditions), more than a hundred simultaneous users can be supported.

The key disadvantage is that starvation may occur. Transmitters transmitting strong signals may use the channel almost continuously denying access to weak transmitted signals. This problem may be overcome by ensuring that each of the users within the group of ten users (or "pod") is on a separate channel. This should ensure that the strong signals in close proximity do not interfere with each other. It is expected that each pod will be a large distance apart, relative to the distance between users within the pod, and the distance over which the data is being transmitted. One method of overcoming this problem may be by implementing starvation controls in higher level protocols. Another approach may be to ration the transmission slots.

8.4 Time Division Multiple Access (TDMA)

TDMA involves each transmitter being assigned a time slot in which they are able to transmit. The key advantage of this technique is that all users get equal access to the communication channel.

The disadvantages of this method are that all transmitters need to be synchronized to ensure that they only transmit in their given timeslot. This is not easy to implement in a closed wireless system, and a small error in timing could render the entire system useless. The time slot allocation is static, hence the number of users in the system is fixed. This results in all of the available bandwidth not being utilized if the number of users is less than the initial maximum and the initial maximum number of users cannot be increased. This problem can be overcome using dynamic time slot allocation but the complexity of the system is greatly increased.

8.5 Frequency Hopping Spread Spectrum (FHSS)

This technique involves changing the frequency used every given time period in a quasi-random pattern. Frequency hopping controls the interference between cells by averaging the inter-user-interference across a wide frequency band.

The major advantages of this technique is that if unauthorized users try to intercept the data, by the time the signal is intercepted it has changed frequency making the part of the message intercepted relatively useless.

Frequency Hopping can be combined with TDMA, providing TDMA with the benefits spread spectrum systems (Cramer *et al.* 1998) (see also Section 8.2). For this technique to be applied to systems of this type, it would need to be combined with TDMA or CDMA to support multiple users.

8.6 Code Division Multiple Access (CDMA)

CDMA is a pure spread spectrum technique. Each transmitter is assigned a unique orthogonal code. Each receiver listens for signals that contain the code word.

CDMA has a number of major advantages. 4.4 trillion codes (WaveLAN 1999) are available so large numbers of users can be supported on a single frequency band, with such large numbers of codes available it should be relatively easy to find one hundred good codes. The transfer of signals is asynchronous. The Quality of Service (QoS) degrades gracefully as more users are added up to a point where unacceptably high Bit Error Rates (BER) occur. It also has much lower power levels than other systems and a greater capacity. For example CDMA handsets typically transmit power levels 1/25 to 1/1000 of AMPS and TDMA (WaveLAN 1999) and have 10 times the capacity of AMPS, and 3 times capacity of TDMA (WaveLAN 1999). A comparative study pertaining to these techniques is presented in Table 2.

Table 2. Comparative study of various techniques.

	Capacity	QoS	Data Rate
CSMA/ DSSS	13 channels, x users per channel	Exponential degradation as user numbers increase	11 Mbps down to 1 Mbps depending on channel quality
TDMA/ FHSS or TDMA	y frequencies, z users per frequency	Fixed for fixed number of users	Dependent on slot size and number of users per frequency
DS/ CDMA	4.4 billion orthogonal codes, choose w good codes	Graceful degradation as user numbers increase	64 kbps to 1.25 – 2.5 Mbps depending on chipset

9 Video System

9.1 Real-Time Video Compression Techniques

An NTSC video signal at a resolution of 352 × 240 pixels, in 8-bit color and at 29.97 frames per second (fps) has an uncompressed data rate of approximately 19.3 Mbps (1 Mega bit is defined as 2^{20} bits). Clearly this data rate is unachievable using existing digital wireless technology, so real-time video compression is required to reduce the transmitted data rate. Some possible techniques include:

- MPEG (Motion Picture Experts Group)
- Wavelet Compression
- Knowledge Based Techniques

This compression needs to be performed in real-time. Some of the requirements of this are as follows

- Compression and de-compression time
- Tx and Rx buffer sizes required
- Frame rate
- Resolution
- Jitter – standard deviation in frame rate

9.2 Motion Picture Experts Group (MPEG)

Using real-time MPEG-1 encoding on a PC based card (ZDNET 1999), an NTSC signal at a resolution of 352 × 240 (SIF – standard interchange format) can be compressed to between 1 Mbps to 3 Mbps at 29.97 frames per second. By decreasing the resolution to QSIF (Quarter SIF, 176 × 112) the compressed data rate can be reduced to between 56 kbps and 1 Mbps. These levels of compression are becoming close to the data rates achievable over wireless communications for high quality real-time video.

With the emergence of object oriented video streams like MPEG-4 (rather than traditional linear streams), compressed data rates are likely to be further reduced. MPEG-4 supports scalable content for varying data rates depending on the quality of the communication available (Raytheon 1999). This makes it particularly suitable for wireless environments.

9.3 Wavelet Compression

Wavelet compression offers higher compression rates than traditional DCT-based (Discrete Cosine Transform) encoders like JPEG and MPEG-1 (Texas Instruments 1997).

Wavelet codecs can compress a 20 frames per second video signal in CIF (common interchange format) to 33 kbps (Texas Instruments 1997), making it suitable for most existing wireless communication equipment.

9.4 Knowledge-Based Techniques

Cramer *et al.* (1998) describe the use of a Random Neural Network (RNN) for video compression. They tested their method on an ATM (asynchronous transfer mode) network at speeds of 64k bytes/s and 128k bytes/s. Their conclusion was that the RNN based system was several times faster than MPEG based systems.

These asynchronous data transfer rates should be achievable with the wireless communications available.

10 Implementation Conclusions and Prototype

The aim of this development prototype system is to provide a short-term solution system largely using off the shelf components. This system will be used to demonstrate the feasibility of a scaleable and robust system. Moreover, the use of video compression over the air will be verified from the quality perspective. This system can then be used to evaluate various aspects of the system and provide a baseline for the evaluation of future improvements.

Some of the performance parameters that will be evaluated are:
Video:

- Frame rate
- Resolution
- Jitter (standard deviation of the time between frames)
- Artifacts

Communication:

- Effective data rate
- Bit error rate (BER) and Packet loss
- Latency

The structure of the system is shown in Figure 6.

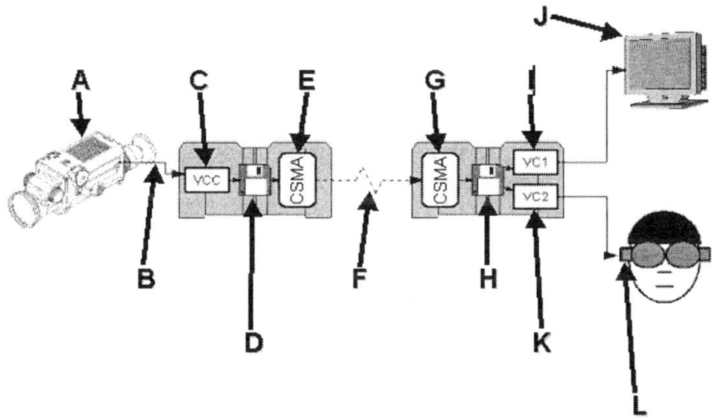

Figure 6. Structure of CSMA based system (Daniels *et al.* 1999).

The video signal is input from A transmitted across F and displayed on the HMD (F) or a VGA monitor (J). This system uses wavelet software video compression (D) and decompression (H). It was chosen over MPEG based compression as it provides lower data rates (Imatex 1999). The particular codec chosen is also easily implemented on a DSP (Digital Signal Processor) (Imatex 1999). The software implementation was chosen at this stage as it is more easily modified in the development stages.

This system uses direct sequence spread spectrum modulation and CSMA/CA (E) and (G). Each soldier within section has their own channel. A limitation of CSMA/CA, is that when two soldiers from different sections are on the same channel (see Figure 7) the effective throughput of both links will decrease, and hence decrease the video quality received. This is because of the carrier sense and collision avoidance mechanisms used in CSMA/CA.

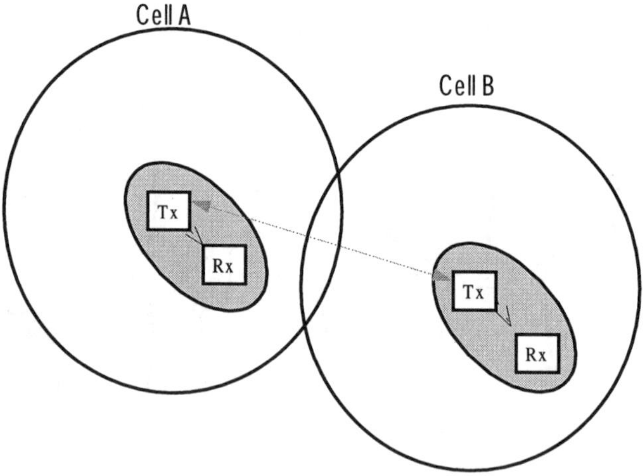

Figure 7. Adjacent links on the same channel.

Such deterioration of the video should not occur frequently as we expect that the company will typically occupy an area of approximately 50 square kilometers, hence the distance between units on the same channel will usually be much greater, and therefore the desired signal for the link, than the distance between the transmitter and receiver.

The similar performance degradation will occur when two soldiers on the same channel's cells are adjacent. Some cross interference will occur but it should not greatly affect the data rate/video quality greatly because the strong signals transmitted between the transmitter Tx and receiver Rx, (shown in gray in Figure 7), distance is still less than the distance between the Tx/Rx pairs. The strong signal between the local transmitter and receiver will not be greatly affected by the weaker interference signal (Katz 1996) – this issue needs further investigation.

(E) and (G) are PCMCIA cards operating at 2.4 GHz (WaveLAN 1999). In future 5.8 GHz (ZDNET 1999) may be tested, as the higher frequency should be more interference resistant. 11 Mbps data transfer rate is achievable (WaveLAN 1999) – making them far superior in speed to other wireless communication equipment, for example CDMA (2.5 Mbps max soon (Qualcomm 1999b)) and Frequency Hopping systems (38.4 kbps (Wi-LAN 1999)). The issues of frequency band selection needs to be further explored considering its defense use. This is because these bands are unlicensed in most countries and it is undesirable to be competing for bandwidth with other non-military users in critical missions.

These PCMCIA cards support 13 channels in the bandwidth between 2.4 GHz and 2.4835 GHz (WaveLAN 1999). So to support 100 users, each soldier within each section will have a different channel. Hence 10 out of 13 channels will be used within each section. We envisage that we will be able to take advantage of short range of the link compared to the range of the distance between, and hence maximizing the bandwidth utilization (Katz 1996). It may be required that we need to use narrow beam antenna to restrict the signal spread.

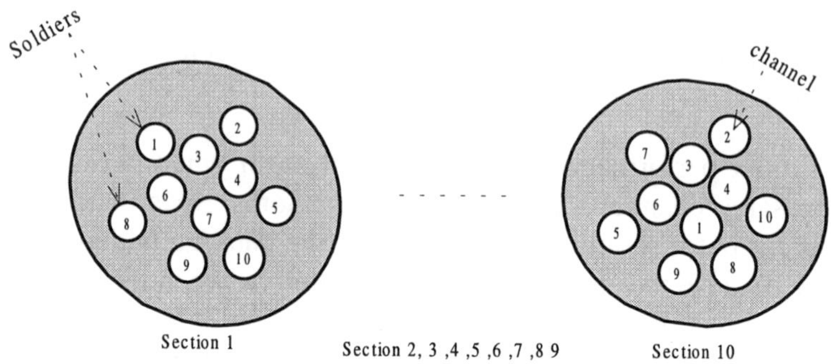

Figure 8. CSMA frequency allocation.

A variety of communication protocols with suitable modifications can be considered including TCP/IP (Transmission Control Protocol/Internet Protocol – more suited to wired environments), UDP/IP (User Datagram Packets/Internet Protocol), RTSP (Real Time Streaming Protocol), and WAP (Wireless Application Protocol).

VGA display is preferred (J), (L) as it does not require conversion to NTSC which adds extra equipment and increases the power usage.

11 Conclusion

The present system provides a short-term solution largely from off the shelf components. This provides a baseline for evaluating future developments. The long-term system aims to provide a more realistic system for field use. This system is more robust and has a small mass.

The long term aim is to use Code Division Multiple Access (CDMA) as a communication mechanism. This is more suited to the outdoor environment, is more interference resistant and also more secure (Qualcomm 1999a).

CDMA can use a large number of unique orthogonal codes (Qualcomm 1999a) that can be used for modulation, so each transmitter and receiver can be assigned a unique code. This avoids the multiple users on each channel associated with CSMA, and FHSS, and TDMA type systems.

It also has the advantage over other techniques in that it allows users to be added dynamically and the quality of communication degrades gracefully (Qualcomm 1999a). It is asynchronous and overcomes the problems associated with systems that need to have the receiver and transmitter synchronized, like TDMA.

Using CDMA requires custom boards be built using 2.5 Mbps chipsets operating at 2.4 GHz (Qualcomm 1999b). This provides a more customized system, including the control of power levels.

The present software implementation of the video compression will be replaced with hardware to provide faster compression/decompression resulting in better quality video.

Further interference suppression techniques will also be investigated.

The CDMA based system will provide a wireless-link enabling wireless transmission from the TWS to the soldier's HMD. It provides quality in terms of communication quality in outdoor environments and supports the required data rates.

New and emerging techniques and technologies for communication, video compression – like neural network based compression techniques, and object oriented video streams will provide impetus to make the Wireless video in weapon applications a reality.

Acknowledgments

Thanks are to due to Vicente Barrientos for his contribution in the early stages of this project. This work was performed under a research contract with financial support from the Defence Science and Technology Organisation. Photo in Figure 1 courtesy of Wayne Hobbs, DSTO Salisbury.

Acronyms, Abbreviations and Definitions

ALOHA – A dynamic channel allocation system developed in the 1970s by Norman Abramson and colleagues at the University of Hawaii
AMPS – Advanced Mobile Phone System
BER – Bit Error Rate
CA – Collision Avoidance
CD – Collision Detection
CDMA – Code Division Multiple Access
Cell – defined as a single transmitter/receiver pair
CIF – Common Intermediate Form
Codec – COder and DECoder
CSMA – Carrier Sense Multiple Access
DCT – Discrete Cosine Transform
DSSS – Direct Sequence Spread Spectrum
DSTO – Defence Science and Technology Organisation
DV – Digital Video (SD mode)
TWS – Thermal Weapon Sight
FHSS – Frequency Hopping Spread Spectrum
Frame – a group of slots
GSM – Global Systems for Mobile communications (uses narrrowband TDMA allowing 8 simultaneous calls on the same frequency)
GPS – Global Positioning System
HMD – Helmet Mounted Display

IEEE – The Institute of Electrical and Electronics Engineers
IP – Internet Protocol
Packet – Transmissions are broken down into packets (discrete amounts of data + control information)
QoS – Quality of Service
QPSK - Quadrature Phase Shift Keying
RF – Radio Frequency
Slot – slotted systems like slotted ALOHA use slots with one packet per slot
SVGA – Super VGA
TCP – Transmission Control Protocol
TDMA – Time Division Multiple Access
TWS – Thermal Weapon Sight
VGA – Video Graphics Array
VLC – Variable Length Coding

References

Camera Interface Guide (1999), [Accessed 11 Feb. 1999 at http://www.ael.co.kr/inst/New/Interface/camera_guide.htm]

Cramer, C., Gelenbe, E., and Gelenbe, P. (1998), "Image and video compression – their underlying principles," *IEEE Potentials*, February/March.

Daniels, W.M., Jain L.C., Mahajan, A., Forbes, S., and Puri, V. (1999), "Real-time multi-user wireless video link," *Third International Conference on Knowledge-Based Intelligent Information Engineering Systems*, IEEE Press USA.

Imatex Communications Inc. (1999), W1 Codec. [Accessed 23 Mar. 1999 at http://www.imatex.com/e_codec.html]

Katz, R.H. (1996), "CS 294-7: Media Access – TDMA and CDMA," CS Division, University of California, Berkeley. [http://www.cs.berkeley.edu/~randy/Courses/CS294.S96/TdmaCdma.pdf]

Qualcomm (1999a), "CDMA Digital Technology: Features and Benefits." [Accessed 1 Feb. 1999 at http://www.qualcomm.com/cdma/phones/whatiscdma/features.html]

Qualcomm (1999b), "IS-95-HDR." [Accessed 14 Feb. 1999 at http://www.qualcomm.com/IR/SH98/cdma_dnp/cdma_dnp6.shtml]

Raytheon (1999), "AN/PAS-13 Thermal Weapon Sight." [Accessed 2 Feb. 1999 at http://www.raytheon.com/rsc/ses/spr/spr_013/spr_013.htm]

Texas Instruments (1997), "Nightsight Thermal Imaging Solutions: Portable Handheld/Weapon," Installation and Operational Manual, p. 2-2.

WaveLAN (1999), "WaveLAN Products." [Accessed 18 Jan. 1999 at http://www.wavelan.com/products/productdetail.html?id=20]

Wi-LAN (1999), "Technology in Action." [Accessed 23 Jan. 1999 at http://www.wi-lan.com/tech/desc.html]

ZDNET Australia (1999), "RadioLAN Mobilink 140 Review." [Accessed 2 Aug. 1999 at http://www.zdnet.com.au/products/stories/products_display/au0000465.html]

Chapter 6

Agents of Change: the Impact of Intelligent Agent Technology on the Analysis of Air Operations

**C. Heinze, M. Cross, S. Goss, T. Josefsson,
I. Lloyd, G. Murray, M. Papasimeon, and M. Turner**

From the earlier 1990s the Air Operations Division of Australia's Defence Science and Technology Organisation has invested heavily in Intelligent Agent research and development to support a strong requirement for simulating human decision. The first operational system, SWARMM, entered service in the mid 1990s and was used daily to support analysis of military fighter operations. Since that time there has been a maturing of the technology and experience gained with appropriate methods. Today most of the simulations in use take advantage of the advances in intelligent agent development to model a wide variety of personnel. This chapter describes the systems that employ intelligent agents and provides insights into some of the designs, problems, lessons, and research that have surrounded the decade of development.

1 A Challenging Application Domain

All technology tends toward obsolescence. The process is accelerated when maintenance, update, and upgrade, are not fully addressed. By the early 1990s it was clear that maintaining the large simulation codes[1] was untenable. The Royal Australian Air Force (RAAF), as the primary customer, often requires rapid answers to questions. Response times to requests that required even superficial modification of the software had blown out to several months. The brick wall of technology obsolescence had been hit. It was simply not possible to manage the complexity of the existing models. This was due to the old technology computer languages in which they were written and to the lack of professional software engineering practices[2]. Redressing this situation required a paradigm shift in tools, languages, and practices. The technology shift was precipitated by a decision to transition some aspects of the modeling and simulation activities toward what was then an embryonic emerging technology – intelligent agents.

At the time intelligent agent technologies were being examined for the utility in modeling situation awareness but it was soon apparent that they were appropriate for the entire spectrum of military decision-making.

Modeling air operations in a military Operations Research (OR) context presents a number of significant challenges (McIlroy *et al.* 1997b). The cost to acquire, upgrade, and operate modern military systems creates multi-billion dollar decisions for defense organizations. Air Operations Division is an important *advice provider* during the acquisition and through-life support of Air Force aircraft and as such is exposed to the pressure to provide informed impartial advice in response to firm dead-lines.

[1] Air Operations Division makes use of simulations of air combat to study scenarios of interest and to gain insights into the operational effectiveness of aircraft and their subsystems. The simulations were developed through the 1980s and were largely coded with procedural languages such as FORTRAN.

[2] Core professional skills lay in a culture of "research" rather than code production. Since that time work practices have recognised the importance of software engineering and processes have matured.

The following account presents an overview of intelligent agent research and development within Air Operations Division (AOD) from 1990 to the present day. This research is characterized by the innovative employment of new technology and its success has been recognized internationally (Heinze *et al.* 2001, Tidhar *et al.* 1999). Not only has the investment in agents resulted in order-of-magnitude improvements in simulation modification times but many other benefits, some of which were unexpected, have followed.

1.1 Air Operational Analysis

The modeling and simulation activities of the Division require that models of fighter pilots be developed. These virtual pilots fly fighter aircraft inside constructive simulations. Data collected from this virtual combat informs decisions about acquisition of new aircraft or subsystems, or the employment of extant aircraft and their subsystems.

AOD provides timely operational advice to the Australian Defence Force (ADF) about questions that might arise within the ADF specific to air operations and systems. This means that there must be a capability to answer a variety of questions and to analyze a wide range of scenarios. AOD may be called upon to model, to a high degree of sophistication the physical systems (aircraft, radars, missiles, etc.) but must also be able to use these models within a scenario that is capable of simulating their tactical use and the interactions between them. A simulation will typically involve many aircraft, ground-based radars (GBR), missiles, electronic counter measures (ECM), and a variety of other systems. A critical requirement is that the simulations also model the pilots who fly the aircraft, air defense controllers who assist those pilots, and other human operators within the scenario. The models of the human operator should be capable of being modified rapidly whilst the behavior that they exhibit must be explainable in terms that can be understood and accepted as valid by those interested in the output of the work. Within this context analysts provide advice with respect to:

1. Capability analyses in support of the ADF's acquisition process.
2. Assistance with tactics and doctrine development for the effective employment of ADF assets.
3. Advice for the development of training procedures for pilots and controllers at the operational level of command.

1.2 Modeling and Simulation

The modeling and simulation undertaken for the operations analysis activities of the Division are largely constructive – there is no human interactivity. This distinguishes it from the view of military simulation as the human-in-the-loop (HIL) simulator providing high fidelity for a pilot.

Constructive simulation places requirements on the intelligent agent development that are quite different from the computer generated forces observed in training simulators (Tambe *et al.* 1994). There are two quite different categories of military simulation systems. The distinction will become important in characterizing the requirements and the resulting designs in later sections. The vision of military simulators that most people have is that of the graphically intensive human-in-the-loop simulator of the type shown in Figure 8. These simulators are most commonly utilized for training, but also provide research environments for exploring human factors issues. Examples include the monitoring of hand and eye movements or the evaluating the ergonomics of proposed systems. These simulators provide the human participant with an immersive visual and auditory experience that allows them to practice procedures and tactics. This reduces the need for, and cost of, training on the real hardware. Agents have been used to provide both friend and foe in these simulators (Tambe *et al.* 1995, McIlroy *et al.* 1997a).

Military operational analysts also use simulation as a means of evaluating the effectiveness, performance, and tactical employment of military systems prior to acquisition or upgrade. The information that results from these studies guides acquisition and can be used to evaluate successive designs, thereby mitigating some of the risks associated with the physical development of modern defense systems. These simulations are most often constructive – that is they make use of no new human input during execution. All of the human aspects of these simulations must be computationally modeled. This type of simulation can also provide data for the evaluation of tactical options for situations that cannot easily be explored by conventional training.

The requirements associated with these two different types of military simulation show some marked differences. An obvious one is that the human-in-the-loop (HIL) simulations are normally configured to run as real-time systems. Some military simulations have a capacity to run much faster than real time. This allows the scenario to be fast-forwarded through the areas that offer little training benefit and then returned to real-time when human interaction is required. A similar facility is common on many PC based computer games. There is normally no need for OR simulations to run in real time. Commonly the Monte Carlo style simulation they perform requires them to explore large parameter spaces resulting in many hundreds or thousand of runs of the scenarios. These simulations run many times faster than real time[3] in order to provide timely information.

A more subtle difference in the requirements concerns the modeling of the simulation entities. In HIL simulation the quality of the outcomes is frequently judged by the fidelity of the images pre-

[3] For instance: a typical study considering 10 scenarios each of which lasts approximately one hour of simulated time, and randomises 4 parameters over particular ranges, running at 100 times faster than real time will still require about a week of compute time.

sented to the human participant. As long as the entities appear to behave correctly the detail of the internal modeling is not questioned. This is not true of OR simulations where the result is the accumulation of knowledge about the functioning of the hardware or the efficacy of particular tactics. In these cases it is necessary to develop a detailed understanding of the reasons for observed behaviors and a more explicit representation of the human component, and the tactics in particular (Heinze *et al.* 1998).

1.3 Intelligent Agents

Agents are, despite the flourishing research interest, relatively immature in industrial applications. The innovations that agents bring to software development require engineering methodologies that deal with concepts and techniques that differ from existing approaches. The immaturity of agents is reflected both in the small number of commercially available agent languages and the lack of supporting software engineering methodologies.

Defining agency is not a simple task. Strict definitions are not particularly helpful and general definitions often so inclusive as to be uninformative. Agents are expected to perform tasks for their user and to do this with a certain amount of *autonomy*. This autonomy is usually taken to mean that the agent will initiate action by itself if required and does not need continual instruction. Agents are thought of as residing in *environments*. Thus the agent perceives the environment and can decide to carry out certain actions that effect the environment. Agents then, are not necessarily a strong attempt to build human intelligence but rather an approach to building software that leverages some human-like characteristics and simplifies the visualizing, describing, and designing of software.

Since the 1980s when the first agent technologies emerged from distributed AI and computational cognitive modeling there have been several strong but different threads that have characterized the applications pull and the technology push that drives research and development in the area.

Perhaps the strongest application area for intelligent agent technologies is the Internet. Information agents are flourishing and crawlers, spiders, and bots move over the web gathering information for the human users or software systems that deployed them. Information agents are not typically modeled directly on human intelligence but they act as surrogates for us, searching the web and returning when they have found the requested information.

The agents that are described in this chapter are more strongly 'human-like'. Builders of military simulations use agents as computational models of pilots and other military personnel to inhabit the simulated battle-space and to fly or control the virtual aircraft. These types of agents are strongly influenced by cognitive science, psychology, and artificial intelligence because they are representations of actual human decision-making.

There are few commercially available agent languages. The ABE (Agent Building Environment) and AGLETS are two examples of lighter-weight languages aimed at the web-market. 'Heavy-weight' languages that implement some human cognitive model are rarer. Perhaps the most notable family of languages in this class are those classified as BDI (belief-desire-intention) agents. Languages in this class include dMARS, JACK, PRS, C-PRS, and JAM. These languages form the basis of the examples in this chapter. Some details of these languages are provided throughout the chapter as are pointers to further information

2 Innovative and Successful Application Deployment

The first intelligent agent project was a technology demonstrator that utilized the Procedural Reasoning System (PRS) from Stanford Research Institute (SRI). This technology demonstrator, exhibited two qualities that proved convincing in deciding to transition the demonstrator into an operational system. The PRS programming language allowed coding of tactical plans in a fashion that supported rapid responses to suggested modifications. It was comparatively easy to develop and modify quite sophisticated tactics. Perhaps even more telling was the graphical nature of the plan language. PRS plans are graphical flow-chart like representations of decisions. With careful design they can be made to resemble the types of decision trees that pilots would draw when explaining their evaluation of the tactical situation. Clearly there were benefits to be had beyond simple software engineering maintainability that went to the very core of the promise of agent technology. These agents appeared to offer an appropriate level of abstraction and a set of modeling constructs that were ideally suited to the requirements. The following sections provide some details of some of AOD's projects that have made use of intelligent agents.

2.1 Fighter Combat

Air combat is a highly dynamic environment where decisions must be made rapidly in response to incomplete knowledge (Shaw 1985). Pilots, air-defense controllers, and commanders must respond quickly to changing situations. Being a military domain in which pilots act in teams with the assistance of fighter-controllers in other aircraft and air-defense commanders on the ground, there is fairly strict adherence to command and control structures that have been trained for and developed over time. Standard operating procedures (SOP) are often mandated for specific situations. Pilots and fighter controllers must communicate with other team members and must coordinate their actions over time in order to achieve their goals. Pilots must operate many complex systems simultaneously; they must fly the aircraft, operate the radar, the radios and datalinks, and launch weapons and countermeasures. They must communicate with other aircraft and act in a way that maximizes their own survivability and the chances of successfully completing their mission. Tactics must be selected in response to enemy activity. Furthermore enemy tactics must be pre-empted and counter tactics selected for possible eventualities in advance. Long-term goals must be complemented with responses to unexpected eventualities. All of this decision-making must occur rapidly, in a dangerous and highly dynamic environment. Perhaps the most critical factor of all is that decisions are often made with incomplete (or at least uncertain) knowledge about the true state of the world.

The above description is the day-to-day work environment of fighter pilots and controllers. If simulations are to be credible they must be capable of reflecting the scope and detail of this behavior. Furthermore the agents that model human decision making should not only be functionally accurate but must also be designed in a way that allows for validation by domain experts. These agents must not only function correctly; they must be seen to function correctly.

2.2 SWARMM

SWARMM was developed by DSTO in co-operation with the (former) Australian Artificial Intelligence Institute (AAII) within the Co-operative Research Centre for Intelligent Decision Systems. The design concept for SWARMM was to retain existing software representing physical systems performance from the PACAUS constructive simulation and to marry this to agent-oriented software for tactical reasoning and tactics selection.

The initial SWARMM project was successfully completed in June 1996 and further developments have matured its use as a tool for operational analysis.

Planned as a progressive development project, SWARMM was heavily reliant on the content of previously developed models of aircraft and their systems. These models of aircraft and engine performance, weapons, countermeasures and environment comprise validated code that was developed by many staff over a period spanning more than a decade. The design concept of SWARMM foresaw the next stage of development being a move to an object-oriented (OO) environment for the physical systems modeling. The combination of agent-oriented tactical reasoning and OO physical code was achieved in the production of the BattleModel, discussed elsewhere in this chapter.

SWARMM was designed to alleviate, or avoid altogether, many of the problems associated with developing pilot tactics using traditional scientific constructive simulation code. To this end, the language used for simulation of tactical reasoning was dMARS (distributed Multi-Agent Reasoning System) and, in part, the SWARMM requirements specification influenced the development of dMARS as a C++ refinement of its LISP based precursor, PRS. SWARMM's dMARS employment of Beliefs, Desires and Intentions (BDI) agent formalism was an apt method of encapsulating the descriptions of human reasoning required in the domain of military air operations. The BDI approach enables the operational analyst to consider agent behavior at the tactical level and to map analyst operational experience into an appropriate set of tactical dMARS plans.

The tactics are written with respect to individual agents, or teams of agents. Teams are broken into lead and wing elements and corresponding team plans recurse down to the individual lead and wing aircraft.

SWARMM's design concept expanded considerably on PACAUS and specified that the simulation should feature many other tactical roles, such as ground attack aircraft, fighter escorts, refueling aircraft, air defense surveillance and command and control entities. In defining a scenario, the analyst is able to assign high level mission plans to the appropriate agent team on the basis of their tactical role and mission assignment. Without any stimulating trigger the agent flies the mission as briefed, just as in real life.

Figure 1. A scene from the SWARMM graphical replay utility.

In addition to the high level mission plans, the agents are assigned a suite of tactical plans commensurate with their role and mission in the scenario. Deviation from the pre-briefed mission behavior is part of the central interpretive reasoning that considers the current state of the environment within the tactical context of the current mission phase. The plans are divided into a number of groups, of which the most significant is that dealing with maneuvers. Other categories of tactics are those handling the operation of sensors, communicating, and weapons and countermeasures employment. Agents respond to events in the environment with tactics selected from one or more of these groups. SWARMM's tactical repertoire has been steadily developed within operational study requirements, in line with the original scope to provide a comprehensive suite of tactics commensurate with existing Air Force procedures.

2.3 BattleModel

The emphasis of operational analysis studies in support the Airborne Early Warning and Control (AEW&C) projects was initially on the evaluation of comparative hardware solutions and required only simple agents with shallow command and control structures. The behavior required of these first agents was merely the operation of the hardware in a reasonable manner consistent with the possibilities afforded by the hardware. As the hardware progressively becomes fixed in an aircraft acquisition program, the emphasis of these studies turns to the development of operational concepts and tactical procedures. These studies require better fidelity representation of human behavior, both as individuals and in teams.

The crew of an AEW&C aircraft are required to perform a large number of disparate but complementary tasks. An AEW&C team is an information gathering, interpretation and dissemination center as well as fulfilling a battle management role.

Developing a picture of the battle space and passing it up and down the line is only part of the story. The AEW&C is a weapon system, and is capable of employing its weapons in offensive action and to defend itself and friendly assets. The difference between the AEW&C and most other weapons systems is that it delegates responsibility for direct hostile action. For example, it uses fighter aircraft in the way that fighter aircraft, in turn, use ‘smart’ missiles. Constructing an agent-based representation of an AEW&C crew is quite a challenge and is still being developed. The early agents were only required to perform sub-tasks, sufficient to achieve proper operation of the equipment under evaluation, in the way test pilots are used in trials of hardware systems. Relationships between agents were very formalized and simplified. Simulations are now moving to multi-agent systems that have depth in both command and control structure and in time. The depth of command and control structure requires that agents be capable of packaging and delegating tasks to those agents that have the capability to understand the requirements, read the situation, and carry out the intention of the delegating agent. These delegating agents must in turn be capable of understanding the intention of the agent above them in the command control structure in the tasks delegated to them. Another requirement is multi-agent systems that have depth in time. By 'depth in time' we mean the ability of an agent to progressively gather an understanding of the situation over a period of time, infer the probable intention of those forces that are under its direct control and project the situation forward in time.

2.4 F-111 Stand-Off Weapon Employment

The Australian Defence Force continually evaluates its capabilities and upgrades or replaces them if appropriate. One such project is the acquisition of long range smart weapons for its fleet of F-111 strike aircraft. Due to their long range, these weapons are often referred to as stand off weapons, as they are designed to strike high value targets outside the reach of any air defenses protecting those targets. Hence the Stand-Off Strike project investigated tactics for employment of these weapons by F-111 aircraft. The studies conducted were interested in looking at finding the optimal tactics the maximized the effectiveness of the weapon against the selected target, while at the same time minimized the risk which the F-111 aircrew faced from hostile ground based air defenses such as surface to air missiles.

The BattleModel simulation architecture used in for the AEW&C evaluation was also used in this project. The flexibility and object oriented nature of this architecture allowed models for different scenarios to be plugged in. These varied from different physical models (aircraft, weapons, environment) to different reasoning models some of which were agent based others which were not. A typical scenario might involve an F-111 strike aircraft flying in low making use of a terrain following radar, popping up to a required altitude, launching the weapon against the target and guiding it in and then returning to base. At the same time the target will be protected by a system of surveillance and tracking radars to detect and track enemy aircraft and launchers with surface to air missiles to try and shoot them down.

The crew of the strike aircraft (pilot and navigator) and the operators of the ground based air defense system (commanders, radar operators and fire control operators) can be modeled as agents, as we can ascribe notions of agency on them, as they perceive, reason and act in their world. However, the behavioral requirements for the agent models in this project were less complicated than projects such as SWARMM. Therefore the agent models did not the sophisticated capabilities of a heavyweight BDI agent language such as dMARS. The activities which the operators could perform were firstly defined.– representing the high level goals which the agents could have. Also, a set of conditions were defined based on each agent's perceptual inputs and own internal knowledge about the world for going from one goal to another. Each agent was implemented as an object oriented "State" pattern, with the high level goals being states that each agent could be in, with conditions from changing from one state to another. Hence the agent implementation is represented as a simple finite state machine, which is quite common for representing agent entities in computer games.

An interesting aspect of this project was that an agent oriented analysis and design of the system was performed, however at implementation an object oriented solution was chosen. The system was implemented in the BattleModel architecture which allowed simple finite state machine agents to interact in the same environment as behaviorally more complex agents using the BDI model.

This demonstrates that an architecture like BattleModel can be used as a test-bed for different agent behaviors and interactions of agents of different types. Furthermore, it shows that a component based architecture where models of systems (sensors, platforms) or models of humans (FSM agents, BDI agents), interact is scalable, because it allows agents to be de-coupled from one another and allows different models of agency to interact in the same environment. As the behavioral requirements change, and more complex behaviors are required, the simple agent models can be migrated to more complex agent models without any change in the architecture or how the different models interact via the architecture.

Figure 2 shows an example visualization from an F-111 strike mission simulation.

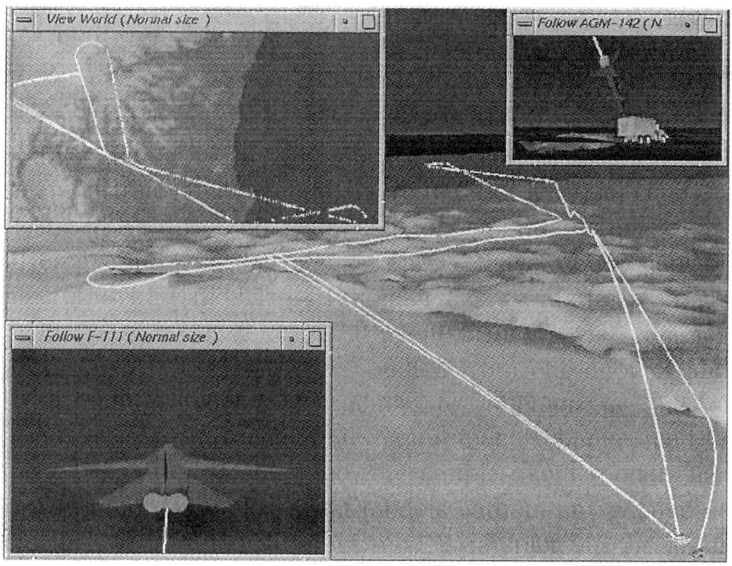

Figure 2. 3D visualization of a simulated F-111 strike mission. A variety of views and visualization tools are available to assist the analyst in understanding the scenario and the implications of particular tactical decisions.

2.5 The Virtual Air Environment

The Virtual Air Environment (VAE) concept integrates real assets and virtual simulations (comprising Human-in-the-Loop (HiL) and computer generated entities) in one synthetic environment to create a virtual world for training and other purposes. A number of new capabilities are due to enter RAAF service in the next few years, including the Airborne Early Warning and Control (AEW&C) and Jindalee Operational Radar Network (JORN). Current training practices, which rely heavily on the use of live assets, are unlikely to meet projected training demands cost-effectively. Modeling and simulation can be used to provide a training system which will overcome many of the projected shortcomings of the current training systems by providing a synthetic environment in which live assets can be stimulated by, and interact with, networked simulators and/or computer generated forces (CGFs). This is the basis of the VAE. To determine the viability of this concept, an Initial Development Phase (IDP) is under-way to investigate the technical and operational issues associated with this proposal. There are two aspects to this work:

1. linking HiL simulators based in geographically separate locations with the real (operational) system in order to stimulate and interact with the operational system; and
2. stimulating the operational system with computer generated entities from a constructive simulation or other scenario generator.

It is in this second area of work for the VAE that the BattleModel is being applied. This represents an application of the BattleModel and agents in the domain of Modeling and Simulation for training as distinct from the OR domain.

A number of aspects have been developed to enable BattleModel to provide CGFs for the VAE. A Distributed Interactive Simulation (DIS) interface has been development to enable BattleModel entities to interact with other DIS entities in the VAE. Intelligent agents, developed using dMARS, are being employed to control the behaviors of the CGFs.

The intelligent agents controlling the BattleModel entities respond to commands but also provide a high degree of autonomous behavior. The agents are used to model pilots of the various simulated entities (fighter and strike aircraft for friendly and hostile forces) and they respond appropriately to commands issued through the GUI described above. Their autonomous behavior includes flying waypoints, flying in formation, engaging opposing force aircraft, and firing missiles in an attempt to destroy opposition forces. The CGFs required for these scenarios are required to exhibit potentially complex and tactically credible behaviors.

How well the current agent technology will scale to a fully developed VAE is yet to be established. There are also likely to be issues particular to the HiL simulation domain which are not encountered in the OR domain in which the base technology has been developed.

It has already been found that inserting humans as surrogate agents in an OR simulations is considerably easier than retrofitting existing HiL simulators to incorporate agents.

2.6 AP3-C Orion Operating Procedure Update

This project supports the RAAF Maritime Patrol Group in developing new tactics and 'concepts of operation' for the upgraded AP-3C Orion Maritime Patrol Aircraft. These are used in peacetime for maritime search, surveillance and intelligence gathering operations in and around Australian territorial waters.

While the recently upgraded sensors and avionics significantly improve the capability of the aircraft. BattleModel is used to baseline the expected mission performance of the aircraft in typical mission profiles and scenarios, and also to develop new, integrated flying and sensor employment policies that allow the aircraft to function at its full potential.

The requirement was to report on the effectiveness of different flying tactics and sensor employment policies. Therefore a model of the tactical decision making process on board the aircraft that was capable of representing the actual human operator and crew workload, the sensor data-fusion process and chain of command was needed. The tactical decision making model also had to be flexible and robust enough to allow timely modification to investigate different operational procedures and tactics. Individual intelligent agents were used to model the tactical decision making for each crew member being simulated on board the AP-3C. These agents were implemented with the dMARS language.

Intelligent agents were chosen because of the requirement to model decision making based on a degree of awareness of the environment or tactical situation the aircraft finds itself in. Maritime surveillance tactics, as with almost all tactical decision making, rely on making an assessment of the current situation based on fusing data from different sensors and also on making some assessment about the intent of other entities in the environment. The Beliefs, Desires, Intention formulation of the dMARS agents lends itself to this type of modeling. Crewmembers were modeled to the extent that the type of information, the amount of information and the work responsibilities of the operators were accurately represented. As is the case on board the real aircraft, communication between the agents was modeled via a radio intercom link.

The simulation can operate in one of two modes; constructive or 'crew in the loop'. In the constructive mode, missions are run faster than real time, hundreds of times using Monte-Carlo techniques analyzing the effectiveness of different tactics using statistical techniques. In this case, the tactical decision-making is made by the agents. In the 'crew-in-loop' mode, the agents are removed from the simulation and are replaced with user interfaces which are controlled by actual crew members who make the tactical decisions (see Figure 4). The interactive or 'crew-in-loop' mode is used to test and evaluate new tactics in a realistic environment. In this mode, the tactical picture is projected onto a large screen showing the current sensor information which is superimposed on to a geographical map of the region. This allows the crew to focus on developing and evaluating higher-level tactical procedures rather than on low-level interactions with individual controls.

Figure 3 illustrates the tactical development cycle used by the RAAF and where BattleModel fits into the process. Initial tactics, mission profile and measures of effectiveness are supplied to AOD in the form of scenarios describing proposed new concepts of operations (CONOPS). AOD then obtains appropriate models to represent all the components of the scenarios. This is followed by the analysis stage where the constructive version of the BattleModel is used to obtain statistical data.

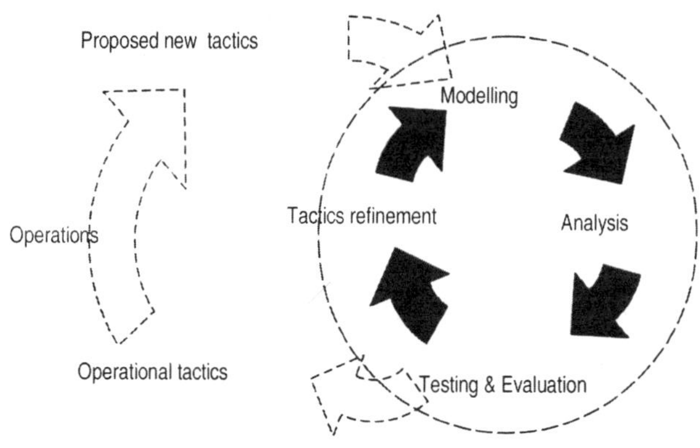

Figure 3. The tactical development cycle. The processes inside the circle are under occur "in the laboratory" and those outside reflect the day-to-day operations of the Air Force.

The results of the analysis are presented to the RAAF and a testing and evaluation stage is then conducted with RAAF personnel using the crew in loop version of the BattleModel. This stage results in some tactics refinement, and the cycle begins again. After an appropriate amount of simulation based development operational tactics are developed that are then employed on board the actual aircraft. The implementation of the tactics development cycle, and the use of the constructive version of the BattleModel (which used the dMARS agents) and the crew in the loop version of the BattleModel has proven to be an effective means of developing tactics.

Figure 4. RAAF P-3 crew members from Maritime Patrol Group participating in a simulated mission using the BattleModel interactive crew-in-loop mode to evaluate proposed maritime search and classification tactics.

3 Leading Edge Agent Technology Development

From the early 1990s AOD pioneered the employment of intelligent agent technologies in military simulators. Through a series of successful collaborations with local academic institutions and technology vendors AOD has been able to influence the direction of technology growth through involvement at all levels of agent technology research and development. A strong research and development program is continuing.

3.1 dMARS and JACK

AOD's demanding requirements have driven the development of agent technology. AOD played a role in the development of dMARS both as an alpha and beta tester of the language and as a co-developer of methodologies, tools and language extensions. As a user AOD played a larger role: developing some of the largest and most complex agent applications yet deployed.

dMARS is a multi agent architecture that implements a BDI model of agency based on the concepts of intentions, plans and practical reasoning developed by Bratman (1987). Further information about the underlying formalism of dMARS is available (d'Inverno *et al.* 1997, Georgeff and Lansky 1986). Every dMARS agent comprises a set of beliefs, desires (goals), plans and intentions.

The beliefs of an agent are stored in a relational database and contains information about the nature of the world. These beliefs may refer to hard physical data or to more abstract concepts. The goals of an agent are descriptions of required behaviors or desired outcomes, whilst the plans are declarative procedures specifying actions to take to accomplish these goals. The intentions are instances of plans that have been selected for processing to achieve some goal. They represent commitments by the agent to the achieving of a goal through the course of action specified by the plans. The plans represent the most visible part of a dMARS agent. They constitute the procedures that an agent will use to deal with situations as they arise. The plans are graphical in nature (see Figure 5) and are capable of being displayed during the simulation through the dMARS Control Interface (DCI). This feature is an attribute of dMARS and not of agents per se; it allows the plans adopted by the agent to be displayed and the current state of the agent reasoning evaluated during the simulation. With careful de-evaluated during the simulation. With careful design these plans can be read and understood by lay-people with little or no additional explanation. A detailed description of the dMARS system as it pertains to air-combat modeling can be found in a paper by McIlroy and Heinze (1996). A more recently developed alternative to dMARS is JACK (Howden *et al.* 2001). JACK is based on JAVA and like dMARS implements a BDI model and provides many of the same programming constructs[4]. AOD is actively cooperating in the future development of extensions and supporting infra-structure for JACK.

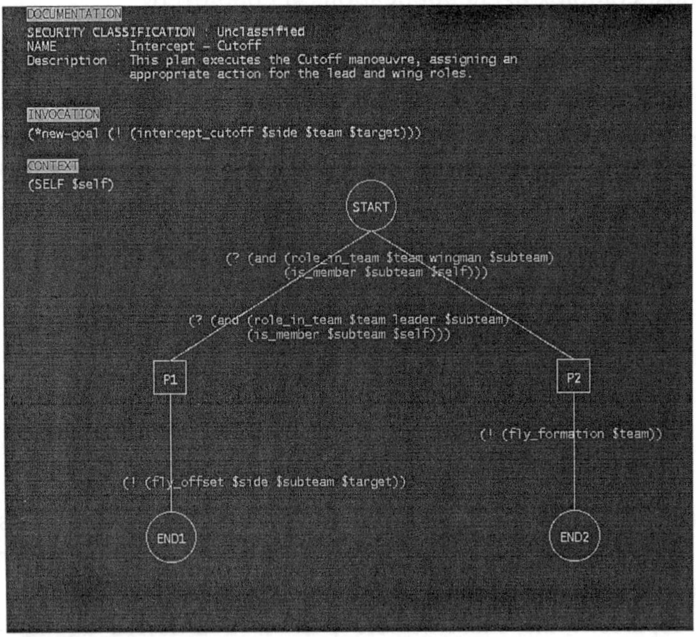

Figure 5. An example of a dMARS plan represented in the graphical plan language.

[4] Details on JACK can be obtained from www.agent-software.com.

3.2 The Agent Architecture

In the course of defining the agents it was a specified design objective that they should reason in a way that was accepted as intuitive by pilots and other lay-people. The architectural design (see Figure 6) summarizes the processes that the agent goes through when making decisions. The process is more complex than the discussion here may suggest but the detail is outside the scope of this chapter. More information about the design of these agents may be obtained from a paper by Tidhar *et al.* (1995). This design has proven to be a good framework about which to build agents. It provides a balance between allowing the flexibility to capture the complexity and richness of human decision-making and the simplicity and modularity required to keep the agents simple enough to understand and easy to modify.

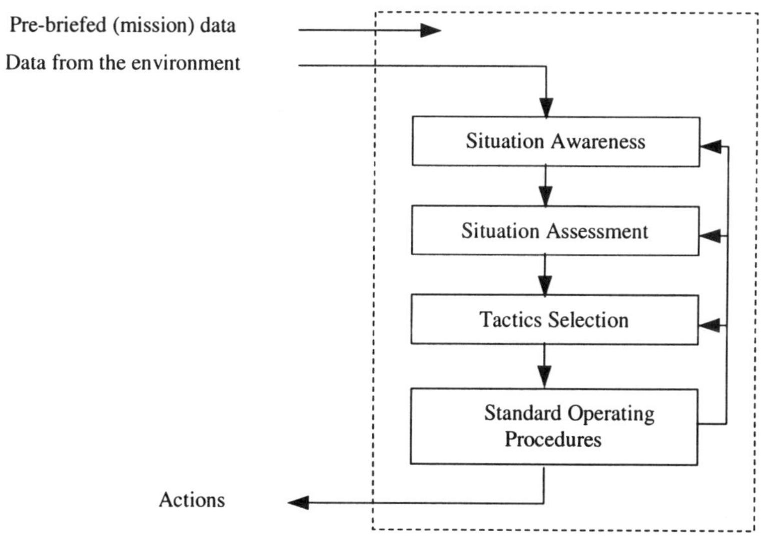

Figure 6. The agent architecture used in Air Operations Division, DSTO Teams and Command and Control.

The four modules displayed here are general enough to span the breadth of the reasoning required for our simulations but at the same time are specific enough such that pilots report a familiarity and an understanding of the architecture chosen. All software, indeed any engineering project, is a compromise and in this case an attempt has been made to strike a balance between software engineering utility, simplicity and transparency, and cognitive credibility. This formulation has been able to achieve this balance by avoiding becoming to focused upon a single design criterion but allows a continual updating of input from software engineers, pilots, and psychologists.

The model described above is a computational implementation of the OODA loop model widely known within military circles (Boyd 1987).

4 Ongoing Research and Development

AOD's commitment to the long term use of agent technology has seen a broad based research and development program address several of the pertinent issues facing the developers and users of intelligent agent technology.

Modeling teams and command and control has been identified as an important pre-requisite for many of the future simulation based studies that are likely to be required. Increasingly, in military circles there is an emphasis on 'joint' activities where all three forces cooperative to achieve more effective and efficient operations. With modern systems[5] there is increasingly a need for the commander to understand the possible arrangements of assets that are available in order to assemble the best possible fighting force for a particular mission. Simulations in support of the innovative em-

[5] Datalinks are fitted to many military surface, sea, and air platforms. This allows them to share information and to relay important tactical data. In this environment large coordinated operations are possible.

ployment of teams and the flexible use of command and control must provide facilities for the explicit modeling of these concepts (Hodgson *et al.* 1999, Tidhar 1999, Tidhar *et al.* 1998).

4.1 Perception and Recognition of Intention

In some of the earliest recorded descriptions of combative human behavior it was noted that recognizing the intent of an opponent was vital for success[6]. Two and a half thousand years later intention recognition is still one of the most important aspects of military decision making and is significant in many other competitive and cooperative human endeavors. The simplest explanation for the need for recognition of intention lies in an inability or unwillingness to communicate. If someone is both willing and able to communicate fully and honestly about their intent then there is little need for recognition. If communication is unavailable, unreliable, or in the case of combative or competitive environment, undesirable then a mechanism for deducing the intent of others is required. In virtual environments recognition of intention is potentially as important as it is in the real world. This is due to both the proliferation of heterogeneous computer systems, networks, and agents that speak different languages, or because of security or resource constraints are unable or unwilling to communicate and there are applications that might benefit from computer systems capable of inferring human intent. The Division is undertaking research into the provision of intention recognition capability for agents in simulation (Heinze *et al.* 1999).

[6] Five hundred years BC the oft-quoted Chinese military strategist Sun Tzu wrote: "His plans are calm and deeply hidden, so no one can figure them out. He changes his actions and revises his plans, so that people will not recognize them. He changes his abode and goes by a circuitous route so people cannot anticipate him. When people never understand what your intention is, then you win."

4.2 Software Engineering

As mentioned elsewhere, the employment of agent technology in SWARMM was one of the earliest steps in improved software engineering practices in model construction as it decoupled the reasoning models from the models representing physical systems. The BattleModel architecture took this a step further enabling all models to be completely decoupled from each other, each model being highly cohesive, in an object oriented framework. Basically this meant that each model whether it be a radar, platform, missile, or reasoning model was independent and communicated with each other via the BattleModel architecture, enabling a plug-and-play type of simulation. Furthermore, it allowed for agent reasoning models of different fidelities and types to interact in the same simulation. For example, heavyweight BDI agents written in dMARS could interact with lighter weight reasoning models implementing a variety of techniques written in languages such as C++ and Java.

The move to an agent friendly architecture led to research into how agent oriented systems should be engineered. To date this has involved looking at a number of different approaches including research into how existing object oriented methodologies such as the Unified Modeling Language (UML) (Booch *et al.* 1999) could be adapted or modified for agent oriented systems. Whereas agent researchers such as Odell and others looked at issues representing agent communication in UML through the use of AUML (Agent-UML) (Odell *et al.* 2000), the research conducted by AOD looked requirements specification, analysis and detailed design of agent systems. This included looking at how to extend the UML's concept of use cases for requirements specification of a system from a user's perspective, to one of behavioral specification of an agent system from an agent's perspective (Heinze *et al.* 2000). Furthermore, with the adoption of newer BDI agent languages such as Jack, design methodologies to support these new languages are required. Again an approach taken was to look at what existing techniques such as the UML could offer. Since Jack is a superset of the Java programming language with agent oriented extensions, and UML can be used to design Java programs, a logical approach would be to extend the UML to enable the modeling of Jack agents (Papasimeon and Heinze 2001). An example of a Jack agent designed in UML with Jack oriented extensions is shown in the figure in Figure 7.

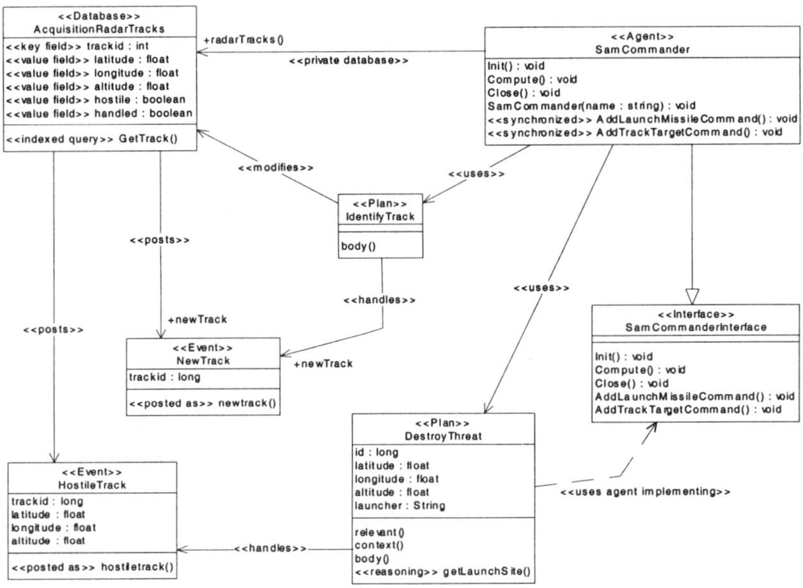

Figure 7. A Jack agent designed in UML with Jack extensions

Although there has been a lot of work over the past decade in agent systems, the diversity of agent systems means that research into agent oriented software engineering is still a very active area for a number of years to come.

4.3 Computer Generated Forces and Other Applications

The work of the Division is largely concerned with constructive simulation but for certain applications above there are good reasons for involving flight crews directly in the simulation (the AP3-C development described above). There is the potential for intelligent agents to become an integral part of the human-in-the-loop facilities and are a promising part of studies into distributed mission training and distributed simulation generally.

Throughout the world intelligent agents have been used as computer generated forces in HIL facilities and insights into the problems that beset them have been gained from the experiences with agents in constructive simulation.

Many of the issues associated with situating agents in simulations can be addressed in applications outside of the military domain. BattleModel has been applied to RoboRescue research in the form of a bush-fire simulation. By examining the requirements on the environments that suitably support agent activity insights for military simulation development are gained.

Another thread of research is exploring the addition of the principles of naturalistic decision making (NDM) to agents (Norling and Heinze 2000). NDM is a field of research concerned with analyzing the decision making of experts. The term was first coined in 1989 and marked an emergence of a new stream of decision research that emphasized the importance of experience, expertise, and situation awareness. Previous decision research had considered decision making in terms of generating options for action and then selecting from among them. NDM stresses the importance of building and maintaining situation awareness within the context of actual field settings. NDM claims that laboratory settings often remove critical contextual factors that influence the development of situational awareness. NDM and related research is appearing everywhere. Much of it is associated with military domains and it is all concerned with the decision-making performance of experts in field settings.

Figure 8. A Human-In-The-Loop F/A-18 Hornet Flight Simulator.

4.4 Graphical Interfaces for Agent Development

From the first technology demonstrator the importance of user interfaces in providing clear insights into the operation of the simulations has been recognized. The Division is currently collaborating with industry partners in the design of appropriate development environments for engineering intelligent agent systems. Whether or not this will result in GUIs with the look and feel of dMARS is still an open research question but it is likely that influences from other development environments will influence the development. For example AOD has independently described a process for specifying and coding Jack agents using the UML and tools such as Rational Rose.

5 Discussion

The impact of intelligent agents within the Division can be measured in the successful deployment of a number of innovative systems, the associated productivity improvements that have resulted directly from the technology and indirectly from the associated adoption of improved software engineering practices, and the quality of the research that has been fostered in and around the development. In summarizing the advantages of agents it is worth first cautioning that the particular adopted technology matched neatly with the requirements of the Division and though the application development has been successful it has not been without risk, steep learning curves, and cost. As Wooldridge and Jennings (1998) caution, "there are a number of good reasons for supposing that agent technologies will enhance the ability of software engineers to develop complex distributed applications… But agents are not a magical problem solving paradigm."

5.1 Rapid Development

One of the first requirements to be articulated during the system requirements specification for SWARMM was that it should allow for the rapid modification of tactical behaviors. This requirement stemmed from the lack of availability of combat pilots. Fighter pilots have large demands placed upon their time. It is not possible for modelers and analysts to obtain the assistance of a fighter pilot in building a model of their decision processes except for a few days spread throughout the year. When a pilot is available for active participation in the model development process it is essential that the modifications and additions to the model be made as quickly as possible to allow rapid trialing and feedback of suggested modifications.

AODs charter is to provide timely advice to the ADF. One of the bottle-necks in the process of conducting a study is that the analyst's time is spent in tailoring the tactics for that study; again it is essential that the models are capable of rapid modification to supply results in the minimum time.

By utilizing a modeling system based on the intuitive abstraction of: BDI, a variety of graphical tools, an architecture that is both sophisticated yet simple to understand, and a well developed graphical language such as dMARS it has been possible to reduce the gap between the domain expert and the software engineer. By including pilots and fighter-controllers in the design, coding, and testing it has been possible to dramatically reduce the time taken to develop and validate the systems. The result is an environment where operational personnel have confidence in the delivered results because of their participation in the whole process. The scientists and engineers responsible for building and using these systems are thus able to spend more time concentrating on the operations research aspects of the simulation and relatively less time attempting to model the complexities of human decision making. The result is a suite of simulation and modeling tools which are flexible, easily modifiable, and above all capable of providing answers to questions in an environment of demanding time schedules.

5.2 Graphical Interfaces

The dMARS language is graphical and a number of interfaces allow the plan-language to be viewed and edited by a graphical user interface (GUI). Furthermore the dMARS control interface (DCI) allows for the graphical viewing of the dMARS plans at run time.

Thus it is possible to view the plans that an agent has selected for execution as the simulation progresses along side detailed 3D visualizations of the physical models (see Figure 1). Traditional simulation visualization techniques have concentrated on viewing the physical world. In the air-combat domain this visualization is vital but while watching a simulation unfold the question often posed by pilots is "Why did that aircraft do that?" By incorporating a graphical representation of the reasoning processes we are able to answer that question as the simulation runs.

5.3 Appropriate Abstraction

One of the advantages of the agents that have been developed by AOD is that they have the quality of appropriate abstraction - the system used to model the human reasoning matches neatly with the design objectives of the human reasoning models. When modeling human reasoning it is advantageous to make use of a paradigm and a development environment that has its origins firmly rooted in the research of intentionality and practical reasoning. That dMARS (a BDI system) is well suited to this task is no accident. Work by Rao and Georgeff, bound the science of intentional systems to the practical application of dMARS in a way that aids the modeling of human reasoning (Rao and Georgeff 1995).

Pilots and controllers who regularly interact with AOD simulations report that the notion of a single agent mapping to a single human is an intuitive one. To capitalize on this, some effort has been made to design the agents for ease of understanding by people from a non-computer science background. This was a requirement because of the difficulties associated with the transfer of knowledge from the domain expert to the software-engineer. It was considered vital that models of human reasoning should be easily understood by pilots and that they could comment on, and suggest modifications to, these models. For this reason the level of abstraction afforded by a BDI model and a language such as dMARS is a good solution. The result is a system whose architecture and agent design is a natural, almost obvious, choice. The decision to design the agents in this way is in keeping with the folk-psychological roots of BDI (Dennett 1987). Indeed it may be argued that attempting to construct a model of human reasoning more in keeping with traditional cognitive psychology with a system rooted heavily in BDI may be asking for trouble. Whatever the relative merits of different cognitive models may be it appears that BDI, and in particular the dMARS implementation of it, combine very nicely with the present design requirements.

5.4 Enforced Modularity

Agents are by most popular definitions autonomous. This autonomy is enforced within dMARS and is a feature of AOD's most recent analysis models. Autonomy of operation means that models of reasoning are completely modular with well-defined inputs and outputs, yielding significant software engineering advantages by allowing the agents to be reused across a number of simulations.

The modularity of the agents has meant that they tend to be built as stand-alone models that can be easily interfaced to many simulations. The notion of agents as portable models of reasoning which may be integrated quickly into a simulation has allowed the reuse our fighter-pilot agents in at least two different simulations yielding a significant time-saving.

5.5 Distributed Processing

The implementation of dMARS is such that a distributed architecture for our agents can be obtained. Not only does this allow maximum throughput of the simulations by distributing the processing load of many networked machines but it also allows different models to be run on different machines. This gives an advantage when utilizing models that are largely operating system dependent. There is no requirement to port every model to a specific machine, each model can be run as a distributed process on the most appropriate machine. It is important to highlight the double-edged nature of distributed processing. Whilst it allows for faster throughput by utilizing all of the available hardware there are challenges posed in any distributed system. Whilst the simulations described here have all been tested as distributed processes and can successfully function there are issues of timing synchronization, process control, debugging, and logging output data that make distributing a process difficult. A current research activity at AOD is addressing these and other very real problems so that distributed simulations can be more easily implemented.

5.6 Improved Interactions

By creating an environment where tactics and standard operating procedures are more faithfully and explicitly represented it has been possible to improve the interactions between analysts and military personnel. The abstract graphical representation of plans provides the flight crews with visibility into the simulation increasing their confidence in the modeling. It allows tighter faster validation of tactical plans by inspection of crews and therefore improves the confidence that analysts have in the models.

Transitioning to agent technology shifted the focus from a systems view of military operations to a human centered view. This allowed problems to be viewed in radically different ways, for new research threads to be explored and for collaboration with psychologists, physiologists, and human factors experts to explore modeling options.

5.7 Software Engineering

Improved software engineering was a forced side-effect of adopting the intelligent agent technology. Mainly due to the collaborative nature of the early development it was essential that clear specifications, designs, and documentation were available to communicate ideas to geographically separated staff. This alone had the effect of improving procedures. In addition the professional staff of AAII were experienced software developers and there was a transfer of software engineering methodology techniques and tools that accompanied the agent technology transition.

6 Conclusions

The acquisition of intelligent agent technology has resulted in large gains in productivity. Problems that were previously intractable have been opened up and explored while other problems are now addressed more effectively and efficiently.

The particular implementations and technologies described here are clearly not suitable for all problems in all domains and care must be exercised when selecting novel technologies.

The particular domain and challenges facing the Division meshed ideally with the adopted agent approaches and the significant risks associated with new technologies were mitigated through iterative development, careful tool acquisition and a firm research base.

The success of a decade of intelligent agent research has resulted in quality outcomes for the Australian Defence Force, both in timeliness and quality, and an internationally recognized research program for DSTO.

Acknowledgments

The authors would like to thank the many engineers and scientists who have led their support and time to the various projects described here.

As a part of the strong local agent community the authors recognize the valuable insights and input from the local academic institutions, particularly the Intelligent Agent Lab at the University of Melbourne, and RMIT University.

A number of contractors have collaborated with these projects. In the 1990s employees from AAII played a significant role in the development of SWARMM and many of those engineers and scientists are now with Agent Oriented Software or KESEM International are still actively involved in collaborative research and development activities.

References

Booch, G., Rumbaugh, J., and Jacobsen, I. (1999), *The Unified Language User Guide*, Addison Wesley, Reading, MA.

Boyd, J.R. (1987), "A Discourse on Wining and Losing," unpublished set of briefing slides available at Air University Library, Maxwell Air Force Base, Alabama.

Bratman, M. (1987), *Intentions, Plans, and Practical Reason*, Harvard University Press, Boston, MA.

Dennett, D. (1987), *The Intentional Stance*, MIT Press/Bradford Books, Cambridge, MA.

d'Inverno, M., Kinny, D., Luck, M., and Wooldridge, M. (1997), "A formal specification of dMARS," in Singh, M.P., Rao, A., and Wooldridge, M.J. (eds.), *Intelligent Agents IV*, LNAI Vol. 1365, Springer-Verlag: Berlin, Germany, pp. 155-176.

Georgeff, M.P. and Lansky, A.L. (1986), "Procedural knowledge," *Proceedings of the IEEE Special Issue on Knowledge Representation*, vol. 74, pp. 1383-1398.

Halford, G.S., Wilson, W.H., and Phillips, S. (2002), "Processing capacity defined by relational complexity: implications for comparative, developmental, and cognitive psychology," *Behavioral and Brain Sciences*. Accepted as target article.

Heinze, C. and Goss, S. (2000), "Human performance modelling in a BDI agent system," *Proceedings of OZCHI*, Sydney, Australia.

Heinze, C., Smith, B., and Cross, M. (1998), "Thinking quickly: agents for modeling air warfare," *Proceedings of Australian Joint Conference on Artificial Intelligence (AI '98)*, Brisbane, Australia.

Heinze, C., Goss, S., and Pearce, A. (1999), "Plan recognition in military simulation: incorporating machine learning with intelligent agents," *Proceedings of Workshop on Team Behavior and Plan Recognition IJCAI'99*, Stockholm, Sweden.

Heinze, C., Papasimeon, M., and Goss, S. (2000), "Specifying agent behaviour with use cases," *Proceedings of Pacific Rim Workshop on Multi-Agents.*

Heinze, C., Goss, S., Josefsson, T., Bennett, K., Waugh, S., Lloyd, I., Murray, G., and Oldfield, J. (2001), "Interchanging agents and humans in military simulation," *Proceedings of Thirteenth Innovative Applications of Artificial Intelligence Conference*, Seattle, Washington. (Deployed application case study paper.)

Hodgson, R., Ronnquist, R., and Busetta, P. (1999), "Specification of coordinated agent behavior (the SimpleTeam approach)," *Workshop on Team Behavior and Plan Recognition, International Joint Conference on Artificial Intelligence*, Sweden.

Howden, N., Ronnquist, R., Hodgson, A., and Lucas, A. (2001), "JACK intelligent agents: summary of an infrastructure," *Pro-*

ceedings of the 5th International Conference on Autonomous Agents.

Jones, R.M. and Laird, J.E. (1997), "Constraints on the design of a high-level model of cognition," *Proceedings of Nineteenth Annual Conference of Cognitive Science*, pp. 124-132.

McIlroy, D. and Heinze, C. (1996), "Air combat tactics in the Smart Whole AiR Mission Model," *Proceedings of First International Simulation Technology and Training Conference (SimTecT '96)*, Melbourne, Australia.

McIlroy, D., Heinze, C., Appla, D., Busetta, P., Tidhar, G., and Rao, A. (1997a), "Towards credible computer generated forces," *Proceedings of Second International Simulation Technology and Training Conference (SimTecT '97)*, Melbourne, Australia.

McIlroy, D., Smith, B., Heinze, C., and Turner, M. (1997b), "Air defence operational analysis using the SWARMM model," *Proceedings of Asia Pacific Operations Research Symposium.*

Norling, E. and Heinze, C. (2000), "Naturalistic decision making and agent oriented cognitive modelling," *Proceedings of Fifth Australasian Cognitive Science Conference*, Melbourne.

Odell, J., Parunak, H.V.D., and Bauer, B. (2000), "Extending UML for agents," *Proceedings of the Agent-Oriented Information Systems Workshop at the 17th National conference on Artificial Intelligence.*

Papasimeon, M. and Heinze, C. (2001), "Extensions to the UML for designing Jack agents," *Proceedings of the Australian Software Engineering Conference (ASWEC)*, Canberra, Australia.

Rao, A.S. and Georgeff, M.P. (1995), "BDI agents: from theory to practice," *Proceedings of the First International Conference on Multiagent Systems.*

Shaw, R.L. (1985), *Fighter Combat – Tactics and Maneuvering*, 6th edition, US Naval Institute Press.

Tambe, M., Jones, R.M., Laird, J.E., Rosenbloom, P.S., and Schwamb, K. (1994), "Building believable agents for simulation environments: extended abstract," *Collected Papers of the SOAR/IFOR Project*, Information Sciences Institute, University of Southern California, Marina del Ray, CA, pp. 78-81.

Tambe, M., Schwamb, K., and Rosenbloom, K.S. (1995), "Building intelligent pilots for simulated rotary wing aircraft," *Proceedings of the Fifth Conference on Computer Generated Forces and Behavioral Representation*, pp. 39-44.

Tidhar, G. (1999), *Organization-Oriented Systems: Theory and Practice, in Department of Computer Science and Software Engineering*, University of Melbourne: Melbourne, p. 278.

Tidhar, G., Selvestrel, M., and Heinze, C. (1995), "Modelling teams and team tactics in whole air mission modelling," *Proceedings of Eighth International Conference on Industrial and Engineering Applications of Artificial Intelligence and Expert Systems (iea-aie '95)*, Melbourne, Australia.

Tidhar, G., Heinze, C., and Selvestrel, M. (1998), "Flying together: modelling air mission teams," *Applied Intelligence*, vol. 8, pp. 195-218.

Tidhar, G., Heinze, C., Goss, S., Murray, G., Appla, D., and Lloyd, I. (1999), "Using intelligent agents in military simulation or 'using agents intelligently,'" *Proceedings of Eleventh Innovative*

Applications of Artificial Intelligence Conference, Orlando, Florida. (Deployed application case study paper.)

Wooldrige, M.J. and Jennings, N.R. (1998), "Pitfalls of agent-oriented development," *Proceedings of the 2nd International Conference on Autonomous Agents (Agents-98)*, Minneapolis, USA, pp. 385-391.

Chapter 7

Agent-Enabled Sensor Network for Target Observation

R.S. Bowyer and R.E. Bogner

Target observation is a problem where the application of multiple cooperating sensors can improve the probability of detection and observation of the target. One way to organize multiple sensors is as a network, where the nodes are the sensors and the links are the channels of communication between the sensors. The structure of a network can vary greatly and we employ team formation as the means to grow a dynamic network structure. The sensors, which we shall refer to as agents, are situated in an area of interest with the goal of observing a moving target. We apply a team approach to this problem, which combines the strengths of individual agents into a cohesive entity – the team. In autonomous systems, the mechanisms that underlie the formation of a team are of interest. Team membership can be determined by an external authority, or it can emerge as the result of internal self-organizing (SO) mechanisms. We propose that existing sensors require augmentation with a process to facilitate this SO behavior. Internally motivated mechanisms are particularly challenging, but offer the benefit of being unsupervised, an important quality for groups of autonomous cooperating machines. This is the focus of our research. By studying natural systems such as colonies of ants, we obtain insight into these mechanisms of self-organization. We propose that the team is an expression of a distributed agent-self, and that a particular realization of the agent-self exists, whilst the environmental conditions are conducive to that existence. We describe an algorithm for agent team formation that is inspired by the self-organizing behavior of ants, and describe some simulation results for team formation for various configurations of sensors.

1 Introduction

1.1 The Target Observation Problem

In this chapter, we examine the problem of target observation using a network of sensors whose organization is inspired by biological ecosystems. The sensors are located on platforms such as ships, and are distributed throughout a geographic space that we call variously the surveillance region, the region of interest (ROI), or battle space as illustrated in Figure 1. The sensors are connected together by communications links, thus forming a network of sensors. It is important to realize, however, that not every sensor is connected to every other sensor. A communications barrier is shown that serves to block direct communication between sensors in the north and sensors in the south east of the region of interest. Some sensors are shown with detection regions, whilst other sensors may be inactive or faulty. Typically, the number of operational sensors within the ROI varies over time as some sensors move across the ROI boundary or may become faulty. Likewise, the number of targets may vary as they move through the ROI.

Specifically, we are interested in networks that comprise Electronic Support (ES) systems, which are passive sensors, meaning that they do not radiate energy at a target. Examples of ES sensors include infrared and ultraviolet detectors, CCD imaging cameras, and wideband and superhet microwave receivers. Each of these sensor systems typically generates digital data that describes the emissions they intercept.

When information from sensors is shared between the sensor users via the communications links, then each user gains the advantage of enhanced situation awareness. Our objective is to explore how the collective behavior (Wolpert and Tumer 1998) of the network can be self-controlled so that we can observe a target that moves within the region of interest.

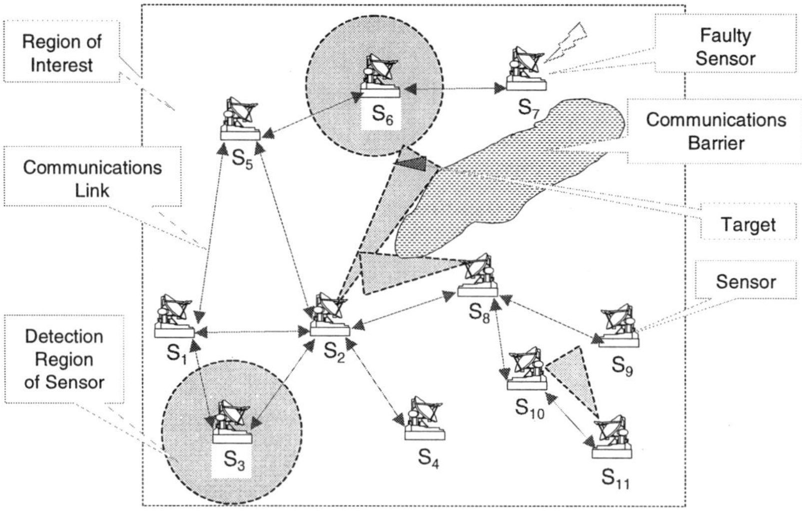

Figure 1. The sensor network.

Each of the sensors in the network possesses specific capabilities to observe and report on electromagnetic emissions within the environment. Such a sensor network is often called a network-centric (Alberts *et al.* 1999) surveillance system as distinct from a platform-centric system. Of course, a network of interconnected sensors is not of itself of any greater inherent value than many independent sensors, unless the sensors and their associated effectors can be scheduled to operate in a cooperative manner. An effector is the device that acts upon the environment, for example a flare or chaff dispenser. We want to explore methods that will support the self-organization of the distributed sensors into cooperative sensor teams, to provide the best possible situation awareness of the ROI. We want the sensors within the network to autonomously organize themselves and cooperate so that the information about the target can be maximized. Integration of appropriate sensors within the network into temporary teams is an important problem in multisensor management (Roy and Bosse 1998).

The amount of information that is generated by an ES sensor (Curtis 1999) can be quite large, and the interpretation of that information in real-time is an even larger challenge. We aim to put in place a system that makes intelligent choices about what information to collect, and from what sensor. Such a system needs to be autonomous, react to its environment, and regulate its own functions, similar to a multi-cellular organism.

The target observation problem presents many opportunities to explore the application of autonomy. However, for the sensors to engage in autonomous behavior first requires that each sensor is equipped to interact with another sensor, and secondly, requires a framework in which they can organize the development of their interaction. We shall develop this concept later.

As an example, position estimation of targets is an important task that requires cooperation between sensors, particularly if the sensors are passive and can only provide bearing information. A single sensor that receives all the target reports from all the sensors in the network is clearly faced with a large data association and fusion problem. It is better to send a sensor only the information it requires, when it is needed in a cooperative exchange, thus limiting the data traffic on the network.

Cooperation between sensors effectively generates virtual sensors, whose capabilities are well beyond those of any single sensor. An example is to extend sensing range, where the field of view of one sensor is partially obstructed by an object such as a hill, but the field of view of another cooperative sensor is located in a position such that the occluded region is within its field of view. Clearly, if both sensors engage in cooperative behavior with each other they gain from the complementary nature of the relationship, with both sensors receiving the benefit derived from shared situation awareness. In this way, the network can generate virtual sensors, an example of which is stereo-vision, where depth perception is possible when two cameras, or two eyes, provide complimentary information.

The important ingredient in this scenario is cooperation. If we increase the number of sensors that are situated within the ROI, we are able to generate more sensor reports about the area under surveillance. More reports, however, do not necessarily translate into more and better knowledge. We are considering the problem of forming an understanding, or situation picture, of a geographic region, within which there may be a target, and to achieve this objective, we have at our disposal a number of sensors.

To help us set the scene, consider the following scenario, which is illustrated in Figure 1, where a platform is monitoring a region of interest. A target enters a region of interest far beyond the detection range of your optical sensors. The ES sensors aboard this platform, which incorporate high sensitivity receivers, detect the presence of electromagnetic radiation and classify the source as a threat tracking radar. The ES system can only provide the commander of the platform with an approximate compass bearing to the object: it is too far away for the radar to track. Additional information is required from other sensors in the ROI, which could then be fused to generate a situation picture that would identify the threat location. If the sensors could reach out beyond their individual ranges, combining the capabilities and information from neighboring sensors, then the location of the threat, and possibly its identity and intent could be more precisely determined. Once the identity is known, a suitable response can be generated from the cooperating collective of sensors and their associated platforms.

Other friendly sensors may be distributed about the platform's position, some closer to the object and some with better sensitivity than the sensors, which belong to other command structures and are all busy with their own tasks. We need some way for the sensors to cooperate, to share in a common observation task, without the need for human intervention.

An autonomous system, that is a system that senses, reasons, and acts upon its environment independently of external command structures, such as a robotic aircraft or ground vehicle, requires the knowledge of specific objectives and a set of guiding principles for it to be able to function in a safe manner in an environment. We want to be able to deploy such autonomous systems by providing them with a mission statement and little else: allowing them to engage the environment and each other to achieve their mission objectives. The deployment philosophy is stated as; *go over there, fix the problem, and don't do any harm.*

1.2 Agent-Enabled Sensors

In our view of the multi-sensor target observation problem, we refer to agent-enabled sensors, which as the name suggests are sensors that are augmented with an agent process. This concept is illustrated in Figure 2 where, in practice, an agent process is connected to each sensor's data and control bus. An agent is an autonomous software process that interfaces to the real world via sensors, reasons about its perceptions, and takes action by using its effectors.

The objective of this approach is to enable a complimentary information-processing network to coexist with the original sensor network. The agent will not interfere with the normal operations of the sensor, but will act in a symbiotic mode (Bowyer and Bogner 1999) to generate additional target information by interacting with neighboring agents.

A single sensor operating in isolation has a limited capacity to provide complete surveillance coverage of the region of interest. Clearly, if the range of this sensor is limited to less than the boundaries of the ROI, and the sensor is unable to relocate itself within a short time frame, then it cannot accomplish complete area surveillance. Likewise, additional processing and detection measurements would be required to perform geolocation of a target. We may add more sensors to improve the coverage, but if the behavior of each sensor is not coordinated with that of its neighbor then considerable time and energy will be wasted and the expected sensing performance will not be realized.

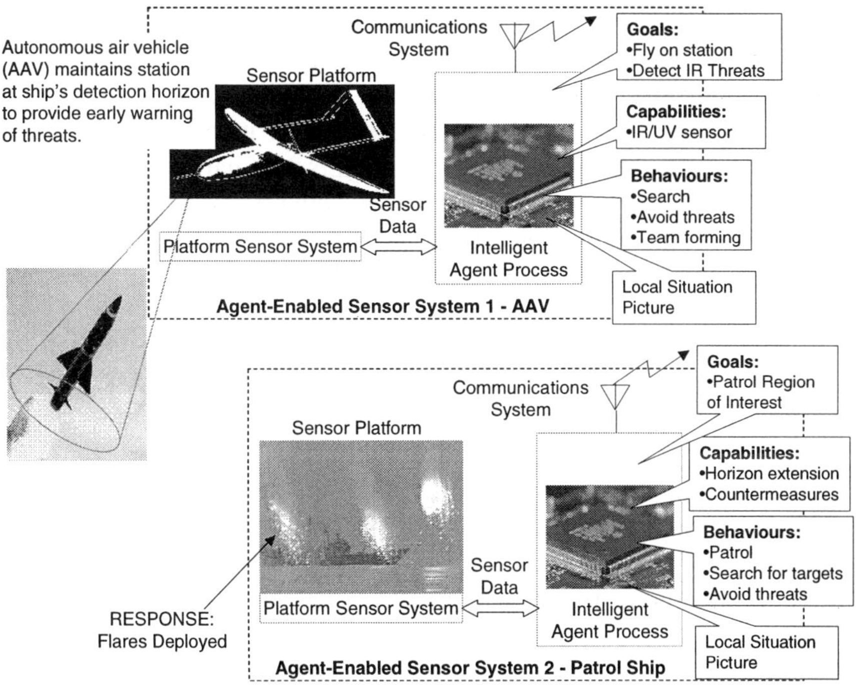

Figure 2. Agent-enabled sensors communicate with each other as if they were members of a community. This illustration shows an autonomous air vehicle providing extended sensing horizon to a ship. The ship is able to deploy counter-measures when a threat is detected by its counter-part.

When there are multiple sensors in the region of interest and each has the ability to communicate over a particular range, we can expect a performance improvement in accomplishing the surveillance and target observation tasks. Two sensors looking in different directions and sharing their information form a virtual sensor. When more sensors are added, we are faced with allocating sensors into teams, where each team focuses its efforts on solving one section of the overall task. Of course, this problem can be solved in an optimal manner if all sensor capabilities and locations are known prior to commencement of the task: we perform a search that optimizes the coverage of the region of interest, and allocate sensors accordingly to exhibit particular behaviors.

We want to know when a target is present in the region of interest, together with the target's position and behavior. The behavior of a target describes its spatial dynamics such as its velocity and maneuvers, and by deriving these parameters, we can build up a picture of the environment and improve situation awareness for the operators of the sensor assets. However, we shall leave behavior and intent analysis for another time.

This problem facilitates the study of many related issues, which include:

1. individual sensor performance measures,
2. cooperative sensor performance measures,
3. mechanisms to coordinate individual sensor behavior,
4. ensuring that autonomous sensor systems behave in a safe manner, and
5. mechanisms for fusing sensor data.

From this point, we shall understand the term *sensor* to refer to the physical device that intercepts and processes the signal from a target. The term *agent* shall be used to refer to an autonomous process that can sense, reason and act pro-actively. It engages in the networked social interaction and has available to it the information generated by the sensor.

1.3 Centralized versus Decentralized Control in Multisensor Systems

A multisensor data fusion system combines the output information from multiple sensors to generate situation picture. A common approach to forming a situation picture, based on multiple reports from geographically distributed sensors, is to fuse all the information at a central processing facility as illustrated in Figure 3. Sensor data fusion (Anthony 1994, Hall and Llinas 2001, McMichael *et al.* 1997, Thomopoulos 1994, Waltz 1998) is the process of combining the data that is generated from multiple sensors, which are measuring phenomena within a region of interest, into a coherent situation picture. The main steps of data collection, dissemination and processing must occur within a period that is commensurate with the possible actions that may be taken by the end-user of the data. The next step involves allocating the finite sensor resources, in the light of what has been discovered, so that a particular goal may be realized. Of course, evaluating the integrity and accuracy of the processed information and the effectiveness of any actions taken also requires considerable effort on the part of the fusion system.

We want to develop a suitable mechanism whereby geographically distributed sensors can intelligently interact with each other, make decisions at a local level, and provide a robust and fault tolerant sensor network. In a centralized architecture like that illustrated in Figure 3, all sensors report to a central fusion node where the user is able to interpret all the data to generate the global situation picture and apply control back to the sensors. Considerable processing capacity is required to collect, align and correctly associate related reports from different sensors. This method also relies on secure and stable communication links between the sensors and the central processor. In addition, if the central processor is damaged, then the performance of the entire sensor network is compromised.

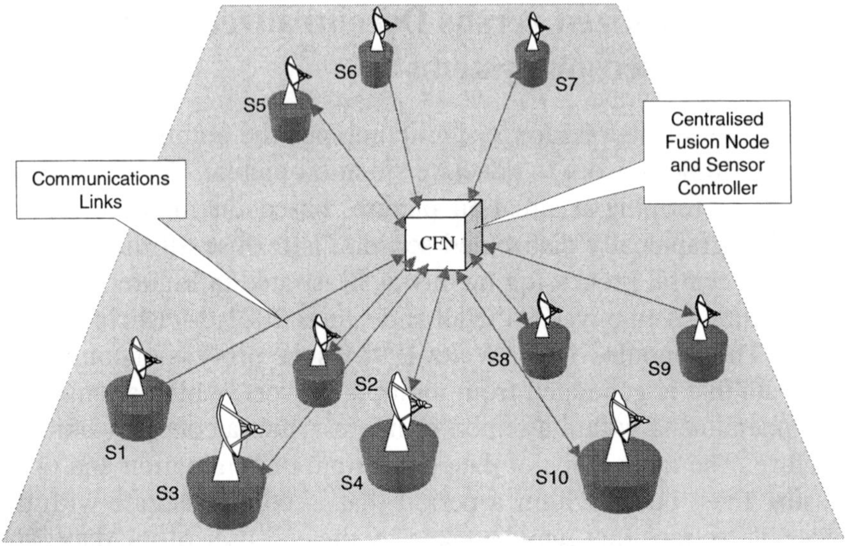

Figure 3. Centralized architecture: sensors may be organized and tasked by a central controller with centralized information fusion.

Information relevance and response speed are two key factors in distributed sensing and control systems. When operating at a local level, the end-user of information, whether human or machine, requires real-time situation awareness within their context region. If a central platform is receiving sensor reports from all of the sensors in the region of interest and attempting to fuse all of the information, then the platform's fusion system will be placed under considerable processing load. The fused information is then transmitted to every sensor platform in the ROI.

If we now consider a single sensor platform within the ROI, it will receive information from distant sensors that is possibly not relevant to its own context. Further, sending non-essential information over the links represents an undesirable load on the communications links and processing algorithms that run on each sensor node. The region within which the platform can impart an affect on, or perceive an affect due to objects in the environment is defined as the context region for that platform.

Increasing the number and variety of sensors, increases the flexibility of the network to fuse data and create virtual sensors. However, just throwing more assets at a problem does not imply more success, as the processing required increases exponentially for a centralized architecture. We are constrained by time, energy, multiple user demands, and the dollar cost of the sensors and infrastructure. What we would like is a method of integrating sensors dynamically, that can account for these constraints and account for the dynamic nature of the environment in which they operate. The number of sensors changes over time because some are subject to failure, new sensors are moving into the area of interest and some are leaving the area of interest. We are interested in the best use of available sensor resources, which change over time, for the task of area surveillance and target observation. A community of cooperating sensors will provide a means to achieve the task. The cooperating sensors share and fuse information to generate local awareness of the environment as illustrated in Figure 4.

Social interaction is the key to forming these relationships that respond to local target activity. To facilitate the social interactions, each sensor requires the ability to act as though it were a member of a society. This is accomplished by installing an agent with each sensor. The agent relationships emerge and evolve as the environment changes, forming sub-networks or teams as illustrated in Figure 5. As a result, information flow is restricted to the relevant agents that belong to the sub-networks. Locally relevant responses are guided by goals common to all agents within a sub-network. This local sensing and action cycle is central to our approach.

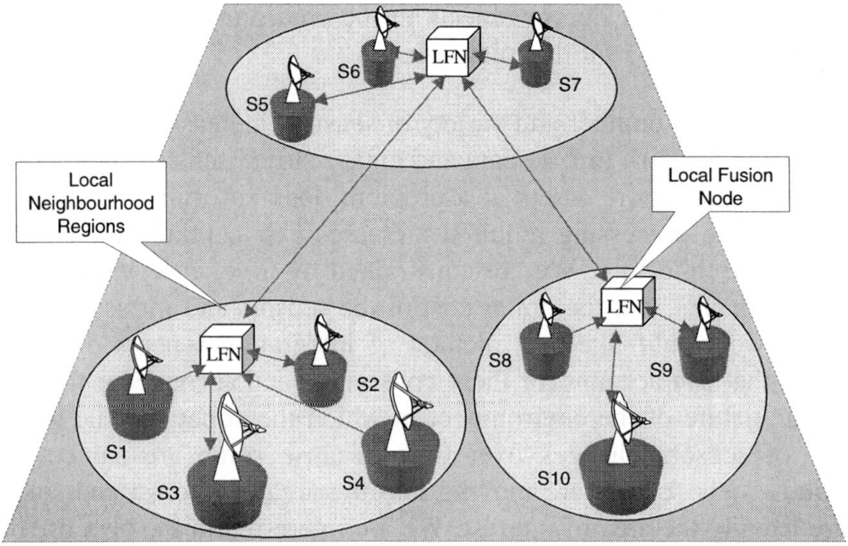

Figure 4. Decentralized architecture: the objective is to have sensors that interact with their local neighbors. The local fusion node may actually reside on one or all of the sensors within the neighborhood.

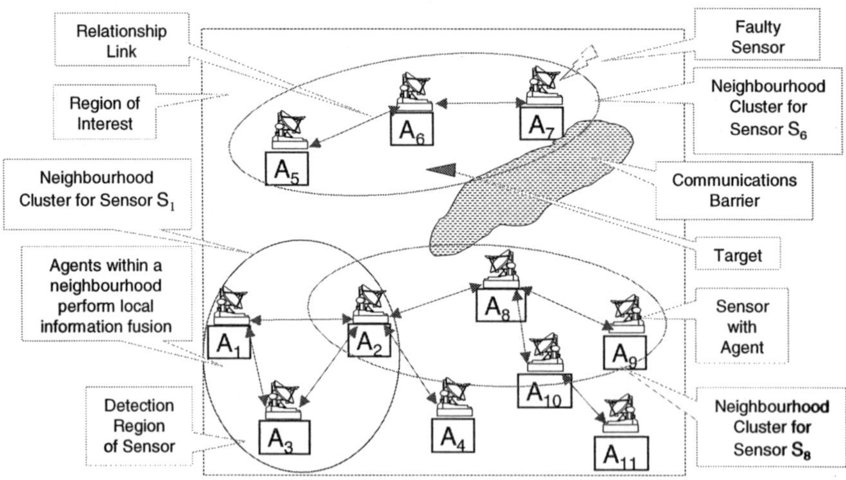

Figure 5. The collection of sensors is now augmented with an intelligent agent that enables the sensors to interact with each other in a social network. Relationship links are established between agents and clusters are formed.

1.4 Motivation for Self-Organizing Sensor Network

Coordinating the behavior of a group of autonomous sensors is an area of considerable interest. To achieve coordinated behavior of the network in response to target activity, the actions of every sensor must be continually monitored and adjusted. The obvious point of inquiry is to determine the mechanism that performs the monitoring and the adjusting.

Sensors in the network may be controlled externally, or they may be controlled by an internal mechanism. The idea that a group of sensors can organize its own collective behavior to meet a particular goal is an appealing one. To impose external control on a network of interconnected machines is a non-linear problem: one that is made even more intractable because the number, type and a myriad of other factors vary with time. The term self-organization brings to mind the concept of possibly getting something for nothing, or at least getting something with minimal effort on the designer's part.

We are motivated to understand how to engineer a self-organizing sensor network because it has the potential to respond more quickly and effectively to events within the region of interest. Further, as sensor systems become more numerous and target dynamics become more advanced, the human operators of such networks are faced with over-whelming quantities of information to interpret before appropriate actions can be taken. The time available for making decisions is also diminishing to the point where only machines can make decisions and execute actions fast enough to keep up with the tempo of the modern sensor/target environment. An autonomous self-organizing network can help to relieve the human element from much of the demanding task of interpreting enormous quantities of information.

The success of the area surveillance task depends on how well the available sensors are allocated to *cover* the region of interest, and take advantage of the complimentary functions that exist within a network. Sensor coverage depends on the individual capabilities of the sensors and the methods by which the sensors are being tasked. Figure 3 shows the situation where various types of sensors report to a centralized controller, where the data is fused and a situation picture is developed. Commands are issued to the sensors from the central controller to adjust their behavior to optimize the situation awareness.

Decentralization and devolution of control decisions to the sensors has the potential payoff of enabling the sensing system as a whole to react and adapt faster to events. This is particularly important when the normal communication links and the command structures break down (Vicente 1999) or are placed under an extreme load. If smaller groups of sensors are enabled to make decisions following strict standing orders, then such a system is more likely to exhibit graceful degradation in the presence of disturbances to the communications infrastructure.

We desire a system comprising many sensors, each of which are capable of social interaction with their neighboring sensors, to exhibit self-organizing behavior that evolves over time and hence be more flexible. The result of this behavior, as observed from outside the system, is the emergence of patterns (Kelso 1999) that are not pre-programmed into any of the individual system agents. We wish to harness this self-organizing behavior to provide an automated means of coordinating sensor behavior and hence reduce the workload on human operators. By decentralizing the control of sensors and devolving the behavioral control to the members of the community, the system is free to organize its topology according to the prevailing environmental conditions.

One approach to this surveillance and control problem is to enable the sensor network to make its own decisions on resource allocation based on the strength of inter-agent relationships. We consider the mechanisms that would enable the sensors to self-organize in order to observe a moving target. Our premise is that the sensors that are located in the neighborhood of each other, and of the target, are in the best position to make decisions about how to cooperate and share target surveillance data. We propose *team formation* as the means to organize the groups of heterogeneous sensors to achieve a coordinated, autonomous observation system.

1.5 Motivation for Team Formation

Team formation is one way to approach the class of problems called resource-allocation problems (Beckerman 1997, Reynolds 1987). That is, given a limited number of heterogeneous sensors, how many sensors and of what type do we allocate to each task that arises? The team approach combines the strengths of individual agents into a cohesive entity – the team. In autonomous systems, the mechanisms that underlie the formation of a team are of particular interest. By studying natural systems such as colonies of ants, we obtain insight into mechanisms for self-organization that are useful for our efforts to achieve a coordinated observation system. We propose the concept of the agent-self, or agent phenotype, which we describe in Section 4.

We shall identify important characteristics of self-organizing systems and consider the membership requirements of the individual elements of such systems. We describe an algorithm for agent team formation that is inspired by the observed behavior of ants, and describe simulation results for team formation amongst a group of networked sensors.

The natural systems that we consider here possess mechanisms for interaction. In the case of ants, we shall see that chemical pheromones play a significant role in the communication between individuals in the colony. Therefore, if we are to adopt similar models of interaction, it is not surprising that our agents must also be equipped with the ability for social interaction, which is why we refer to agent-enabled sensors. Our objective is to develop a distributed control strategy to manage the sensor and effector resource allocation in response to the target activity. Figure 2 illustrates a representative situation where one sensor system, the UAV, detects a possible missile threat to a ship that is within the UAV's context region. The UAV and the ship share a relational bond. The UAV responds by informing the ship of the threat. The ship responds by launching flares to deceive the approaching threat.

Given this approach, we then ask the question, "how do we organize the membership and lifetime of a team?" To answer this question, we consider two approaches. The first approach calls for a single external authority to impose a structural topology. We want to avoid this approach because it often requires knowledge that may not be available to such an authority at the time of selecting the composition of a team. Also, the approach can result in rigid lines of control and communication. The second approach, which we adopt here, relies on the self-organizing behavior of the individual agents within the collective, where appropriate structural topologies emerge as the need arises. We call each topology that emerges a team.

This method is particularly challenging because factors such as cooperation, stability and member suitability must be addressed and are no longer controlled from a single location. However, it does offer the benefit of being autonomous, which is an essential quality for agent systems that need to operate with little or no human intervention. We investigate these factors and specific mechanisms that enable agents to form relationships with their neighbors.

We propose that by studying natural ecosystems such as colonies of ants, we obtain insight into these mechanisms of self-organization. Further, we propose that the team is an expression of a distributed and extended agent-phenotype or *agent self*, and that a particular realization of the agent self exists, for as long as the environmental conditions are conducive to that existence. In Section 8, we describe an algorithm for agent team formation that is inspired by the self-organizing behavior of ants, and describe simulation results for team formation amongst networked agent-enabled sensors.

1.6 Related Work

There are many research areas that relate to self-organizing sensor networks. Multi-sensor target observation is a problem that has inspired many solution approaches, and is particularly prevalent in the data fusion (Anthony 1994, Blackman and Popoli 1999, Dorigo and Gambardella 1997, Hall and Llinas 2001, Harney 1997, McMichael *et al.* 1997, Thomopoulos 1994) and target tracking (Blackman and Popoli 1999) research communities. The mobile robot community has contributed to this area. Teams of cooperating agents can take many shapes, and can arise from the action of a number of mechanisms (Cohen *et al.* 1997, Dorigo and Gambardella 1997). For example, a system may have a Captain process, where the Captain selects team members according to the skill level of each individual. This is a centralized approach. Parker (1996, 1997) uses a behavior-based approach to explore team behavior, defining a framework problem called *Cooperative Multi-robot Observation of Multiple Targets* (CMOMMT).

Much can be learned by studying the behavior of natural ecosystems such as colonies of ants (Anderson and Franks 1999, Dorigo and Gambardella 1997, Resnick 1994), that display surprisingly high levels of organized collective behavior. This collective behavior may be viewed as a form of collective intelligence (Wolpert and Tumer 1998), where the interaction of the parts, the agents, gives rise to the emergence of behavior patterns that optimize a global goal: the observation of the target.

Other relevant work has examined the under-lying mechanisms that can imitate the self-organizing formations observed in a flock of birds (Reynolds 1987). This work proposes a decentralized explanation of the collective behavior of the birds, and is relevant because we can see how machines may be coordinated by the emergent properties of the collective.

Task decomposition is another related area of research. Stone and Veloso (1999) examine issues of team synchronization and formations of agents. Team formation mechanisms are investigated, in which varying degrees of information about team structure is exchanged between robots. Our approach is based on fundamental qualities of each agent within the neighborhood of each agent. Tambe (1998) examines teamwork using a framework called TEAMCORE.

Unattended ground-based sensor networks is an area of work that incorporates many of the problems that we are interested in, such as distributed processing by autonomous sensors which are linked together by time-varying communication networks. In work by Brooks (2001), the point is made that such networks of autonomous sensors must be able to configure itself in response to operational needs. We address this point in our current work.

Robot soccer or RoboCup (www.robocup.org) is an area where team strategies are explored. To achieve this we apply a social ecosystem paradigm, where the teams emerge because of social interaction between agents. This approach provides a flexible, plastic and reactive mechanism, and supports learning. Higher levels of decision-making are possible for generating teams, but our current approach is based on inter-agent attraction due to potential fields.

Self-organization is a desirable property for a colony of agents to possess because (a) a self-organizing colony requires little or no supervision, (b) the colony can potentially perform self-repair, and (c) the colony can respond in an appropriate manner when unusual events occur in its domain of interest. Vicente (1999) states that *the social organization determines the form of the communication among actors.* A corollary to this statement is that the communication mechanisms, the infrastructure, available to the actors, shape the social organization. We shall return to this concept in Section 4.

The chapter is arranged as follows. In Section 2, we introduce the system models and in Section 3, we examine self-organization as a means of network management. Then, in Section 4, we consider the emergence of form and function that we can observe in some natural ecosystems. The mechanisms that are at play in these systems provide us with inspiration to develop algorithms that generate self-organizing behavior in artificial ecosystems. In Section 5, we develop the architecture for our biologically inspired agent. Section 6 describes the method by which the behavior of an agent is directed. Goal-directed behavior is central to the operation of a computational ecosystem comprising agents. The details of team formation are discussed in Section 7, and we introduce our ANTeam algorithm in Section 8. Finally, we present some simulation results for a number of sensor network scenarios in Section 9.

2 System Models

2.1 Characteristics and Assumptions

Having introduced the problem and characteristics of network-centric target observation we now place some bounds upon the problem, which allows us to investigate some of the key concepts by way of simulations. In Table 1, we have listed the assumptions that we employ in this work.

In this section, we present the models for the simulation environment and for the dynamics of the target. We describe the method by which the sensors perceive their environment, and how interactions are represented between the target and sensors. A key concept in this work is the idea of social interaction and we present our model in this section. The models that we use in our simulations have been implemented in MATLAB for convenience.

Table 1. Characteristics of the distributed sensor problem.

1. The number of sensors N_s is dynamic.
2. There is a single target, N_T=1 (for simplicity).
3. A sensor is susceptible to the presence of a target.
4. The time taken for the target to move an appreciable distance is much less than the time required for information processing by any agent.
5. Information has a must-have-by date – *"I need to know now."*
6. Information has a use-by date depending on the context.
7. Each sensor has a particular sensitivity to events to which it is susceptible.
8. Each sensor is under the authority of a local supervisor V.
9. Each agent is capable of communicating with neighboring agents.
10. Each sensor perceives a local neighborhood of sensors via mutual communication.
11. Existence of communication channels between sensors is time varying.
12. Each sensor has a specific field-of-view.
13. Information has a lifetime that decays.
14. Important information may have an infinitely long lifetime – always useful.

2.2 Environment Model

The environment is modeled by a two-dimensional $m \times n$ lattice shown in Figure 6. The lattice is populated with N_s agents that observe the region of interest. An agent i is located at (x_i, y_i) within the region of interest, and it possesses a velocity v_i. A global temperature is defined to provide a method to control the rate of evaporation of the pheromone discussed in Section 5. Each agent in turn inspects each of its neighbors, and reacts according to the situation it perceives. Now we shall consider how the sensors and targets behave within this environment.

2.3 Sensor Models

Let the distance between sensor i and the target T be denoted $d_{i\mathrm{T}}$ and the detection radius of sensor i be denoted $\rho_{\mathrm{det},i}$. We simplify the characteristics of the detection process by declaring a detection by the ith sensor if the target lies within the detection field of view of that sensor, where $r^{\mathrm{S}}_{\mathrm{skip},i} \le d_{\mathrm{T}i} \le r^{\mathrm{S}}_{\mathrm{det},i}$ and $\theta^{\mathrm{S}}_{\mathrm{start},i} \le \theta_{\mathrm{T}i} \le \theta^{\mathrm{S}}_{\mathrm{stop},i}$ (Figure 6). The sample or *look* time of the ith sensor is defined by $T_{\mathrm{S},i}$.

As mentioned in the introduction, electronic support systems comprise passive sensors that are designed to provide a platform commander (platform is used throughout this document to describe the physical support structure on which the sensors are installed) with reliable situation information. Based on this and other sources of data, the commander may gain an early warning of potential threats to the platform. The ES system is usually designed to monitor particular sections of the electromagnetic spectrum (Curtis 1999, Brooks 2001), and through a variety of processing stages, establish the presence of all emitters near the platform, and consequently, the level of threat.

The indications from the ES system are used in conjunction with other intelligence information to establish a local situation picture. The commander refers to the situation picture to make tactical decisions. Active sensors such as radar provide information at the expense of energy, and in the military case, at the expense of remaining undetected. This may well affect the platform survivability. Although we are only considering passive sensors in this study, combinations of active and passive sensor suites should be exploited.

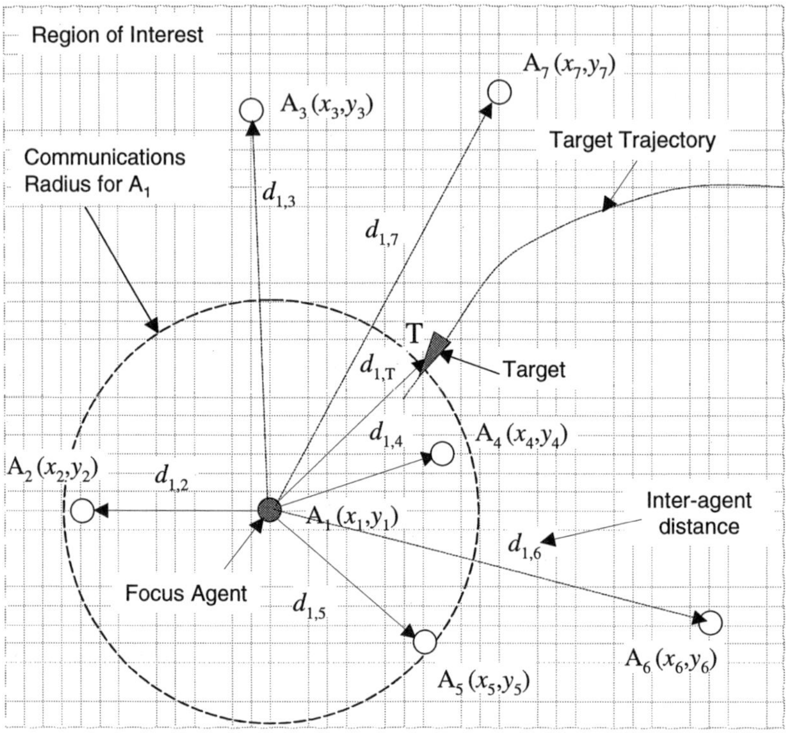

Figure 6. Influence diagram for the object agent A_1 (repeated for each agent in the society). This section of the $m \times n$ lattice shows the influences associated with A_1, which are the moving target T and the Nn_i neighboring agents that can communicate with A_1. We assume each agent has the same communications range. There are no communication links between agent A_1 and agents A_3, A_6 or A_7 and therefore have no influence over the behavior of agent A_1.

2.4 Target Model

The target T is modeled as a simple point-object that exhibits one of four basic behaviors as listed in Table 2. These behaviors were chosen, as they are typical of objects that exist in a variety of natural ecosystems and in electronic warfare scenarios in general. The target moves around within the region of interest and occupies a single lattice site at any given instant in time as shown previously in Figure 6. The time between moves is defined by T_T. The behavior selected by the target is controlled by a transition algorithm. The target detects an active sensor if $d_{Ti}<\rho_{T,det}$, where d_{Ti} is the Euclidean distance from the target to the ith sensor.

Table 2. Target behaviors supported by simulation software.

Behavior	Modes
Patrol	Target moves in one of three types of motion: 1. random walk 2. waypoint-defined 3. orbit about a central point
Avoid	Target moves to avoid sensor position.
Homing	Once the target detects an active sensor, the target homes-in on the sensor – e.g., anti-radiation missile.

2.5 Inter-Agent Communication Model

One agent cannot affect the behavior of another agent, unless there exists a mutual channel of communication between the two agents. A communication link must exist between any two agents for an exchange of information to occur, and hence an influence exerted.

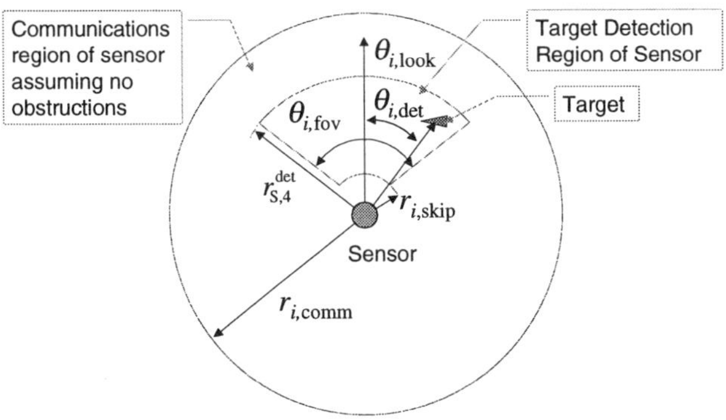

Figure 7. Generalized sensor detection and communication characteristics.

We define inter-agent communication by the exchange of state information between two agents (Figure 8).

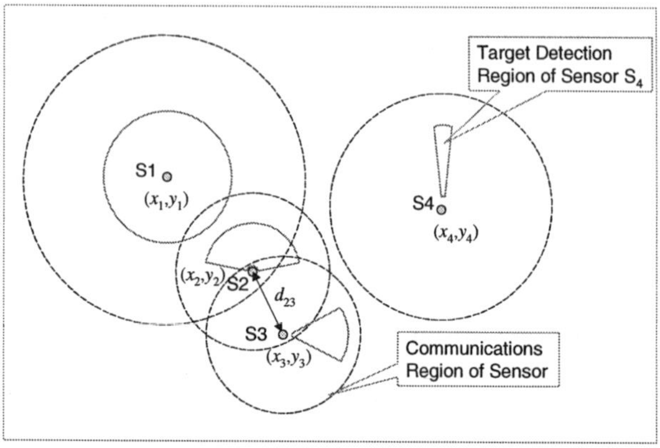

Figure 8. A group of four sensors is clustered together and shows the communication boundaries and detection regions. Sensors S2 and S3 are aware of each other as their communication regions overlap. Sensor S1 is only aware of sensor S2, and sensor S4 is not communicating with any other sensor.

In our simulations, agent A_i is *aware* of the existence of agent A_j if $d_{ij} < \rho_i$, where d_{ij} is the Euclidean distance between the ith and j^{th} sensors, and ρ_i is the communication radius of agent i. The neighborhood of an agent is defined as the set of agents that lie within communication range of the agent. Any two sensors i and j can *communicate* with each other iff $\rho_i > d_{ij}$ and $\rho_j > d_{ij}$, where ρ_i is the communication radius of the ith sensor, ρ_j is the communication radius of the jth sensor, and d_{ij} is the Euclidean distance between the ith and j^{th} sensors.

2.6 Social Aspects of Sensor Networks

The dynamics of a society are what make them interesting and useful. As sensors move into the range of the region of interest, their capabilities are made available to the rest of the community, if a suitable social infrastructure exists (Figure 9). The addition of new capabilities to a system of sensors is only of value, however, if the *collective intelligence* has the corresponding ability to first recognize the new contribution, and second, to exploit the additions.

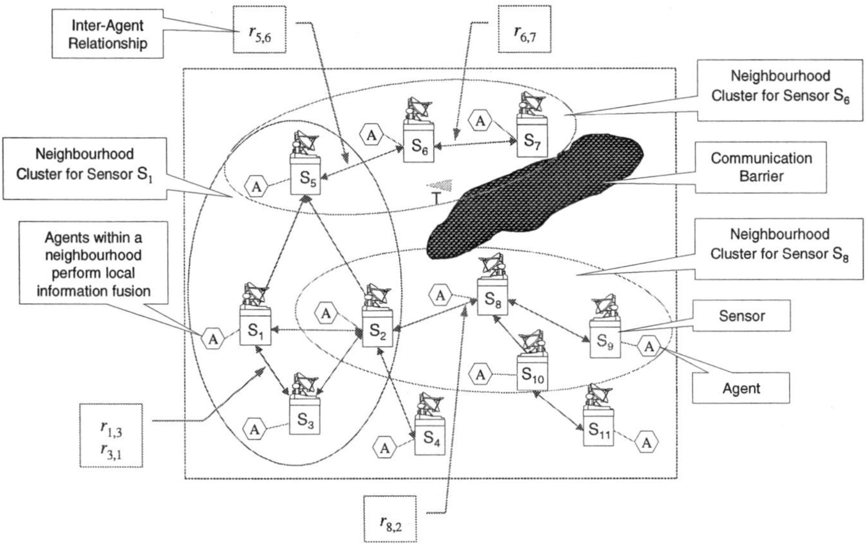

Figure 9. Relationships develop between agents where communication links exist. These relationships are the foundation for team formation (see Section 5).

Conversely, the removal of a capability must also be recognized and taken into account. In this context, the exploitation of resources may be seen as intelligent behavior, distributed among the members of the community. That is, we define intelligence behavior as the *timely,* bounded adaptation to changing circumstances.

A community of *sensors* that exhibits this innate intelligence is desirable in environments where the reliability or life expectancy of a sensor is limited, and where the environment itself exhibits dynamic behavior. The community concept means that in place of a rigid connection circuitry between sensors, there is a *plastic* association, giving rise to a logical social network (Bowyer and Bogner

1999, Kelso 1999) that changes and adapts according to the situation and context.

To formalize the social interactions we apply a terminology familiar to the areas of group theory and social science research. A valued network relation matrix describes the physical relationships between members of a community, and here we apply it to represent the sensors in our network. Note that *connection does not imply that the sensors either are exhibiting or are capable of cooperative behavior.* These relationships are illustrated in the network value graphs in Figures 10a and 10b, where R is called the *relation matrix,* with $R = (r_{ij})$. A relationship is indicated by a positive real number that will be greater than an appropriately chosen threshold value. A number below the threshold indicates an absence of a relationship.

The graph is a dynamic structure that evolves as the system adapts (i.e., its behavior evolves). Matrix R_1 defines a network with five sensors, as in Figure 10a, with no inter-connection, and hence no *direct* cooperation exists. Matrix R_2 describes the network in Figure 10b. These matrices are read as follows: the sensor in row i is said to have a relationship with the sensor in column j, following the direction of the arrows.

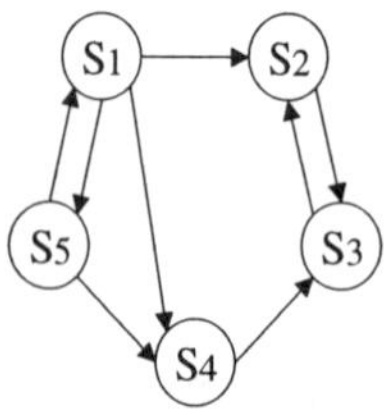

$$R_1=\begin{bmatrix}1&1&0&1&1\\0&1&1&0&0\\0&1&1&0&0\\0&0&1&1&0\\1&0&0&1&1\end{bmatrix}$$

(a)

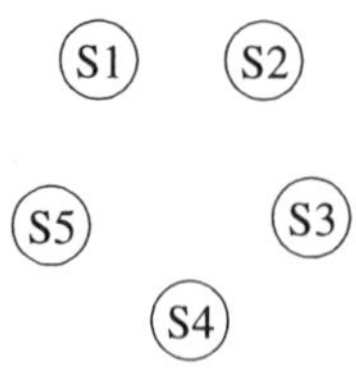

$$R_2=\begin{bmatrix}1&0&0&0&0\\0&1&0&0&0\\0&0&1&0&0\\0&0&0&1&0\\0&0&0&0&1\end{bmatrix}$$

(b)

Figure 10. Two examples of value graphs that represents the logical behavioral relationships between sensors. In (a) there are relationship between agents while in (b) the agents are devoid of relationships.

3 Biological Systems

3.1 Morphogenesis – the Emergence of Form and Function

To understand the interactions that occur when cognitive individuals move into the field of influence of other cognitive individuals, we turn our attention to various physical systems and biological ecosystems in which we observe interesting interactions between the components on the systems, resulting in emergent group behavior. A helpful frame of reference to think about such interaction is the team.

Here we consider the underlying mechanisms that lead to the creation of intricate relationships and resultant behavior. Social organisms have the capacity to self-organize their activities and relationships (Kelso 1999) into teams. Humans are familiar with the requirement to get organized, but for a machine, what are the motivators and mechanisms that allow it, the agent, to alter its current state in the world? What is it about being organized that is so rewarding and what is it about doing it yourself, that is so important?

The word emergence, used in the context of a system, refers to a process of complex behavior developing within the system, over time, due to the interactions between its constituent parts. The parts are usually simple in structure and behavior, but are capable of interaction with each other. The inter-play between simple systems, or simple agents, can give rise to complicated emergent behavior.

The emergence of complex patterns (of behavior) in natural ecosystems is called *morphogenesis* (Kelso 1999). In our work, we focus on the concept of a computational ecology, where the morphogenesis is what we want to be able to direct. We shall say more about this later.

Here we consider some biological systems that exhibit self-organizing behavior. Consider a society of relational sensor agents, where the term *relational* refers to the fact that each agent is the vertex of at least one relationship arc. When a target traverses through the field of view of a sensor, the state of the society of sensors is perturbed: the effect is propagated to the agent neighbors according to the relation links. In this case, the disturbance takes the form of a moving target T, which the agents are capable of detecting, provided the target is within the sensor detection region. A desirable quality for this system to possess is the ability to react in a coordinated fashion and make the necessary inter-agent adjustments to reach an equilibrium state between its members, within an appropriate time.

The internal *self-control* mechanism of each agent must be capable of providing the appropriate governance. That is, we want the society of sensors to be able to self-organize, to gain a better view of the target. Further, we want the relationship structures to emerge over time and to remain plastic so that the agent society may respond to additional changes in the environment. Such emergent flexibility exists in nature, as can be observed in the visual cortex (Beckerman 1997), in human social coordination such as in a crowd of pedestrians (Cohen *et al.* 1997), where people avoid obstacles and form into lines, and in flocking behavior of birds (Reynolds 1987). This approach allows us to consider team formation in terms of a dynamic process rather than a fixed, inflexible structure.

3.2 The Agent Phenotype (Self)

Up to this point, we have been using the word self in conjunction with the word organization, to convey the idea that it is the system that in some way changes its structure to accommodate new circumstances in the environment and in the sensor agent society. This structural organization is not imposed from outside the society, but occurs because of internal mechanisms. We now explore this concept of self in another context, in which an agent perceives and modifies its identity. We introduce a concept of an agent-self, upon which to build our concept of sensor integration.

In Dawkins' books *The Selfish Gene* (1989), and further expounded in *The Extended Phenotype* (1999), a view of the world is described from the perspective of the genes within an organism. The powerful idea that is described in these two books relates to the forces that an organism can exert on its surrounding environment as it extends its phenotype. The organism is no longer confined to the biological boundaries of its body, but extends into the environment in a purposeful way, where the goal is to propagate the most successful gene to the next generation. Dawkins provides the examples of beaver dams, birds' nests, and caddis houses (tubular structures) to illustrate this point.

The extended phenotype concept is one that we now apply to our current problem of sensor integration. The socially augmented sensors are able to extend their structure, through cooperation, to include neighboring agents, reaching beyond their physical boundaries and limitations (Figure 11). This concept provides us with a measure of the level of integration and pervasiveness of an agent's identity within a society of agents. To this end, we introduce the concept of an *agent-self*, the agent's identity, which provides a vehicle for describing the phenotypic effects that social agents may exert on their environment and neighboring agents, in order to fulfill their objectives.

In this context, another example from Dawkins (1989) is that of the spider web. In fact, this is a very good fit for the problem at hand; reaching out with sensors to observe your environment. The spider web, as Dawkins describes it, is a temporary functional extension of her body, "*...a huge extension of the effective catchment area of her predatory organs*". In the same way, we examine how social agents extend themselves, as measured by the agent-self, to create new representations of themselves to fulfill their goals. We consider how an agent may extend its capabilities and awareness by suitable interaction with its neighbors, and consider how one might represent this extended state of existence and awareness. We refer to this process as *absorption*.

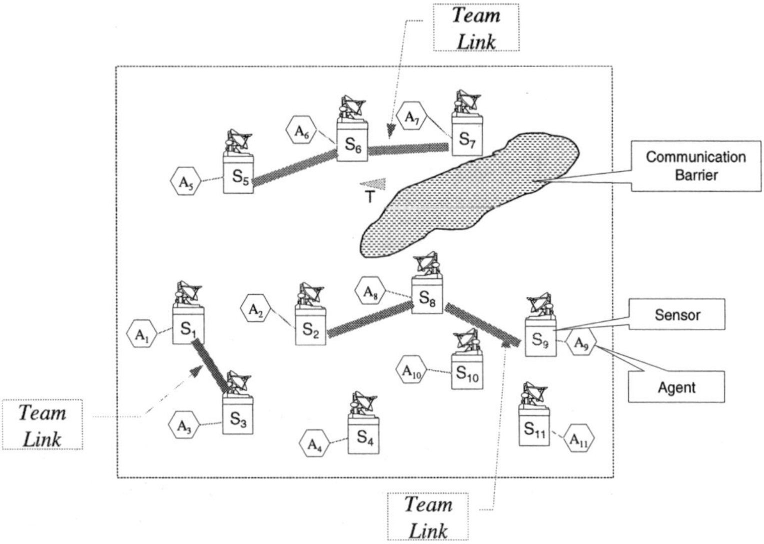

Figure 11. Relationships (thick lines) form between agents within the society; leading to extended agents A_1, A_6 and A_8. Other notation from Figure 9 has been omitted for clarity.

As is the case with humans, we are unable to point to that part of the body, for example the brain, as the sole repository of who we are as humans: our identity or inner self. To simplify our viewpoint for intelligent agents, we consider the agent self to be defined by the state vector of the agent. The extended self therefore, is defined by the original measure of self, together with the capabilities of the absorbed neighbor, forming a collective identity. The absorption of other agents is subject to the relationship that exists between the two agents. An absorbing agent will perceive an extension as belonging to itself providing the relationship is above the threshold value. The number of agents in the extended phenotype of agent i is denoted $N^E_i \leq N_S$, where N_S is the total number of agents in the society.

We would like to be able to measure the degree to which an agent is extended among its neighbors. To this end, we define the dimension of the agent-self as the weighted union of the neighboring agents to which agent i is attracted. The measure we use is based on the mean relationship value of the neighbor agents for which $r_{ij} \geq r_{i,th}$.

Definition 1: Let **A** be the universe of all N_S agents in a society, and assume communications exist between all members of **A**. The agent self (phenotype) of an absorbing agent A_i at time t, is denoted the set comprising the neighboring agents $A_j \in \mathbf{A}$, $j = 1..N_n-1$, for which $r_{ij} \geq r_{th,i}$ and $r_{ij} \in R_i$. R_i is the $N_i^n \times N_i^n$ relation matrix for agent i, where N_i^n is the number of neighboring agents and r_{ij} is the measure of the relationship between agent A_i and agent A_j. The parameter $r_{th,i}$ is the threshold value of agent A_i, above which a relationship is deemed to be significant and support cooperative information exchange.

Corollary 1: The agent self cannot exceed the boundaries of the agent society $\boldsymbol{A}$.

3.3 Lessons Learned from Nature

Ants use chemical markers to build a corporate or social memory of important facts, such as the location of food. The ant paradigm has been applied to optimization problems such as the Traveling Salesman Problem (Beckerman 1997), and the Quadratic Assignment Problem (Dorigo and Di Caro 1999). We address these issues in the ANTeam algorithm. As the name suggests, the algorithm is inspired by the collective behavior of ants (Anderson and Franks 1999, Bonabeau *et al.* 2000, Dorigo and Gambardella 1997, Resnick 1994).

Another example of self-organized behavior is that of army ants which have encountered some prey (Resnick 1994). Neighboring ants join in the task of moving the immobilized prey if they are motivated to do so. The motivation derives from the fact that the prey is stationary in the stream of ants. In other words, it presents an obstacle to the traffic flow. The standard rule that has been demonstrated regarding the ant's behavior is that ants will join the moving task until the prey is moving at the standard retrieval speed. The behavior of collecting the prey no longer hinders the flow of traffic if it is moving with the traffic, and so no additional ants join in the effort. This team is an emergent entity of individual action, but there is a predisposition by each ant to act in a social manner. The cooperation that exists is incidental, being communicated to each ant through the workspace.

So what do we learn from studying the behavior of ants and their social structure? We observe that a single ant benefits from the collective efforts of other ants. The deposition of a pheromone trail by ants returning to the nest with food increases the statistical likelihood that other ants, moving in a random manner, will intercept the trail and be guided to the food source. These ants in turn add to the pheromone trail, and hence we have an example of positive feedback or reinforcement learning, not by the individual, but by the colony as a self-organizing entity. The pheromone trail is constantly diminishing due to evaporation, so that when the food source is depleted, this *memory* will be erased, releasing ants to search for other food sources.

The finite state machine shown in Figure 12 defines the behavior of each ant. Figure 13 shows a series of plots from our MATLAB simulation of ant food foraging behavior. This behavior motivates the approach that by leaving cues in the state space, and looking for cues deposited by other agents, assists in achieving corporate objectives. In the case of multiple agents, as an agent reaches its objective (the detection of a target for example) it deposits markers at sites in its state space. This process increases the probability that other agents in the same neighborhood will be attracted to the site, just as more ant agents find their way to the simulated food sources. We have not included recruitment behavior in this simulation.

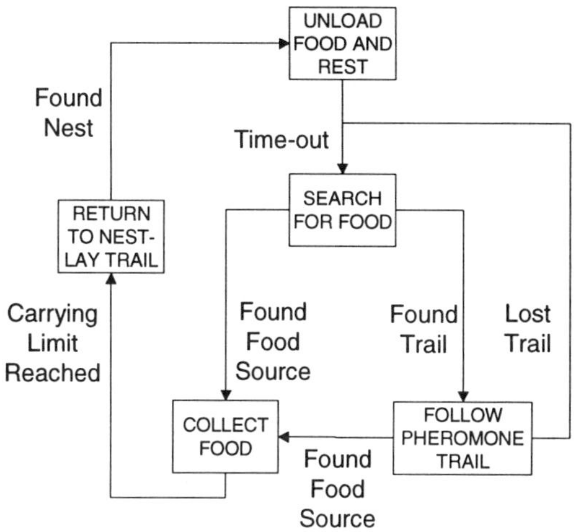

Figure 12. The ant foraging behavior is defined by a finite state machine. The transitions are triggered by sensory information such as the magnitude of the local pheromone concentration, or the discovery of a food source.

4 A Biologically Inspired Agent and the Social Potential

4.1 Agent Goals

In this section, we describe the architecture for our agent. Our design is inspired by the biological ecosystems mentioned in the previous section. We begin by defining some of the key components of an agent. Throughout this next section, we will refer to agent behavior, which describes the actions that the agent exerts on its environment. The relationship between agents is modeled by a social potential field that each agent generates and which each agent can detect provided there exists a communication link between the agents.

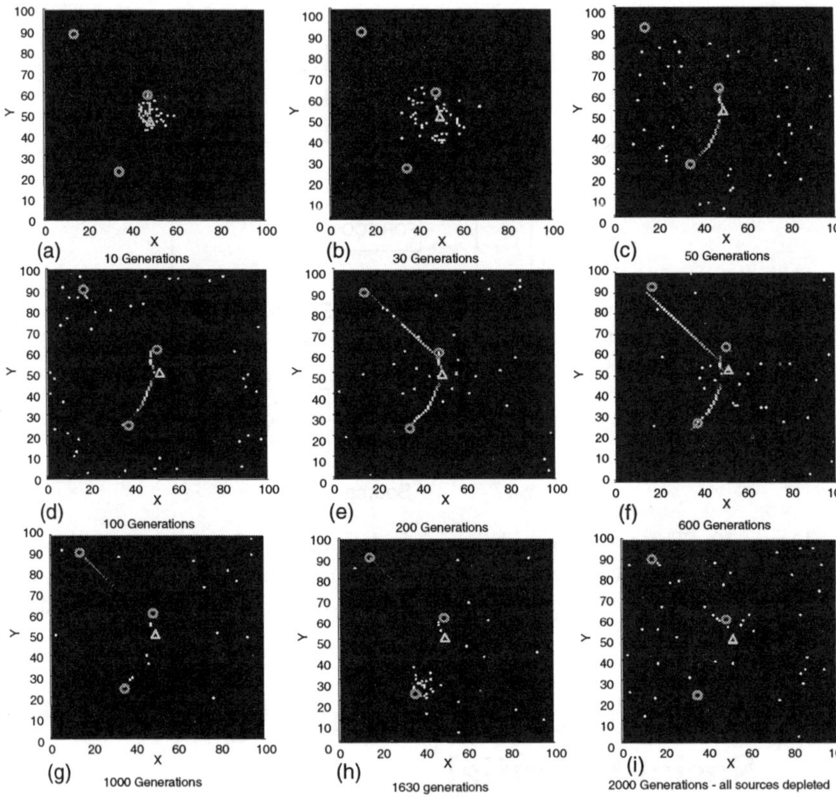

Figure 13. A series of plots showing the evolution of simulated ant foraging behavior. From left to right, (a-b) ants move from the nest (Δ) in a random search pattern. (c-e) When food (o) is detected, the ant returns to the nest with a piece of food, depositing pheromone as it goes. (f-g) Sources closer to the nest are discovered first with a higher probability than sources farther away and depleted first. Trails are formed, which other ants follow. (h) Finally, when all food is exhausted, the pheromone evaporates, and the ants are *released* from the trail following behavior, to (i) begin a random search pattern once again.

Goals are statements of what an agent wants to achieve during its operational lifetime or deployment period T_{deploy}. The goals are programmed prior to deployment of the agent, and are described by the priority-ordered set $G_i=\{g_{i1}, g_{i2}, ..., g_{iNg}\}$, starting with the highest priority. This means that each goal within G_i is accomplished in ordered sequence. Another way to look at the set G_i is to consider it as a statement of the agent's mission, where g_{i1} is accomplished then g_{i2} and so on.

Let us now consider the example depicted in Table 3, where a sensor is required to first move to a designated region of interest. Once on station, satisfying goal g_1, the sensor begins to search for targets. When a target is detected, goal g_2 is satisfied and the agent moves to g_3, which seeks to determine the coordinates of the target.

Table 3. Goals for target observation.

g_1 =	Physical deployment of sensor: Move to position (x,y) within ROI
g_2 =	Surveillance: Perform search for targets within the ROI.
g_3 =	Knowledge refinement: Determine position of targets within the ROI.

Clearly, the agent needs some method of determining whether its current goal has been achieved. We define utility functions for each goal, which measure the degree of success according to a specific criterion. Each of these goals will cause the agent to behave in different ways.

We have defined time-varying goals because, in general, the sensor systems we are considering are deployed in dynamic environments, and may need to adapt to changing priorities. The sensors are subject to commands generated by human operators, and by electronic sub-systems, such as radar warning receivers (RWR), which serve to affect the agent's perceived goal priority. In the case of the

RWR, the detection of specific types of emitters produces a high priority interrupt into the information processing of the sensor system to adopt a specific threat relevant behavior. To simplify our analysis, we are assuming that no such messages occur and that the set $\mathcal{G}_i$ remains constant for a fixed deployment time. For all the simulations, we shall assume the single goal g_1 from Table 3.

4.2 Agent Capabilities

The capabilities, denoted cap_i, that an agent A_i possesses tell us what that agent is equipped to do. We denote the set of capabilities as the set (1):

$$Cap(A_i)=\{cap_1, cap_2, ..., cap_{N_c}\} \tag{1}$$

That is, if the agent is controlling an ES surveillance receiver, the time taken for that receiver to tune from one frequency to another frequency, or its detection bandwidth, are considered capabilities of the sensor. Some typical capabilities are listed in Table 4. In a society of agents, we assume that all agents can be categorized by the same set of capabilities. Then, we define a vector $\Gamma_i(k)$ with entries between 0 and 1. The degree of capability of each of the N_γ elements of Cap(A_i), at time k is recorded in the vector $\Gamma_i(k)$. This vector is described by (2):

$$\Gamma_i(k) = [\gamma_{i1}, \gamma_{i2}, \ldots, \gamma_{iN\gamma}]^{\mathsf{T}} : \gamma_{in} \in [0,1] \tag{2}$$

A value of agent capability $\gamma_{in} = 0$ indicates zero capability in the particular function (e.g., the sensor cannot scan its antenna) and a value of $\gamma_{in} = 1$ represents the highest subjective capability. Each agent in the society is pre-classified against a common standard. For example, for an ES sensor suite, we may define the capabilities of agent i as described in Table 4. The choice of values for the γ_i is up to the designer and requires knowledge of the systems that are likely to be present in the sensor network.

Table 4. Typical capabilities Γ_i for an ES sensor agent.

Parameter	Capability Description: Cap(A_i)
γ_1	Detection bandwidth
γ_2	Tuning settling time
γ_3	Detection Range $\rho_{det,I}$ (sensitivity)
γ_4	Beamwidth bw_i
γ_5	Processing Gain
γ_6	Scan Velocity ω_i (rad/s)
γ_7	Social Quotient ('cooperativity')
γ_8	Communication Range

The capabilities of the agent define the *species* of the agent. Agents that have identical capabilities are deemed to belong to the same species. The way we measure species kinship amongst neighboring agents is by the correlation between Γ_i and Γ_j, denoted by R^S. A suitable formulation for calculating the similarity between vectors is:

$$R^S\left(\Gamma_i,\Gamma_j\right)=\frac{\sum_{k=1}^{N}\gamma_{ik}*\gamma_{jk}}{\sqrt{\sum_{k=1}^{N}\left(\gamma_{ik}\right)^2*\sum_{k=1}^{N}\left(\gamma_{jk}\right)^2}} \tag{3}$$

(We have a similar formulation for behaviors.) Some examples of different species types are:
1) an infrared sensing agent,
2) a wide-band microwave receiver agent, and
3) a narrow-band microwave superhet receiver agent.

A wideband (non-integrating) receiver generally has less sensitivity and provides poor frequency resolution, but has reasonably good probability of intercept performance for signals with typical signal to noise ratio. A superhet receiver however provides relatively poor probability of intercept because it is tuned to a small part of the frequency band, but provides good frequency resolution and high sensitivity. These sensors can complement each other if they act as a team.

To assist in evaluating the overall capability of a sensor agent, and enable comparison between agents, we define a capability factor Ω_i, which is a somewhat arbitrary measure of the usefulness of the agent. We simply form the average of the γ_i as follows:

$$\Omega_i \triangleq \frac{1}{N_\gamma} \sum_{n=1}^{N_\gamma} \gamma_n \tag{4}$$

4.3 Agent Behaviors

Each of our agents is capable of expressing N_b behaviors defined by the set $\mathcal{B}_i = \{b_{i1}, b_{i2}, ..., b_{iNb}\}$. The behavior that agent A_i exhibits at time k is $b_i(k)$ and is a function of the influences due to each of its neighboring agents $\phi_{n,j}$, the influence due to the target $\phi_{i,T}(k)$ and the current state $S_i(k)$of the agent:

$$b_i(k) = F_{\mathrm{sel},i}(\phi_{n,j}(k), \phi_{i,T}(k), S_i(k)) \tag{5}$$

where $F_{sel,i}$ is the selection function of agent i.

Consider the case of an electronic support sensor suite, where the goal is to monitor a region of interest for the presence of moving objects (targets). In this case, the set of behaviors may include: $\mathcal{B}_i = \{$*standby*, *search*, *stare*, *track*, *report*$\}$. By employing an agent-enabled ES system, we can add team forming to this set. We employ only a few of the many possible behaviors here, with $\mathcal{B}_i = \{$*standby*, *search*, *team-forming*$\}$.

5 Agent Architecture

We now consider which properties the agents need so they can behave in a social manner. The behavior that an agent exhibits is a function of its internal state $X_i(k)$, which is a statement of its awareness of its environment and itself, and comprises the following items:

1. the agent-self σ_i, which records agent phenotype,
2. behaviors B_i that the agent can exhibit,
3. the local pheromone concentration η_i that indicates local detection activity,
4. the capability set Γ_i, which describes the capabilities of the agent,
5. the goal set G_i of the agent.

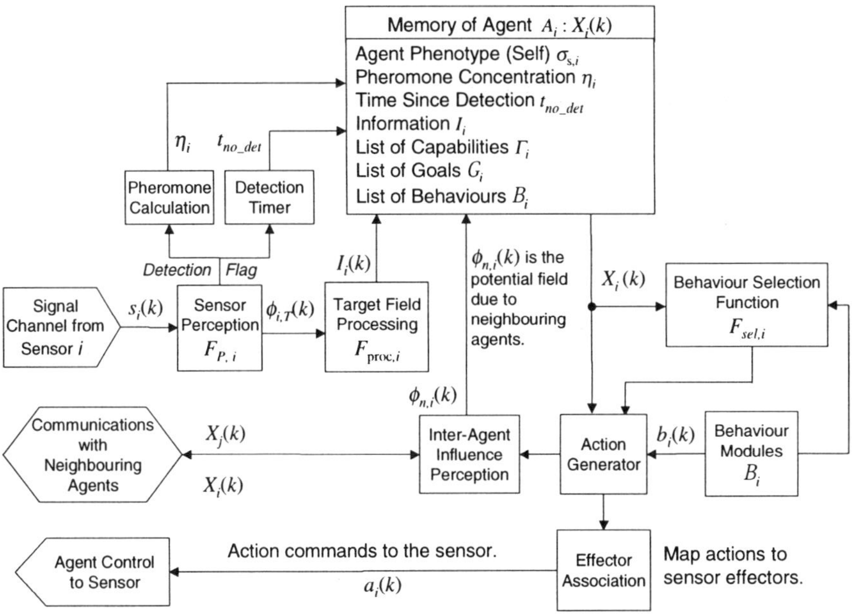

Figure 14. Diagram of the agent architecture showing influences due to targets and neighboring agents. The behavior $b_i(k)$ is selected from set B_i by the selection function. Agent states X_i are shared among neighboring agents.

Biological systems provide us with inspiration to develop models of agents that are suitable for particular niche applications. The emergent patterns of behaviors that may be observed in ant colonies, in flocks of birds, and within human pedestrian and automobile traffic are testimony to a rich diversity of purposeful social interactions. To explore the concepts of self-organizing systems we have developed the agent architecture shown in Figure 14. It is inspired by ant physiology and behavior (Anderson and Franks 1999, Dorigo and Gambardella 1997), and cognitive models of human memory, perception and action selection (Solso 1998). The component parts are shown in the figure, along with the data flows, which are described by the equations in the following sub-sections. (We shall not be describing the functional blocks of the architecture in this chapter.)

6 Building Relationships

6.1 Information

We make the simple assumption that with each observation of the target by a sensor, the information regarding the target increases. We define a local information parameter I_i for the ith sensor, which represents a measure of its situation awareness. A large value of I_i is preferable to a small value.

Let $I_{i,T}(k)$ and I_0 denote the information value that agent i has derived from the target signal $\phi_{i,T}$ and the *a priori* information about the target respectively,

$$I_{iT}(k) = F_{\mathrm{proc},i}(\phi_{i,T} + n) \tag{6}$$

where $F_{\mathrm{proc},i}$ is the processing transfer function of the ith sensor and n is additive noise, which we ignore for simplicity. Given an input signal, this function determines the value of the information derived from that signal. The information present at time t is:

$$I_i(k) = I_0 + I_{i,T}(k) \tag{7}$$

Some important considerations regarding information include:
1. information has a must-have-by date (the information is required now),
2. information has a use-by date depending on the context,
3. information has a lifetime that decays with time,
4. some information may have an indefinite lifetime – it is always useful.

We incorporate the information use-by date into our simulation by decaying the information value I_i of each sensor. In a highly dynamic environment where sensors are changing modes and targets are changing their position, information that was collected in the past soon becomes less valuable.

6.2 Calculating Sensor Relationships

We now consider the mechanism that generates the inter-agent relationships described in Section 4. The approach taken is to assume that each sensor has something to gain by entering into a relationship with a neighbor. Each sensor is an attractor, and effectively generates its own scalar potential field. The factors that make a neighboring sensor attractive to the object sensor combine are represented by a scalar potential field emanating from each of the neighboring sensors. The potential field is a spatially continuous distribution generated by a source. The source is a target or another sensor. The geometry of the radiation pattern of the field depends upon the source and the effects of the environment through which the field propagates.

Each agent perceives the presence of other agents through communication links, and targets via sensors. In the simulation software, we model the effects of the environment on the target signal. The signal radiating from the target is $s_T(k)$ is attenuated according to the propagation characteristics of the environment. Equation (8) describes the signal that is incident on the agent due to the target $s_{i,T}(k)$:

$$s_{iT}(k)=F^{\text{prop}}\left(s_{T}(k),d_{iT}\right) \tag{8}$$

An agent situated within the field experiences a motivational force, which is a function of the distance from the source, the properties of the medium, and the perception of the agent. We use the concept of a field to describe the uniformly distributed nature of the effects that targets and sensors exert on one another. Attractive and repulsive fields (Parker 1997, Reynolds 1987) are often applied to the problem of navigation of mobile robots, in particular for collision avoidance, and computer simulations of animal motion. The field equations describe the influences exerted upon a sensor due to surrounding field sources.

We apply potential fields to guide the growth of relationships between field sources, which are the sensor agents. We adopt a mean field (Beckerman 1997) formulation to calculate the influences exerted on each sensor in the society. This formulation is described by Equation (9) and accounts for the N_P time-varying scalar attractor properties V_i and the N_C scalar repulsive properties K_{ij} that each agent possesses. K_{ij} modulates the sum of the attractors:

$$\phi_{ij}(k)=K_{ij}(k)\sum_{i=1}^{N_P}V_i(k) \tag{9}$$

An important consideration in any social structure is the cost associated with a transaction between two parties. The cost that agent i pays to associate with agent j is denoted c_{ij}. The parameters c_{ij} and $d_{i,j}$ serve to moderate the attraction between two agents. We apply superposition to each of the attractor properties V_i to determine the net attractor potential. The agent self-identity grows (or contracts) according to the effects of its neighbors, and each neighbor generates an influence that is the result of the superposition of several attractors.

We now introduce two key concepts which impact upon how each agent perceives its agent-neighborhood. These are:

1. the *agent behavior correlation* R_{ij}^{b}, and
2. the *agent species correlation* R_{ij}^{s}, and example of which is given in Equation (3).

In the following discussion, we shall refer to the agent of current interest as the *object agent*. Consider a population of agents as depicted in Figure 4. The *agent behavior correlation* is a measure of how well the behavior of a neighboring agent matches the desired behavior of the object agent. Similarly, the *agent species correlation* is a measure of how well the capabilities of a neighboring agent matches the desired capabilities of the object agent. A high correlation indicates similar agent capabilities and a low value indicates diverse capabilities.

The information necessary to calculate these correlation factors is obtained by mutual communication between agents. The object agent perceives the behavior of its neighbor via the communication between them: if there were no communication then there could only be indirect communication via the sensors of each agent. A high correlation indicates similar (homogeneous) agent behaviors and a low value indicates diverse (heterogeneous) behaviors.

In this formulation, we define a maximum cost per transaction $c_{\max,i}$ that agent i is able to pay to establish a relationship link with a neighboring agent j. If the cost per transaction c_{ij} exceeds the maximum then no link is established. This maximum cost parameter serves to model the fact that a component of a system is subject to constraints, which take various forms. Among these constraints is energy, such as battery capacity, and information processing ability, where more relationships imply more data processing time. The constraints modulate the proliferation of linkages within a society; not all agents within a society need to be connected. The cost factor c_{ij}^{a} is defined as

$$c_{ij}^{a} = \begin{cases} \dfrac{c_{\max,i} - c_{ij}}{c_{\max,i}} & : \quad c_{ij} \leq c_{\max,i} \\ 0 & : \quad c_{ij} > c_{\max,i} \end{cases} \tag{10}$$

We now return to Equation (9) and write the next stage of this equation that includes a learning mechanism based on pheromone deposition and evaporation. From Equation (9) the K_{ij} is replaced by the cost factor from Equation (10) and the summation of the V_i is replaced by a scalar sum of parameter values.

The parameters for Equation (11) are listed in Table 5. The values of these parameters are chosen for each agent in accordance with the goal of the agent. In the next section, we explain how the weighting factors $u_i^{\eta}, u_i^{I}, u_i^{\Omega}, u_i^{s}, u_i^{b}$ and u_i^{det} are chosen by the current goal (Figure 15).

Table 5. Parameters for the field equation.

Parameter	Description
d_{ij}	Euclidean distance between agent and neighbor.
$c_{\max,i}$	Maximum cost agent i is able to pay to associate with another agent.
η_i	Pheromone concentration at agent i: large η_i implies a high rate of target detection.
I_i	Information content of agent i: large I_i implies greater value.
$I_{\max,j}$	Maximum information content of agent i (e.g., buffer length of memory).
Ω_j	Capability of a neighbor (a highly capable neighbor agent is an attractive asset).
$R^s(\Gamma_i,\Gamma_j)$	Species correlation.
$R^b(b_i,b_j)$	Behavior correlation.
$F_s(\cdot)$	Goal dependent species function: weights the desirability of either a heterogeneous or a homogeneous group.
$F_b(\cdot)$	Goal dependent behavior function: weights the desirability of either a heterogeneous or a homogeneous group.

We are now in a position to define the normalized scalar field $\phi_{ij}(k)$ that agent i experiences due to its neighbor j, thus:

$$\phi_{ij}(k)=\frac{c_{ij}^{a}(k)D_{ij}}{N_{\omega}}\left(w_i^{\eta}\frac{\eta_j(k)}{\eta_{\max}}+w_i^{I}\frac{I_j(k)}{I_{\max,j}}+w_i^{\Omega}\Omega_j+w_i^{s}\mathrm{F}_s\left(\mathrm{R}^{s}\left(\Gamma_i,\Gamma_j\right)\right)+w_i^{b_b}\mathrm{F}_b\left(\mathrm{R}^{b}\left(b_i,b_j\right)\right)\right), \tag{11}$$

where the distance factor D_{ij} is defined by

$$D_{ij}=w_i^{\det}\left(\frac{r_i^{\det}-d_{ij}}{r_i^{\det}}\right) \tag{12}$$

If an agent goal calls for diversity in its phenotype, then the functions $\mathrm{F}_s(\cdot)$ and $\mathrm{F}_b(\cdot)$ are the means for the agent to judge correlation values for its neighborhood. In our simulations, we assume that diversity is desirable. We now ask the question, how are the various weighting factors selected? This question is addressed in the following section.

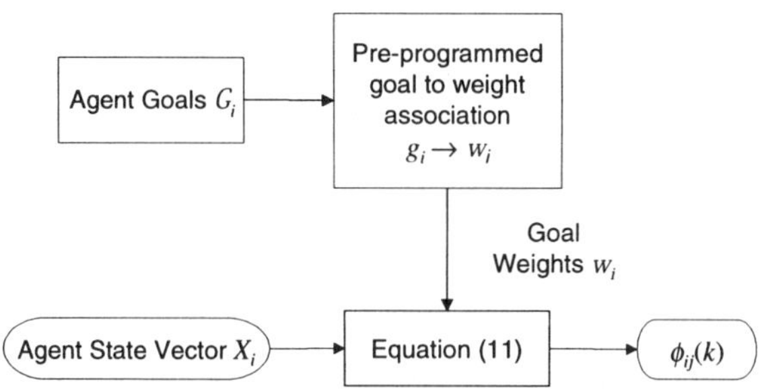

Figure 15. Goal-directed weight assignment: $\mathbf{w}_i = \left[w_i^{\eta}, w_i^{I}, w_i^{\Omega}, w_i^{s}, w_i^{b}, w_i^{\det}\right]^{T}$.

7 Goal-Directed Social Networking

7.1 Setting Goals

We mentioned earlier that we would like to be able to influence the behavior of a social sensor network. We propose that by changing the goal that each sensor is trying to achieve, we can influence the emergence of the relationships that form between agents. Providing all agents in a society with the same goal effectively imposes a bias or constraint upon the behavior of the group as a single entity.

Whether the goals are set by an external authority, or by the agent itself, selected from one of a list of internally stored goals, the effect is to modify the perception of the agent and consequently its singular and cooperative behavior with neighboring agents. The goals affect the weighting of the field function, as shown in Figure 15, for each agent. For example, for area surveillance and target search (target discovery), the characteristics of an agent that improve the sensor coverage and the probability of detection are more highly weighted.

The relationship between the goal and the weights is established off-line by operational experts, namely the humans who make use of the sensor data. These characteristics include sensitivity, false alarm rate, revisit time, digitization level (number of bits), post-processing including de-interleaving.

7.2 Weight Selection

Each agent has a set of goals that determine its behavior. The current goal establishes the relative values of the weighting factors for Equation 11. The nature of the task governs the value of the capabilities of neighboring agents. If, for example, the current goal of the agent is to collect precise frequency information about an observed emitter (target), then the corresponding weights in Equation (11) are set to bias the function towards capabilities that support the goal. In this way, the goal modulates the attractive field strength between agents.

7.3 Inter-Agent Relationships and How They Are Formed

The relationship between agent i and agent j, denoted r_{ij}, is a function of the attractive potential that exists between the agents. The strength of the relationship is calculated by the learning rule

$$r_{ij}(k) \leftarrow \upsilon_i \, r_{ij}(k) + F_{P,i}(\phi_{ij}(k)) \tag{13}$$

where υ_i and $F_{P,i}$ are the learning rate and the perception filter function of agent i respectively. We have previously defined the relation matrix R_i which records the values of r_{ij} for all neighbors of agent i. Now that we have defined the fields that act on an agent, we consider an algorithm (Figure 16) that guides the formation of agent relationships, based on the potential fields and motivated by the behavior of foraging ants.

```
while active
  Get sensor inputs
  Get neighboring agent states
  If target detected calculate information gain:Ii
  Update pheromone concentration
  Calculate the potential fields
  Select Behavior
  Apply evaporation to pheromone
  Apply relaxation to relation matrix
  Apply decay to local information
end
```

Figure 16. Control algorithm for each agent.

8 Team Formation

8.1 Motivation for Team Formation

In general, the motivation for forming a team is to achieve an objective that is beyond the capabilities of a single agent. The concept of the team is very familiar within human society, as it is in the insect world. A football team is an example of an organizational structure whereby the goal, to win the game, is more likely achieved if each member performs as an *integrated part of a whole.* Individual performance, no matter how spectacular, is ultimately no replacement for the synergy developed because of cooperation between team members.

In our football team example, the team's efforts are guided by the captain and by the coach; however, the execution of that guidance is up to the individual who must be suitably equipped with the skills to perform the necessary function. There is no substitute for individual initiative in a collective.

8.2 What Constitutes a Team?

There are many ways to define what constitutes a team. Certainly, a team has at least two members, and these members interact in a coordinated manner to achieve an objective. Consider a flock of birds. The flock exhibits coordinated behavior, but can we say that the birds within the flock constitute a team? It is difficult to study the behavior of a single member of a team, and deduce the overall structure and purpose of the team. Resnick (1994) points out that '*a flock of birds is not a big bird.*' This characteristic seems to be common among team structures. In a flock, there is no concept of the awareness of a global goal, but each bird acts according to its own local circumstances.

We must be careful with our definition to capture those aspects of agent interaction that gives rise to goal directed behavior. A team of agents is not a big agent, but is an expression of the interactions that occur between social agents over time. A team has an existence of its own, supported by the member agents, but not dictated to by any one agent. A team may arise from the result of external explicit influences, or from internal motivating requirements. For the purposes of this work, we shall apply the following definition.

Definition 2: A team is a group of two or more interacting agents, belonging to an agent society, acting under the influence of a time-varying cohesive force. The team exhibits behavior that is the result of local and global influences and goals, with each agent exhibiting coherent behavior directed towards the achievement of those goals.

8.3 What Triggers Team Formation?

As mentioned in the previous section, a team structure provides a means of tackling problems that are beyond the capabilities of individual agents. Therefore, we can assume that one of the primary reasons for a group of cooperative agents to form a team structure is to tackle a complex problem. For area surveillance, which is integral to target observation, it is clear that a single sensor is unlikely to be able to observe the entire region of interest. The goal drives the agent to exhibit team-forming behavior. When the goal of the agent is not being satisfied, this deficiency triggers behavior that seeks to address the situation, such as seeking help from a neighboring agent.

The transition of an agent from one behavior to another behavior is a stochastic process because the events in the environment are unpredictable: hence, teams can form quite suddenly. This is an example of *self-organizing criticality* (Jensen 1998). Specifically, the goal that we set for each agent is to observe the environment, and generate information about detected targets.

We define the following trigger conditions for switching between behavior states:

1. a change of goal – as may be directed from an authority local to the agent or external to it;
2. a target is first detected, triggering a tracking behavior; or
3. a level of anxiety as measured by the time taken to make the first target detection: t_{no_det}.

An agent is motivated by its goals. If an agent cannot satisfy a particular goal, then it must take steps to remedy the situation. In practice, this leads to a rise in the level of anxiety and increases the weighting factor that influences the rate of extension.

A social agent will move from an isolated state, to one where it seeks out neighboring agents with which it can interact. The nature of this interaction is described in the next section. In our approach, a team emerges as the result of the collective attraction between agents. The concept of team that we employ here is external to the society. That is, a team is observed to form and the individual agents are only aware of extending their concept of *self-identity*. However, they do possess mechanisms for transacting the appropriate communication to be influenced by their neighborhood. An agent encounter occurs when at least two agents become aware of each other, via the communication link, and begin a social interaction. Figure 17 shows the behavior transition logic for each agent.

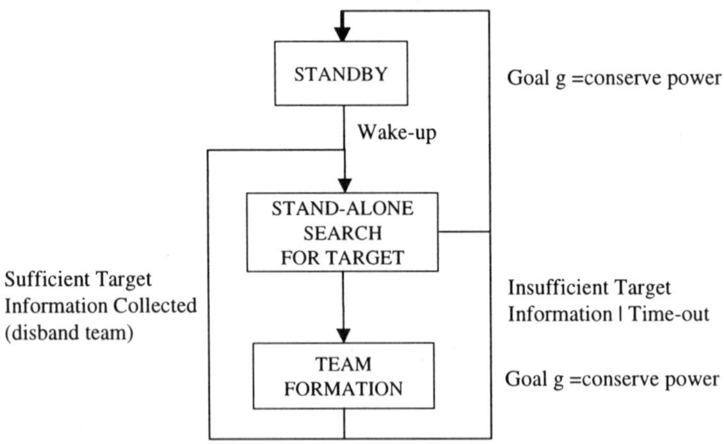

Figure 17. Behavior transition logic diagram for each agent.

The team is an artifact or pattern that emerges due to the effect of potential fields. A *distributed self-identity* is formed and in so doing, the agent attains additional awareness and capabilities by the absorption of other agents, and perceiving them as part of its own phenotype. This mode provides the agent with the capacity to extend the degree of awareness of the environment, and to extend the opportunity to exert its influence upon the environment. Two examples are a spider and its web, and a tennis player wielding his or her racket.

8.4 A Suitable Organizational Structure – Who Gets on the Team?

Consider now which agents join the organizational structure that we call the team. Organizational structures may be thought of as ecologies, where complex relationships exist between the constituent agents. Individual agents may group together according to various authority structures (Thomopoulos 1994). These structures include:

1. autocratic structure, where a single agent takes control of the group,
2. authority heterarchy, which is a chain of command,
3. committee, where decisions are arrived at by voting,
4. anarchy, where each agent is totally anti-social and self-interested, or
5. socially self-organizing, where the formation of a team is the result of attractive forces acting within the society, between the member agents, and not imposed from the outside.

We consider two mechanisms here. The first considers team membership to be determined by *team member selectors*, where the selectors are external to the set of possible members. The selection process is based on a specific performance measure, such as choosing the most capable agent available from the group of candidates. In this case, team membership is determined by a market strategy, whereby the agents that possess the most valuable assets, are deemed suitable team members. The agents buy their way onto the team by virtue of their inherent value. An example is a sporting team.

The second mechanism for deciding who gets on the team is decentralized, and relies on each agent in a society making their own decisions. This is a self-organizing philosophy. An example of such a mechanism is team formation that is based on a distance metric, whereby agents that are sufficiently close to each other (there exist a sufficiently strong attraction) form a team link. The benefit of this mechanism is that it minimizes communication time, and in general, improves reliability, assuming that risk is proportional to the channel length.

The simulations of ant foraging behavior described in Section 4 provide insight into collective behavior. The simulated ants are guided to a food source by the pheromone trail, and hence we see a conformance in behavior and a subsequent team effort emerges. In Section 3.3, we described how army ants group together to carry away a captured prey. We can apply this concept to the observation process, where agents *join the team* until a satisfactory performance is achieved, as measured by the observation status of the target. We remember that the apparent team joining process is an artifact of a self-organizing group of social individuals. For a self-organizing system, we may say that an agent *extends* its phenotype until it is satisfied with the output from its objective function. For example, an agent extends its sensory capabilities through team membership, to maximize the area it can observe.

8.5 What Holds a Team Together?

A team remains together as long as the cohesive force that regulates team behavior remains strong enough. The cohesive force is based on the strength of the relationships that exist between the agents. If a relationship remains relevant to the agent, then that element of the team will continue to exist, though other members may become detached. The inter-agent relationships are based upon mutual attraction to the environmental stimuli, and the cost of inter-agent relationships, giving rise to a fluctuating team membership.

8.6 When Should a Team Disband?

A team will disband when member agents perceive that the cohesive force, that they as individuals experience, falls below a threshold value. Each agent may possess a unique release threshold. To minimize unwanted oscillations in the team forming mechanism, we include a hysteresis mechanism in our model. An agent constantly evaluates its membership status based on the attendant cost and pay-off of remaining on the team. External influences, such as target proximity, affect the relationships that are built within the sensor network. In a self-organizing network, the relationships and hence team links between agents will vary in response to these influences.

9 Team Forming Algorithm

9.1 Algorithm Mechanism

In the ANTeam algorithm, we make use of the corporate learning process mentioned in the previous section, to organize the formation of appropriate teams of agents. During each cycle of the algorithm, every agent evaluates the information derived from target detection, and the potential field due to other agents that lie within its neighborhood. When the attractive potential between agent i and one of its neighbors, agent j, exceeds the threshold, then a team link is established, and agent i perceives agent j as an extended part of its phenotype. Once an agent is attracted to another agent, the new agent experiences a cohesive force that acts to resist team fragmentation. It is at this stage that the agent extends its own internal self-state.

The potential field $\phi_{ij}(k)$ is evaluated at each program cycle and reflects the dynamic trade-off between cost and reward. As long as the team is achieving its objective then the team remains stable. The team structure evolves according to the prevailing conditions. In this case, the agents are stimulated by the presence (within detection range $d_{iT} \leq \rho_{det,i}$)of the target object, and by the interactions between agents that are perceived through their communication channel.

This algorithm relies on a reinforcement strategy for learning and employs the logical equivalent of a pheromone marker at sites within the sensor network, but what is it that is marked? In the biological case, when an ant has collected food from the source, it returns to the nest, marking the surface over which it moves. In our problem, the agents do not move over a physical surface, so we adopt other logical spaces to mark. These are the agent *self-identity* $S_{\sigma,i}$ and the agent *behavior set* B_i. Each agent updates its phenotype or self-state $S_{\sigma,i}$ at each processing cycle. We use a simple finite state machine for the behavior selection function $F_{sel,i}$.

9.2 Corporate Memory and the Role of Agent Pheromone

In our formulation of the problem, when an agent detects a target, the pheromone concentration of that agent increases by an amount α_i, which is proportional to the magnitude of the stimulus $s_{iT}(k)$. Equation (14) describes the pheromone that is *deposited* at agent *i* and is stored in the state vector of the agent. This quantity is subject to evaporation (memory fading), where λ_i is the evaporation constant for agent *i*, *T* is the global temperature, and where $\mu_i \in [0, 1]$ is a scale factor that affects the rate of learning of the ith agent:

$$\eta_i\,(k) \leftarrow (1\text{-}\lambda_i T)\cdot\eta_i(k) + \mu_i s_i\,(k) \qquad (14)$$

We use the term *learning* in the following sense. The concentration of the pheromone increases with each target detection. This has the effect that relationships associated with this agent are enhanced. The agent is, therefore, more likely to be integrated into new team structures in the future. The system as a whole learns which of its members are valuable to the collective of agents.

The evaporation constant is a measure of the susceptibility of the agent to the environmental temperature. Put another way, λ_i is a measure of the insulation properties of the agent. The effect of an increase in η_i is to increase the attraction that other agents experience to merge with this agent. The rational for this statement is that the agent is detecting a target and subsequently deriving information about that target. This represents valuable information to other neighboring agents that may not have detected the target, or which can use the information for data fusion purposes. An agent is naturally attracted to neighbors that are successful in achieving their detection goal. We wish to optimize the probability of achieving the goal of detecting the target within the region of interest. This goal is realized by the self-organization of agents into appropriate teams.

9.3 The Simulation Environment

The pseudo code for the environment and ANTeam algorithms is shown in Figures 18 and 19 respectively. The algorithms have been implemented using MATLAB. The simulation program supports various sensor layouts and target behavior, as listed in Sections 3 and 4. The software supports adjustment of essential parameters, such as number of sensors and temperature. The environment model supports both toroidal (wrap-around) and reflective boundaries. The target motion can be:

1. a random walk,
2. a waypoint-defined track, and
3. an orbit of fixed radius about a fixed center.

We use the second option in the simulations. We define a simple model for the propagation loss, where the signal strength is proportional to the inverse of the distance squared between agent and target d_{iT} and the propagation function: $F_{prop}(x)=x[d_{iT}]^{-2}$.

```
Initialization Target characteristics
Initialize Environment characteristics
Initialize Agent characteristics
while active
  Calculate target dynamics
  Update target position
  for each agent i ∈ society
     Calculate incident signal due to target s_{i,T}(t)
     Call Agent function
  end
end
```

Figure 18. The simulation environment algorithm.

```
Update pheromone concentration (Equation 14)
for each agent A_i ∈ A (set of agents)
  Read pheromone concentration η_j and capabilities Γ_φ
  Calculate species correlation
  Calculate capability factor Ω_j (Equation 4)
  Read the behavior state of neighbor
  Calculate behavior correlation
  Calculate distance to agent A_j
  Read information measure of agent A_j
  Calculate motivation to absorb agent A_j (Eqn. 11)
  Update relation matrix entry r_ij (Equation 13)
  if (r_ij(k) ≥ r_i^th AND A_j ∉ A_i^E)
     Create Team Link and update A_i^E
  elseif (r_ij(k) < r_release,i AND A_j ∈ A_i^E) (hysteresis)
     Delete Team Link and update A_i^E
  end
end
```

Figure 19. The ANTeam algorithm that controls the team forming behavior of each agent. The extended phenotype of agent A_i is denoted A_i^E.

10 Simulations

10.1 Setup

The simulations were programmed with MATLAB. The pheromone evaporation and relation constants were set to the same value for every agent in these simulations. In the first set of simulations the agents only had one goal, to search the ROI. To reduce complexity in the simulations, we set the capabilities and weighting factors of each of the agents to the same values, which produce a homogeneous agent society. The ANTeam algorithm is stimulated by the target model.

In the simulations, we want to investigate how the relationships evolve under varying conditions. First, we consider the case where the sensors are fixed in position, and then we allow the sensors to move. We expect to see a difference in the way an agent extends its phenotype, that is, the membership of the teams should differ from one case to another.

10.2 Results and Discussion

The teams persist while the motivation force is sufficiently high, as shown in Figure 20, where the teams have formed along the path of the target. We believe this is a desirable emergent behavior as it acts to increase the observation area of an agent.

In the first simulation, we distribute the sensors uniformly throughout the ROI and set the detection range of each sensor to the same value of 20 units. This generates the configuration shown in Figure 20a. Initially, the agents form relationships based on the parameters listed in Section 4, and subject to the learning rate. These relationships also fade with time according to the relationship attention factor. In general, the attraction $\phi_{ij}(t)$ does not fall to zero because neighbors still have capabilities that are attractive, even if they are not currently detecting a target, however absorption does not occur unless the relation value exceeds the threshold for that particular agent.

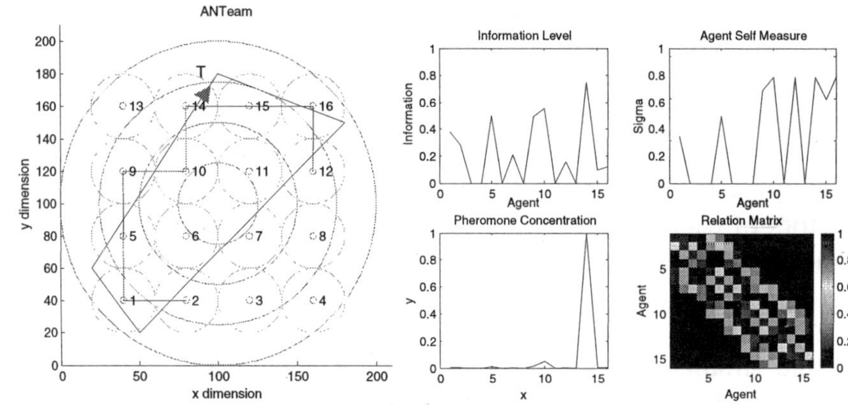

Figure 20. (a) Plot of the situation display, showing a grid placement of stationary sensor agents. Teams are represented by the lines between sensors. The team formations present are: {1,2}, {5,1}, {9,5}, {10,9}, {14,10}, {15,14}, {16,12}. (b) Plot of information, pheromone concentration, self-identity and an example of the relation matrix, for a single agent, after 269 iterations. The information value increases as the target passes through the detection region of a sensor.

When a target enters the detection range of an agent, the relationships for all the neighboring agents will change accordingly. There is a ripple effect across the lattice that is limited due to the thresholds of significance placed on inputs to the agents, and the communication ranges. The simulations revealed that the lifetime of a team is sensitive to the rate of decay of the relation matrix.

In the next simulation, we have distributed 20 sensors randomly throughout the ROI but we have kept the detection range at 20 units, as before. This means that some areas are sparely populated with sensors and others areas have a high degree of overlap giving rise to dense clusters of sensors. In Figure 21a, we see that relationships have developed between sensors that are situated in the dense clusters.

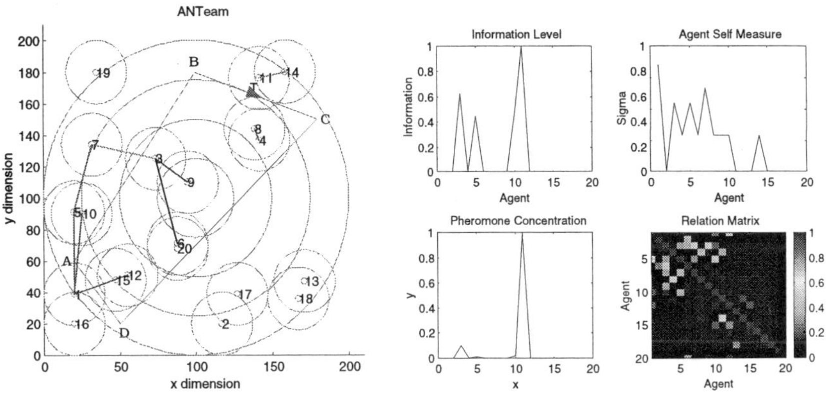

(a) (b)

Figure 21. (a) Plot of the situation display, showing randomly placed stationary sensor agents. Teams are represented by the lines between sensors. This plot shows the following team formations: {1,10,12}, {3,7}, {1,5,10}, {3,5,7}, {4,8}, {3,9}, {5,10}, {11,14}. (b) Plot of information, pheromone concentration, self-identity and the relation matrix, for a single agent, after 50 iterations.

As the target moves through a sensor detection region, the associated agent increases its pheromone concentration. The topology of the agent phenotype is driven by the goal that the agent is programmed with, and can only grow along the communication links between agents. The fading and eventual extinction of a link is modulated by the number of times that link is established. The simulations illustrate the emergence of multiple teams, indicated by the team links.

The team forming behavior generally tracks the target activity, which is a desirable emergent property. That is, the team grows in the direction of the available attractors (nutrient), limited by the cost components of each relationship. One observed emergent behavior was the formation of team links before the target enters the detection range of any sensors. This behavior, which was triggered by the time-out of the no detection period, would be useful for anticipating the appearance of a target within a region (or making periodic surveys when resources permit) of interest to maximize the opportunity of collecting information about the target. The sensitivity of team-formation to parameter selection is of great interest and is to be explored further.

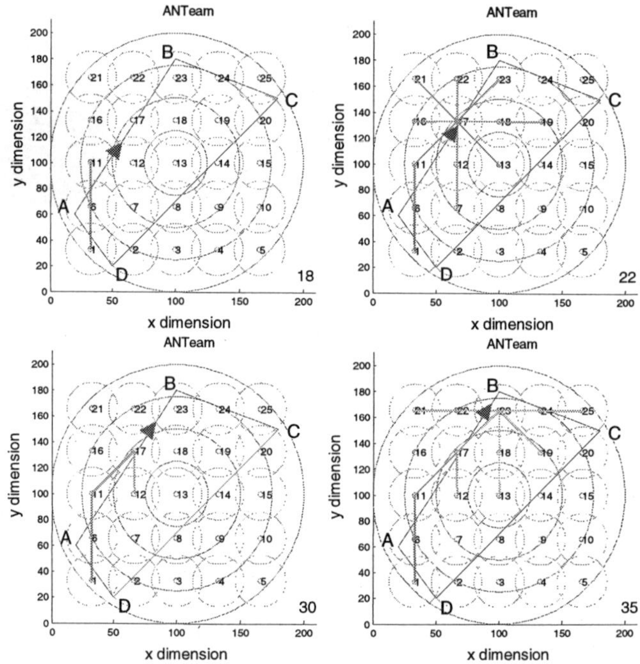

Figure 22. Plot of a sequence of situation displays showing the emergence of teams as the target moves along the path from A to B. The generations are displayed below each display, ranging from 18 to 35.

In Figure 22, we have a sequence of plots showing the development of teams as the target traverses the route defined by the way-points A to D. The first simulation plots show the situation at generation time 18 through to 35. Three distinct teams emerge in this simulation. As the target passes from within the observation range of one team, the next structure is formed and the old structure dissolves away: releasing assets for other duties. Figure 23 shows a simulation run for a sensor network that shows team links together with the target bearing plots for each of the sensors. Notice the rise in pheromone concentration during the periods when the target is within the detection region of each sensor. The simulation ran for one complete circuit of the waypoint route.

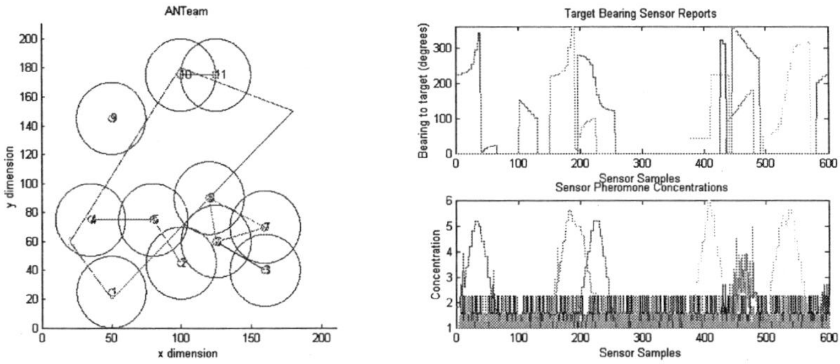

(a) (b)

Figure 23. Plot of a sensor network (a), with goal g_1, showing the team links together with the target bearing plots (b) for each of the sensors. Notice the rise in pheromone concentration during the periods when the target is within the detection region of each sensor. The simulation ran for one complete circuit of the waypoint route.

In the next experiment, each agent was programmed with the first two goals listed in Table 3, and starts with goal g_1. When an agent detects a target, its switches to the second goal g_2. We expected to see the team links change from those which favor detection capability (e.g., sensitivity), to team links that favor position estimation and identification of the target (e.g., small target bearing estimate error and high processing capability).

The change in goal caused a change in the topology of the team links as shown in Figure 24. In the lower half of the ROI there is a dense concentration of sensors with overlapping detection regions, whilst in the upper half there are only a few sensors. One area of current investigation is looking at the affect of sensor density on team behavior. The next phase of this work, which the authors have underway, is to allow the sensors to move in response to target behavior. We are also investigating issues related to agent system stability.

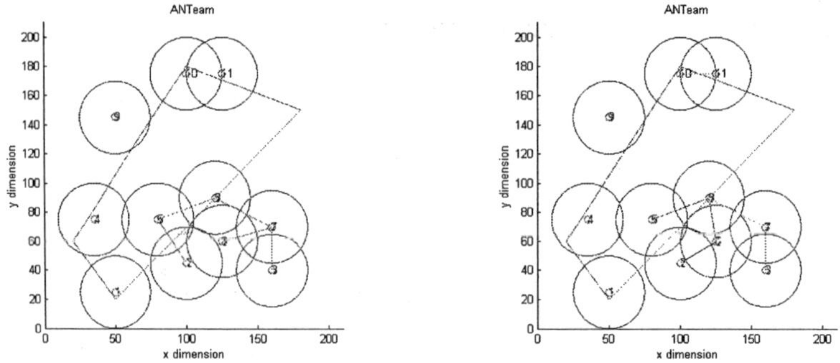

(a) (b)

Figure 24. Plots showing (a) the sensor network with capabilities that increase with the sensor number (from top to bottom), and (b) shows results where the capabilities decrease from a maximum (increase from bottom to top) with sensor number. The target is located at coordinates (98,68). We expect to observe a change in the topology of the network.

10.3 Concluding Remarks

Target observation is one of those ubiquitous problems found in many areas of our world. Whether we are trying to detect an aircraft for the purposes of air traffic control and safety, or trying to identify and avoid an approaching airborne threat such as a missile, or even searching for patterns in some information in a database, we rely on machines to operate in a cooperative manner. Biological systems can provide us with important lessons and inspiration on how to tackle some of these hard problems, so we can design our machines and control algorithms in new ways.

In our case, the collective cooperative behavior of ants has provided this inspiration. Teamwork is an important mechanism for coordinating the actions of multiple autonomous agents and for forming structures that can achieve more than the simple sum of their parts. The ideas in this chapter are based on the paradigm of an extended phenotype or extended self, where the agent perceives its capabilities according to the relationships that it builds with its neighbors. The algorithm presented in this chapter is one method of forming teams between agents, which is reactive to environmental changes and self-organizing.

Finally, a self-organizing sensor network requires more than just mutual communications between sensor nodes. Useful work relies on the ability of the sensors to build relationships between each other, which lead to *structure* that we call a team.

The sensors must be enabled with the means to act in a social framework. In our case, an intelligent agent paradigm has been implemented. Throughout this work, the authors have emphasized the link with social structures such as those observed within biological ecosystems, and the application of behavior-based reasoning. While we seek to operate within our environment under an almost infinite variety of circumstances and situations, our control systems have traditionally been tailored to operate within well defined boundaries: that is, they are niche systems. It is clear that we need to think of the control being woven into the fabric of the systems we create and use. The *intelligent* agent is one tool to enable this approach, and the ecosystems around us remain one of the richest sources of inspiration for developing new algorithms and architectures for information processing, especially in defense-related systems.

Acknowledgment

The authors thank K. Brown for valuable comments on this manuscript. This research is supported by the Electronic Warfare and Radar Division of the Defence Science and Technology Organisation, Edinburgh, South Australia.

Glossary

Symbol	Description
N_S	Number of sensors/agents in the society (S2.1)
N_T	Number of targets in the region of interest for each simulation (S2.1)
i	Sensor/Agent index: $i = 1,2, \ldots, N_S$ (S2.3)
s_i	Sensor i (S2.3)
m	Row index of 2-D lattice (S2.3)
n	Column index of 2-D lattice (S2.3)
A_i	Agent i (S2.3)
N_{ni}	Number of sensors/agents in the neighbourhood of agent i. (S2.3)
(x_i, y_i)	Coordinates of sensor/agent $i : i = 1,2, \ldots, N_S$ (S2.3)
d_{ij}	Euclidean distance between agent and neighbour. (S2.3)
d_{iT}	Euclidean distance between agent i and target T. (S2.3)
$\rho_{det,i}$	Detection range of sensor i. (S2.4)
ρ_i	Communications range of sensor i. (S2.4)
R_i	Relation matrix for agent i (S2.6)
r_{ij}	Relationship strength between agent i and agent j. (S2.6)
k	Generation (time) index (S3.1)
$\boldsymbol{A}$	The universe of all agents making up the agent society. (S3.2)
A_i^E	The extended agent. (S3.2)
N^E_i	Number of agents in the extended phenotype of agent i. (S4.2)
N_g	Number of agents goals. (S4.1)
N_c	Number of agent capabilities. (S4.1)
N_b	Number of agent behaviors. (S4.1)
g_{ij}	The jth goal of the ith agent. (S4.1)
$\mathcal{G}_i$	Goal set for agent i; $\mathcal{G}_i = \{g_{i1}, g_{i2}, \ldots, g_{iN_g}\}$. (S4.1)
$\Gamma_i(k)$	Capability set for agent i; $\Gamma_i(k) = \{\gamma_{i1}, \gamma_{i2}, \ldots, \gamma_{iN_c}\}$. (S4.2)
Ω_i	Capability factor of agent i. (S4.2)
γ_{ij}	The jth capability of agent i. (S4.2)
$R^s(\Gamma_i, \Gamma_j)$	Species correlation function. (S4.2)
$R^b(b_i, b_j)$	Behavior correlation function. (S4.2)
$\boldsymbol{b}_{i,j}$	The jth behavior of agent i. (S4.3)

Symbol	**Description**
$\mathcal{B}_i$	Set of behaviors of agent i: $\mathcal{B}_i=\{b_{i,1}, b_{i,2}, \ldots, b_{i,N_b}\}$. (S4.3)
η_i	Pheromone concentration of agent i. (S5)
$I_i(k)$	Local information of the ith agent at time k (S5.1)
$I_i(0)$	The *a priori* local information of the ith agent. (S5.1)
$x_{i,j}(k)$	The jth element of the ith agent state vector. (S5.4)
w_i^g	Weight for goal j of agent i. (S6.1)
$F_{\text{proc},i}$	The processing function of the ith sensor. (S6.1)
$F_{\text{prop},i}$	The propagation function of the ith sensor (S6.1)
$K_{i,j}(k)$	Cost factor between agent i and agent j. (S6.1)
$V_j(k)$	The scalar attractor properties that agent i experiences due to agent j. (S6.1)
D_{ij}	Distance factor between agent i and agent j – Equation (11). (S6.2)
$X_i(k)$	State vector of agent i at time k: $\mathbf{X}_i(k)=[x_{ij}(k), . . , x_{in}(k)]$. (S6.2)
$\phi_{i,j}(k)$	The scalar field experience by agent i due to agent j. (S6.2)
c_{ij}^a	Cost factor. (S6.2)
w_i^b	Weight factor for behavior j of agent i: $w_i^b = [0...1]$. (S6.2)
w_i^Ω	Weight factor for capabilities of agent i as a discriminant. (S6.2)
w_i^s	Weight factor for species of agent i as a discriminant. (S6.2)
w_i^I	Weight factor for information of agent i as a discriminant. (S6.2)
w_i^b	Weight factor for behavior of agent i as a discriminant. (S6.2)
w_i^η	Weight factor for pheromone concentration of the ith agent as a discriminant. (S6.2)
T	Global temperature used in simulations. (S9.2)
λ_i	Evaporation constant of agent i. (S9.2)
μ_i	Learning rate constant of agent i. (S9.2)
s_i	Stimulus (scalar) perceived by agent i. (S9.2)
r_i^{th}	Relation threshold for agent i. (S9.3)
A_i^E	The extended phenotype of agent i. (S9.3)

Acronyms

ES	Electronic Support (S1.1)
ROI	Region of Interest: the region where our sensors are looking (S1.1)
RWR	Radar Warning Receiver (S5.1)

References

Alberts, D.S., Garstka, J.J., and Stein, F.P. (1999), *Network-Centric Warfare – Developing and Leveraging Information Superiority*, 2nd edition (revised), CCRP.

Anderson, C. and Franks, N.R. (1999), *Teams in Animal Societies*, Duke University, Durham, USA, p. 34.

Anthony, R.T. (1994), *Principles of Data Fusion Automation*, Artech House.

Beckerman, M. (1997), *Adaptive Cooperative Systems*, New York, Wiley.

Blackman, S. and Popoli, R. (1999), *Design and Analysis of Modern Tracking Systems*, Artech House.

Bonabeau, E., Dorigo, M., and Theraulaz G. (2000), "Inspiration for optimization from social insect behaviour," *Nature*, vol. 406, 6 July, pp. 39-42.

Bowyer, R.S. and Bogner, R.E. (1999), "Cooperative behaviour in multi-sensor systems," *Proc. 6th Int. Conf. on Neural Information Processing*, IEEE, Perth, WA Australia, vol. 2, pp. 807-812.

Brooks, R.R. (2001), "Data fusion for a distributed ground-based sensing system," in Hall, D.L. and Llinas, J. (eds.), *Handbook of Multisensor Data Fusion*, CRC Press.

Cohen, P. *et al.* (1997), "On team formation in contemporary action theory," in Hintikka, J. and Tuomela, R. (eds.), *Synthese*.

Curtis, S.D. (1999), *Electronic Warfare in the Information Age*, Artech House.

Dawkins, R. (1989), *The Selfish Gene*, 2nd edition, Oxford University Press.

Dawkins, R. (1999), *The Extended Phenotype*, revised edition, Oxford University Press.

Dorigo, M. and Di Caro, G. (1999), "The ant colony optimization meta-heuristic," in Corne, D., Dorigo, M., and Glover, F. (eds.), *New Ideas in Optimization*, McGraw Hill, pp. 11-50.

Dorigo, M. and Gambardella, J.M. (1997), "Ant colony system: a cooperative learning approach to the travelling salesman problem," *IEEE Transactions on Evolutionary Computation*, vol. 1, no. 1, pp. 53-66.

Hall, D.L. and Llinas, J. (2001), "Multisensor data fusion," in Hall, D.L. and Llinas, J. (eds.), *Handbook of Multisensor Data Fusion*, CRC Press.

Harney, R.C. (1997), "Information-based approach to performance and requirements allocation in multisensor fusion for target recognition," *SPIE*, vol. 36, pp. 789-798.

Jensen, H.J. (1998), *Self-Organized Criticality*, Cambridge University Press.

Kelso, J.A.S. (1999), *Dynamic Patterns – the Self-Organization of Brain and Behavior*, London: The MIT Press.

McMichael, D., Halgamuge, S., Hamlyn, G., Karan, M., and Okello, N. (1997), *Multisensor Data Fusion*, Cooperative Research Centre for Sensor Signal and Information Processing, Australia.

Parker, L.E. (1996), "On the design of behaviour-based multi-robot teams," *Journal of Advanced Robotics*.

Parker, L.E. (1997), "Behavior-based cooperative robots applied to multi-target observation," in Bolles, R., Bunke, H., and Noltemeier, H. (eds.), *Intelligent Robots: Sensing, Modelling, and Planning*, World Scientific.

Resnick, M. (1994), *Termites, Turtles, and Traffic Jams – Explorations in Massively Parallel Microworlds*, MIT Press.

Reynolds, C. (1987), *Flocks, Herds, and Schools: a Distributed Behavioural Model.*

Robocup Website (1998), http://www.robocup.org, 1998-2002.

Roy, J. and Bosse, E. (1998), "Conflict management in the shipboard integration of multiple sensors," *FUSION'98 International Conference*, pp. 381-387.

Solso, R.L. (1998), *Cognitive Psychology*, 5th edition, Allyn and Bacon.

Stone, P. and Veloso, M.M. (1999), "Task decomposition and dynamic role assignment for real time strategic teamwork," *Intelligent Agents V – Proceedings of the Fifth International Workshop on Agent Theories, Architectures, and Languages (ATAL '98)*, Heidelberg, Springer-Verlag.

Tambe, M. (1998), "Implementing agent teams in dynamic multiagent environments," *Applied Artificial Intelligence*, vol. 12.

Thomopoulos, S.C.A. (1994), "Data fusion system design and adaptive control," *SPIE*, vol. 2232, pp. 56-68.

Vicente, K.J. (1999), *Cognitive Work Analysis: towards Safe, Productive, and Healthy Computer-Based Work*, 1st edition, LEA.

Waltz, E. (1998), *Information Warfare – Principles and Operations*, 1st edition: Artech House.

Wolpert, D.H. and Tumer, K. (1998), "An introduction to collective intelligence," in Bradshaw, J.M. (ed.), *Handbook of Agent Technology*, AAAI Press / MIT Press.

Chapter 8

Military Applications of the Disciple Learning Agent

G. Tecuci and M. Boicu

To Mihai Draganescu, on his 75th birthday

Disciple is a theory, methodology, and family of agent shells for the development of intelligent agents by subject matter experts, with limited assistance from knowledge engineers. In the last few years the development of the Disciple approach has been a direct result of its application to three military challenge problems used in the DARPA's "High Performance Knowledge Bases" and "Rapid Knowledge Formation" programs: center of gravity analysis, course of action critiquing, and workaround planning. This chapter introduces each of these innovative military applications of Artificial Intelligence, and the corresponding Disciple agent that was built to address it. Disciple-RKF/COG is used in two courses at the US Army War College. In one course the students use a trained Disciple agent as an aid for learning about center of gravity analysis, and for developing a report containing a case study analysis. In another course the students use personal Disciple agents that were not yet trained, as subject matter experts, teaching them their own problem solving expertise in center of gravity analysis. Disciple-COA critiques military courses of actions with respect to the principles of war and the tenets of army operations. Disciple-Workaround plans how a convoy of enemy vehicles can circumvent or overcome obstacles in their path.

1 Introduction

This chapter presents an overview of the Disciple approach and its military applications. Disciple is a theory, methodology, and family of agent shells for the development of intelligent agents by subject matter experts, with limited assistance from computer scientists or knowledge engineers (Tecuci 1998). A subject matter expert interacts directly with a Disciple agent, to teach it to solve problems, in a way that is similar to how the expert would teach a human apprentice, by giving the agent examples and explanations, as well as by supervising and correcting its behavior. The agent learns from the expert by generalizing the examples and the explanations to build its knowledge base. The Disciple approach integrates methods for mixed-initiative problem solving, teaching, and multistrategy learning, exploiting the complementariness between human and automated reasoning, and creating a synergism between the expert that has the knowledge to be formalized and the agent that knows how to formalize it.

In the last few years the development of the Disciple approach has been a direct result of its application to three military challenge problems used in the DARPA's "High Performance Knowledge Bases" (Gunning and Burke 2002) and "Rapid Knowledge Formation" (Burke 2002) programs:

1. The Workaround challenge problem – planning how a convoy of enemy vehicles can circumvent or overcome obstacles in their path, such as damaged bridges or cratered roads, in order to perform target selection (Tecuci *et al.* 1999).
2. The Course of Action challenge problem – critiquing military courses of actions with respect to the principles of war and the tenets of army operations, in order to assist a military commander to select the best course of action (Boicu *et al.* 2000, Tecuci *et al.* 2000).

3. The Center of Gravity challenge problem – identifying and testing strategic center of gravity candidates in military conflicts (Boicu *et al.* 2001, Tecuci *et al.* 2002).

This chapter introduces each of these innovative military applications of Artificial Intelligence, and the corresponding Disciple agent that was built to address them.

2 Center of Gravity Analysis

2.1 The Center of Gravity Problem

The military literature distinguishes between three levels of conflicts: a strategic level focusing on winning wars, an operational level focusing on winning campaigns, and a tactical level focusing on winning battles (Field Manual 3-0 2001, Joint Publication 3-0 1993). One of the most difficult problems that senior military leaders face at the strategic level is the determination and analysis of the centers of gravity for friendly and opposing forces. Originally introduced by Clausewitz in his classical work "On War" (1832), the center of gravity is now understood as representing "those characteristics, capabilities, or localities from which a military force derives its freedom of action, physical strength, or will to fight" (Field Manual 3-0 2001). It is recognized that if a combatant eliminates or influences the enemy's strategic center of gravity, then the enemy will lose control of its power and resources and will eventually be defeated. Similarly, if the combatant does not adequately protect his own strategic center of gravity, he will be defeated (Giles and Galvin 1996). In spite of the apparently simple definition of the center of gravity, its determination requires a wide range of background knowledge, not only from the military domain, but also from the economic, geographic, political, demographic, historic, international, and other domains. In addition, the situation, the adversaries involved, their goals, and their capabilities can vary in important ways from one scenario to another. When performing this analysis, experts rely on their own professional experience and intuitions, without following a rigorous approach.

Correctly identifying the centers of gravity of the opposing forces is of highest importance in any conflict. Therefore, in the education of strategic leaders at all the US senior military service colleges, there is a great emphasis on the centers of gravity analysis (Strange 1996). Recognizing these difficulties, the Center for Strategic Leadership of the US Army War College started in 1993 an effort to elicit and formalize the knowledge of a number of experts in center of gravity. This research resulted in a COG monograph (Giles and Galvin 1996), which provided a basis for the application of the Disciple approach to this high value application domain, and to the development of the Disciple-RKF/COG instructable agent. Disciple-RKF/COG is used in a sequence of two courses taught regularly at the US Army War College, "Case Studies in Center of Gravity Analysis," and "Military Applications of Artificial Intelligence." In the first course (the COG course), the students become familiar with Disciple-RKF/COG as end-users, using it as an aid for learning about center of gravity analysis, and for developing a report containing a case study analysis. In the second course (the MAAI course), the students use Disciple-RKF/COG as subject matter experts, teaching it their own problem solving expertise in center of gravity analysis. In the next section we will illustrate the use of Disciple in these two courses.

2.2 A Disciple Agent for COG Analysis

In the "Case Studies in Center of Gravity Analysis" course, a personal copy of Disciple guides the student to identify, study and describe the aspects of a scenario (such as the 1945 US invasion of the island of Okinawa) that are relevant for COG analysis. The student-agent interaction takes place as illustrated in Figure 1. The left part of the window is a table of contents, whose elements indicate

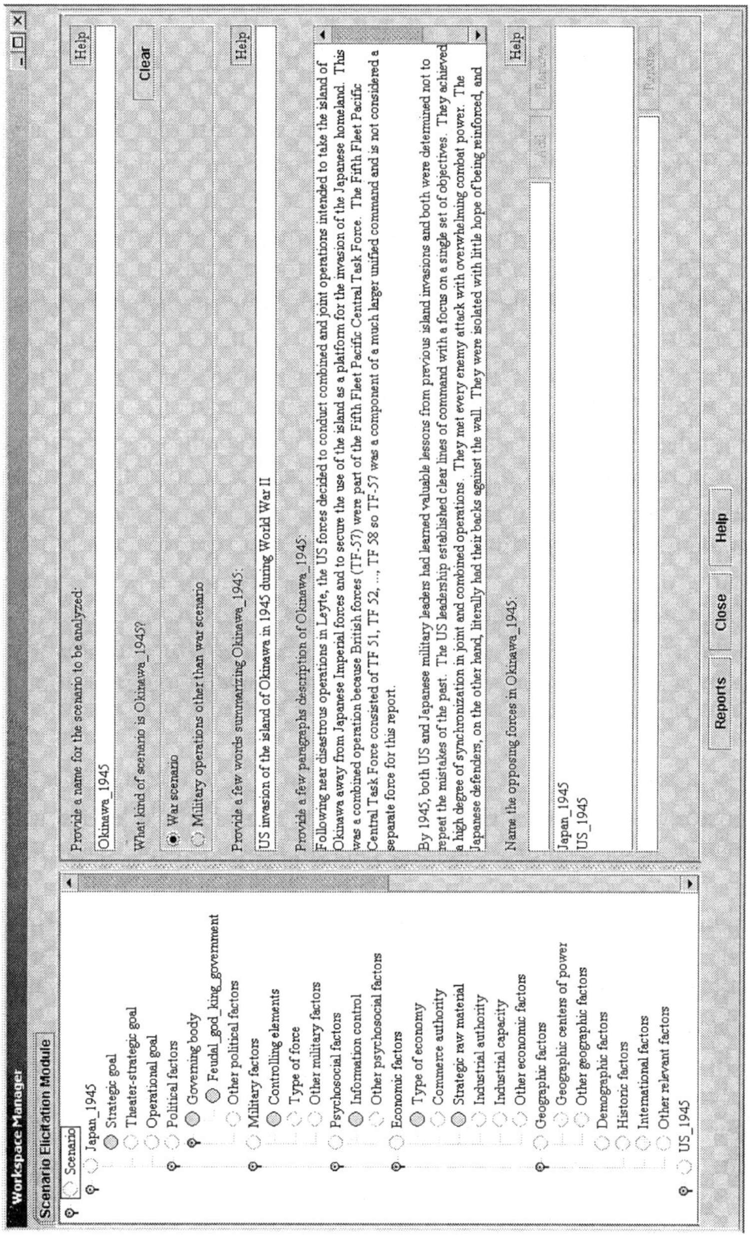

Figure 1. Scenario specification interface.

various aspects of the scenario. When the student selects one such aspect, Disciple asks specific questions intended to acquire from the student a description of that aspect, or to update a previously specified description. All the student's answers are in natural language. Taking the Okinawa_1945 scenario as our example, Disciple starts by asking for a name and a description of the scenario, and then asks for the opposing forces. Once the student indicates Japan_1945 and US_1945 as opposing forces, Disciple includes them in the table of contents, together with their characteristics that the student needs to specify (see the left hand side of Figure 1). Then, the student may click on any of these aspects (e.g., "Industrial capacity" under "Economic factors" of Japan_1945) and the agent guides the student in specifying it. The student's specification may prompt additional questions from Disciple, and a further expansion of the table of contents. An orange, yellow, or white circle marks each title in the table of contents, indicating respectively that all, some, or none of the corresponding questions of Disciple have been answered. The student is not required to answer all the questions and Disciple can be asked, at any time, to identify and test the strategic center of gravity candidates for the current specification of the scenario.

The top part of Figure 2 shows the solution viewer of Disciple that was customized for the COG domain. In the left hand side Disciple lists the strategic center of gravity candidates that it has identified based on the description of the Okinawa_1945 scenario. In the case of Japan_1945, these strategic COG candidates are Emperor Hirohito, Japanese Imperial General Staff, the military of Japan, and the industrial capacity of Japan. When a candidate is selected in the left hand side of the viewer, its (abstract or detailed) justification for identification or for testing can be displayed in the right hand side of the viewer. The top part of Figure 2 shows the abstract justification for the identification of Emperor Hirohito as a strategic COG candidate, and consists of a sequence of questions and answers. Emperor Hirohito is identified as a strategic COG candidate for Japan_1945 in the Okinawa_1945 scenario because Japan is a single-member force and Emperor Hirohito is the main controlling element of its government. After being identified as a candidate, Emperor Hirohito is analyzed based on various elimination tests, but he passes all of them. This testing is shown in bottom part of Figure 2. It has already been established that Emperor Hirohito controls the government of Japan. Because he is the commander in chief of the military, he can also impose his will on the military of Japan. Moreover, Emperor Hirohito could also impose the people of Japan to accept the unconditional surrender of Japan, which is the main strategic goal of the US. Being able to impose his will on the Clausewitz's trinity of power (government, military and people), Emperor Hirohito is very likely to be the strategic center of gravity of Japan in 1945.

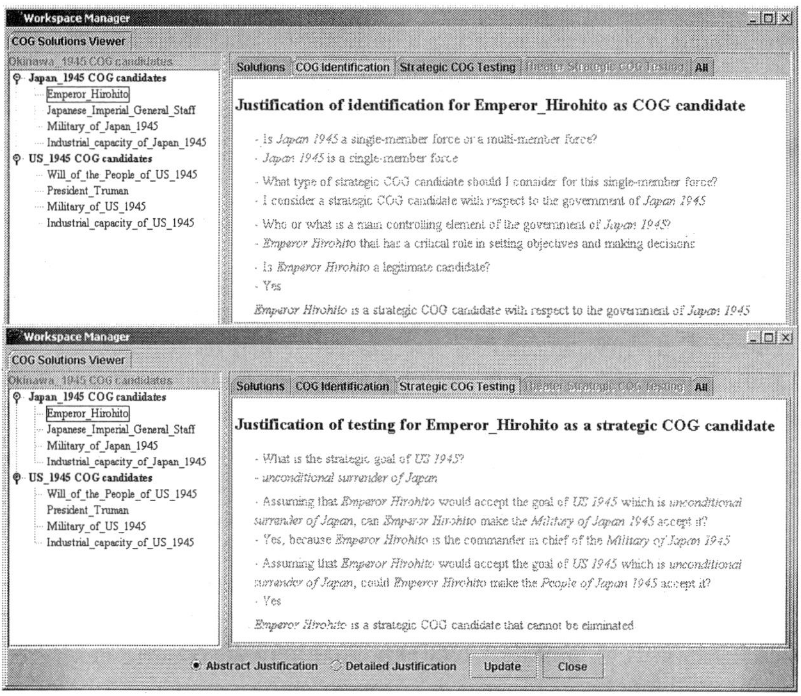

Figure 2. Solution viewer interface.

As another example, consider the industrial capacity of Japan_1945, which is another source of strength, power and resistance because it produces the war materiel and transports of Japan. Disciple, however, eliminates this strategic COG candidate because the military and the people of Japan_1945 are determined to fight to death and not surrender even with diminished war materiel and transports.

In the example scenario portrayed here, Disciple eliminates all but two candidates for Japan – Emperor Hirohito and the Japanese Imperial General Staff – and suggests that the student should select one of them as the strategic Center of Gravity of Japan in 1945. It is important to point out that this example is only a representative approach to the analysis of Japan's center of gravity for the Okinawa campaign. We recognize that subject matter experts often differ in their judgments as to the identification and analysis of center of gravity candidates for any particular scenario. The important point for agent development is that the Disciple agent can accommodate the preferences of the expert that teaches it, as will be discussed in the next section.

As briefly illustrated above, Disciple guides the student to identify, study and describe the relevant aspects of the opposing forces in a particular scenario. Then Disciple identifies and tests the strategic center of gravity candidates, as illustrated in Figure 2. After that Disciple generates a draft analysis report, a fragment of which is shown in Figure 3.

The student needs to finalize the report generated by Disciple. He is required to critically analyze Disciple's logic, correct or complete it, or even reject it and provide an alternative line of reasoning. This is productive for several reasons. First Disciple generates its proposed solutions by applying general reasoning rules and heuristics learned from another COG expert, to a new scenario described by the student. This scenario description incorporates the subjective judgments of the student, and may also be incomplete. Therefore the agent may not generate all the solutions that the expert would have generated, and some of the solutions and justifications provided by the agent may not always be entirely correct or complete. Secondly, as mentioned before, COG analysis is influenced by personal experiences and subjective judgments, and the student (who has unique military experience and biases) may have a different interpretation of certain facts.

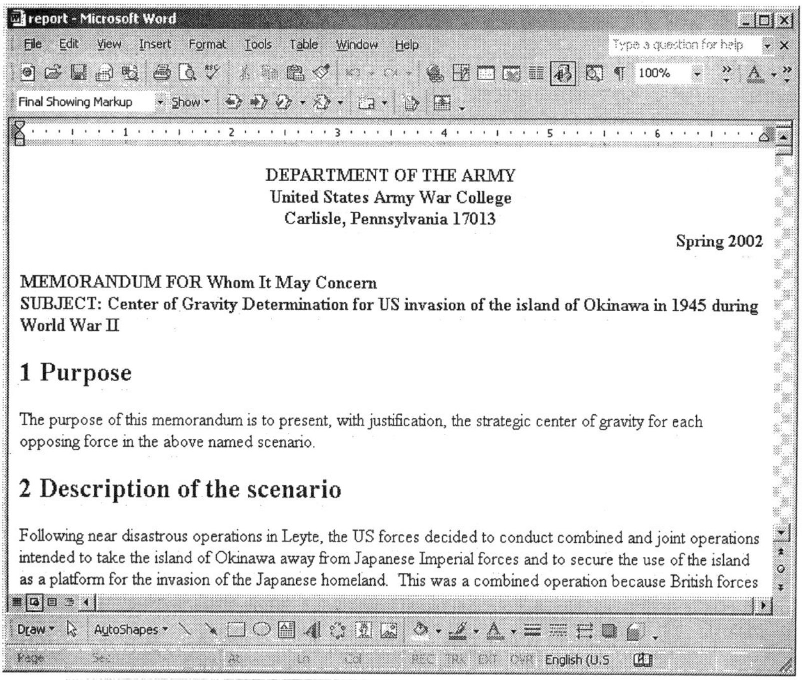

Figure 3. Draft COG analysis report generated by Disciple.

This requirement for the critical analysis of the solutions generated by Disciple is an important educational component of military commanders that mimics military practice. Commanders have to critically investigate several courses of actions proposed by their staff and to make the final decision on which one to use.

During the 2001 and 2002 academic years, Disciple was successfully used in both the Winter and Spring sessions of the COG course. As a result of this initial success, the USAWC decided to continue and expand the integration of Disciple in this course for the next academic year and beyond. At the end of the courses the students complete detailed evaluation forms about Disciple and its modules, addressing a wide range of issues, ranging from judging its usefulness in achieving course's objectives, to judging its methodological approach to problem solving, and to judging the ease of use and other aspects of various modules. For instance, on a 5-point scale, from strongly disagree to strongly agree, 9 students of the Winter-2002 session agreed and the other 4 strongly agreed with the statement "The use of Disciple is an assignment that is well suited to the course's learning objectives." To our knowledge, this is the first time that an intelligent agent for the strategic COG identification and testing has been developed.

The next section presents the use of Disciple in the "Military Applications of Artificial Intelligence" course.

2.3 Agent Development with Disciple-RKF

Several of the students that took the COG course in the Winter 2001 session, together with additional students, took the "Military Applications of Artificial Intelligence" course in the Spring 2001 session. In this course the students were given a general overview of Artificial Intelligence, as well as an introduction to Disciple-RKF. These students used the agent not as end-users, but as subject matter experts charged with developing their own COG agents using Disciple. The students were organized in five two-person teams. Each team was given the project to train a personal Disciple agent according to its own reasoning in COG identification for its historical scenario. All five teams succeeded in developing working agents, with each team addressing one of the following scenarios: 1) the capture of the Leyte Island by the US forces in 1944; 2) the

Inchon landing during the Korean War in 1950; 3) the Falklands war between Argentina and Britain in 1982; 4) the stabilization mission in the Grenada Island in 1983; and 5) the US invasion of Panama in December 1989. In the following we will present how the Disciple-RKF shell is used to develop such an agent.

Generally, a knowledge-based agent includes two main components, a knowledge base and a problem solving engine. The knowledge base contains the data structures representing the entities from the expert's application domain such as objects, relations between objects, classes of objects, laws, actions, processes and procedures. The problem solving engine consists of the programs that manipulate the data structures in the knowledge base in order to solve problems in a way that is similar to how the expert solves them.

Disciple-RKF is a tool for building knowledge-based agents. It consists of a general problem solving engine, an empty knowledge base, and a general learning engine. A subject matter expert can train Disciple-RKF to solve problems in a way that resembles how the expert would teach a student or an apprentice. For instance, the expert defines a specific problem, helps the agent to understand each reasoning step toward the solution, and supervises and corrects the agent's behavior, when it attempts to solve new problems. During such mixed-initiative interactions the agent learns from the expert by employing complementary learning methods, building and refining its knowledge base to represent the problem solving expertise of the human expert. This knowledge base has two main components: an object ontology that defines the concepts from a specific application domain, and a set of problem solving rules expressed with these concepts.

In general, the process of developing a knowledge-based agent with Disciple-RKF consists of two major stages: 1) the development of the object ontology by a knowledge engineer and a subject matter expert, and 2) the training of Disciple by the subject matter expert.

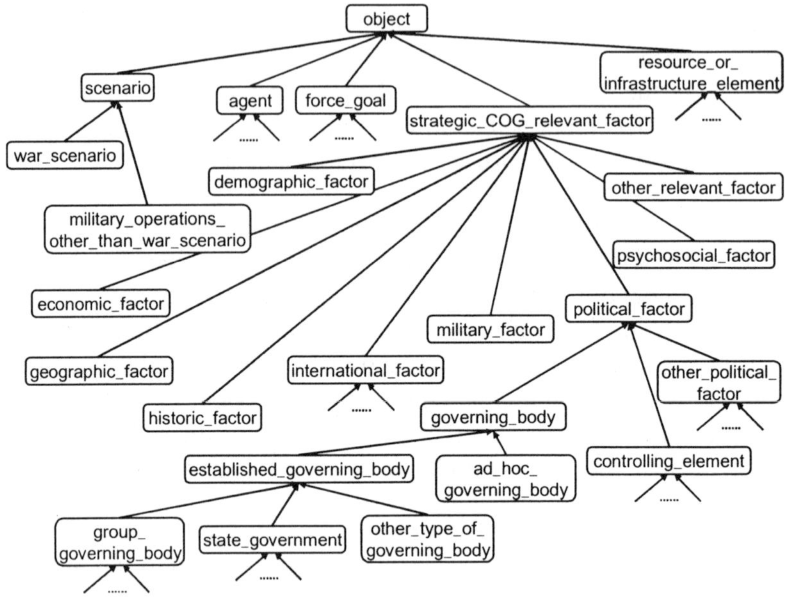

Figure 4. A fragment of the COG object ontology.

In the first development stage a knowledge engineer works with a subject matter expert to specify the type of problems to be solved by the Disciple agent, to clarify how these problems could be solved by Disciple, and to develop an object ontology. A fragment of the object ontology developed for the COG domain is shown in Figure 4. The object ontology consists of hierarchical descriptions of the types of objects (called concepts) from the domain, specifying their properties and relationships. The objects are represented as frames, according to the knowledge model of the Open Knowledge Base Connectivity (OKBC) protocol (Chaudhri *et al.* 1998). Some of the top level objects from this hierarchy are scenario, agent, force_goal, and strategic_COG_relevant_factor. Each of them is the top node of a different hierarchy. Notice, for instance, that immediately under strategic_COG_relevant_factor are various such factors, including political_factor. Under political_factor are various types of political factors, one of them being governing_ body. Under governing_body are the various types of governments, and so on.

Disciple-RKF includes several types of ontology browsers and editors that facilitate the ontology development process. The careful design and development of the object ontology is of utmost importance because it is used by Disciple as its generalization hierarchy for learning, as will be illustrated latter.

The result of the first development stage is a customized Disciple agent. This agent is trained to solve problems by a subject matter expert, with very limited assistance from a knowledge engineer, in the second major stage of agent development. Figure 5 shows the main phases of the agent training process, which starts with a knowledge base that contains only a general object ontology (but no instances, no problem solving tasks, and no task reduction rules), and ends with a knowledge base that incorporates expert problem solving knowledge. In the following we will illustrate these agent training phases.

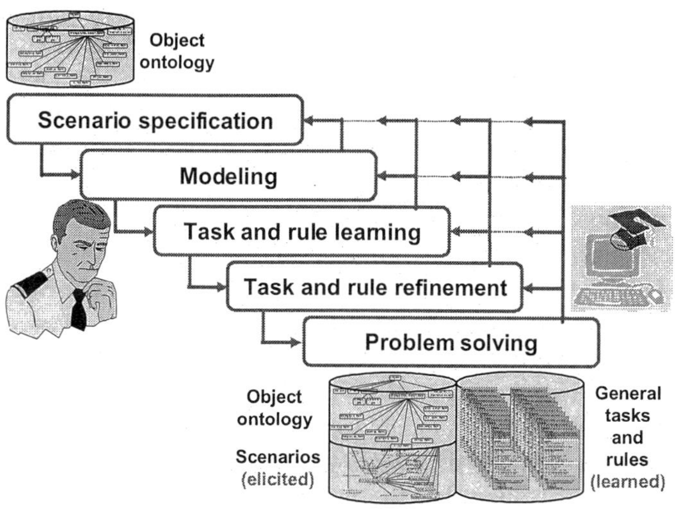

Figure 5. The main phases of the agent training process.

During the ***Scenario specification*** phase, the Scenario Specification module of Disciple guides the expert in describing the objects that define a specific strategic scenario (e.g., the US invasion of the island of Okinawa in 1945). The expert does not work directly with the object ontology in order to specify the scenario. Instead, the expert-agent interaction takes place as presented in Section 2.2 and illustrated in Figure 1. The left hand side of the screen in Figure 1 contains the elements that need to be described by the expert. Notice that for each of the strategic_COG_relevant_factor in Figure 4 (such as political_factor or military_factor) there is a corresponding element in the left hand side of Figure 1 (i.e., Political factor, Military factor). These factors appear both under Japan_1945 and under US_1945. When the expert clicks on one of them in the interface shown in Figure 1, Disciple asks various questions about that factor. For instance, when the expert clicks on the "Political factors" under "US_1945", Disciple asks the expert to indicate what is the type of the government of US_1945. When the expert answers that it is a representative democracy, Disciple asks additional questions about this type of government, such as, who is the head of the government, which is the name of the legislative body, and so on. Based on the answers provided by the expert, Disciple creates formal representations of specific objects that define the Okinawa_ 1945 scenario. These specific objects, such as government_of_US_ 1945, President_Truman, and Congress_of_US, are included by Disciple into the object ontology, as instances of their corresponding concepts. A fragment of the extended ontology is shown in Figure 6. For example, government_of_US_1945 is represented as an instance (or a member) of the concept "representative_democracy." This means that government_of_US_1945 is a representative democracy. Notice also that "representative_democracy" is a subconcept of "democratic_government." This means that any representative democracy is a democratic government. Therefore, because the government_of_US_1945 is a representative democracy, Disciple can conclude that the government_of_ US_1945 is a democratic government. Based on the answers provide by the expert,

Disciple defines also the features of these instances. For example, as one can see in Figure 6, the government_of_US_1945 has as head of government President Truman, and as legislative body Congress_of_US. Experimental results show that the experts can easily interact with the Scenario Specification module to answer Disciple's questions.

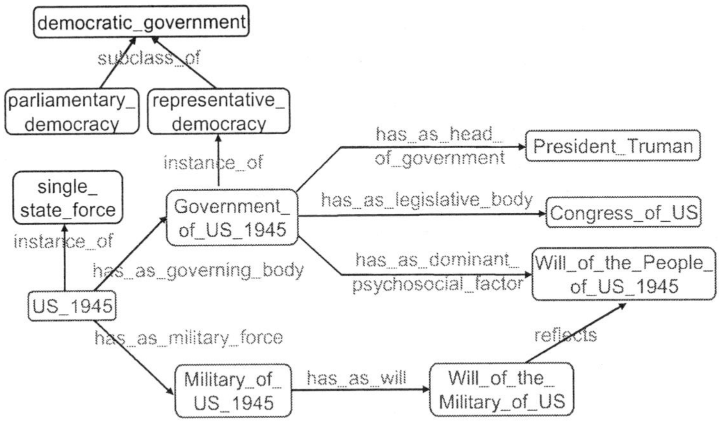

Figure 6. Fragment of the object ontology extended with descriptions of objects.

After the expert has specified the Okinawa_1945 scenario, he can start to teach Disciple how to identify and test the strategic COG candidates for this particular scenario. The expert needs to show Disciple how he solves this problem by using the task reduction paradigm. The use of the task reduction paradigm is necessary because this the general problem solving strategy used by the problem solving engine of Disciple-RKF. According to this paradigm, which is illustrated in Figure 7, a complex problem solving task is performed by successively reducing it to simpler tasks, finding the solutions of the simplest tasks, and successively composing these solutions until the solution of the initial task is obtained. In the illustration from Figure 7, the initial problem solving task T_0 is first reduced to the simpler task T_1. Then T_1 is reduced to the simpler tasks T_{11}, …, T_{1n}. These tasks are simple enough to find their solutions S_{11}, ..., S_{1n}. These solutions are first composed into the solution S_1 of the task T_1. Then the solution S_0 of the initial task T_0 is obtained from S_1.

Task reduction is used in many domains under different names, such as "divide and conquer," or "problem decomposition." In the Disciple approach we have refined this general problem solving strategy by introducing questions and answers that guide the task reduction process, as illustrated in the right hand side of Figure 7. T_0 is the initial problem solving task to be performed. Finding a solution of this task is an iterative process where, at each step, we consider some relevant information that leads us to reduce the current task to a simpler task or to several simpler tasks. The question Q associated with the current task identifies the type of information to be considered. The answer A identifies that piece of information and leads us to the reduction of the current task.

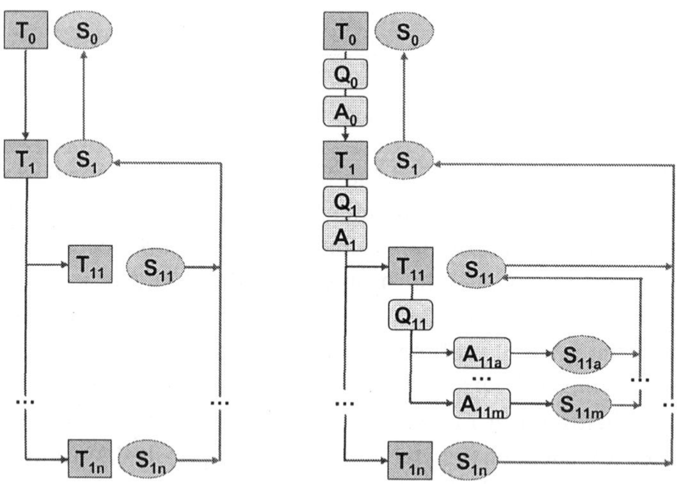

Figure 7. The task reduction paradigm of problem solving.

The expert uses the **Modeling** module of Disciple-RKF to make explicit the way he identifies and tests COG candidates, within the task reduction paradigm. He expresses his reasoning in English, as a sequence of task reduction steps like the ones illustrated in Figure 8. First he formulates the top level problem solving task: "Identify and test a strategic COG candidate for the Okinawa_1945 scenario." To perform this task, the expert asks himself a series of questions.

The answer of each question allows the expert to reduce the current task to a simpler one. This process continues until the expert has enough information to first identify a strategic COG candidate, and then to determine whether it should be eliminated or not. Let us follow the task reduction steps from Figure 8. Through a series of questions and answers the top level task is successively reduced to the task "Identify and test a strategic COG candidate for US_1945 with respect to the People_of_US_1945" This task is then reduced to two simpler tasks:

- "Identify the Will_of_the_People_of_US_1943 as a strategic COG candidate with respect to the People_of_US_1943"
- "Test the Will_of_the_People_of_US_1943 which is a strategic COG candidate with respect to the People_of_US_1943"

Then, each of these tasks is reduced to find their respective solutions:

- "The Will_of_the_People_of_US_1943 is a strategic COG candidate with respect to the People_of_US_1943"
- "The Will_of_the_People_of_US_1943 is a strategic COG candidate that cannot be eliminated"

Notice that this is one of the solutions of the initial task: "Identify and test a strategic COG candidate for the Okinawa_1945 scenario."

The right hand side of Figure 8 shows that Disciple learns a general task reduction rule from each specific problem solving step indicated by the expert. This learning takes place in the ***Task and rule learning*** phase. The learned rules allow the agent to identify and test COG candidates for new scenarios.

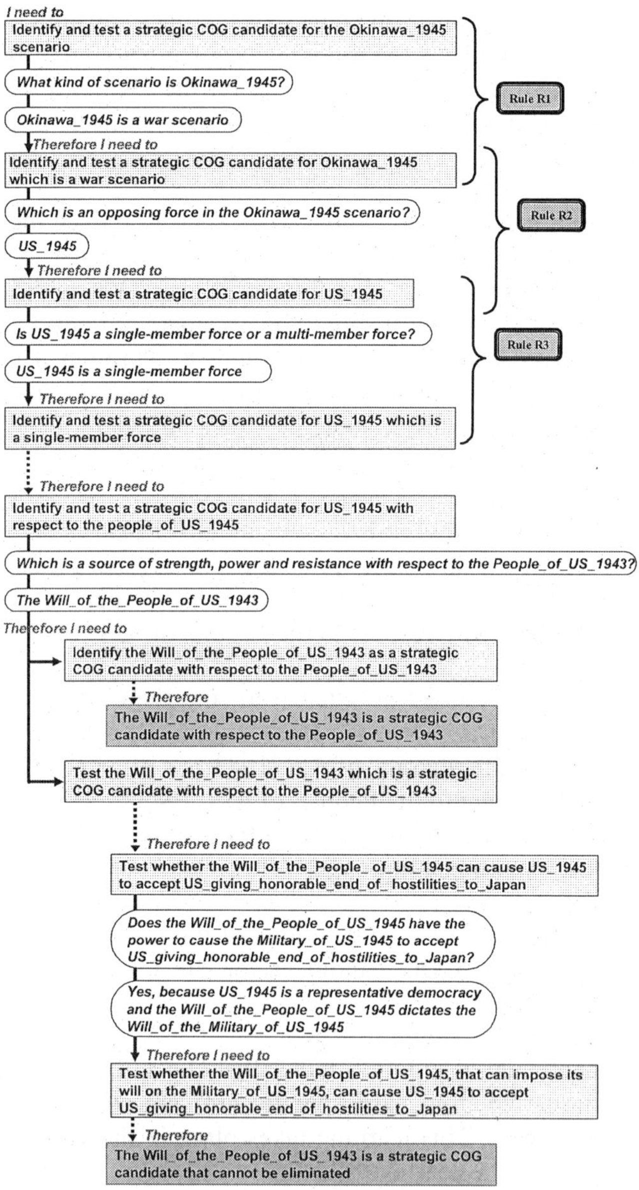

Figure 8. Sample modeling of the expert's problem solving process.

Let us consider, for instance, the task reduction step from the bottom part of Figure 8, which is also shown in the left hand side of Figure 9. The top task is the current task that needs to be reduced. The expert has to define a question that is relevant to the reduction of this task, then answer the question, and then reduce the top task to a simpler one that incorporates the information from the answer. Because all these expressions are in natural language and are not understood by the agent, the expert and the agent have to collaborate to translate them into the formal logical expressions on the right hand side of Figure 9.

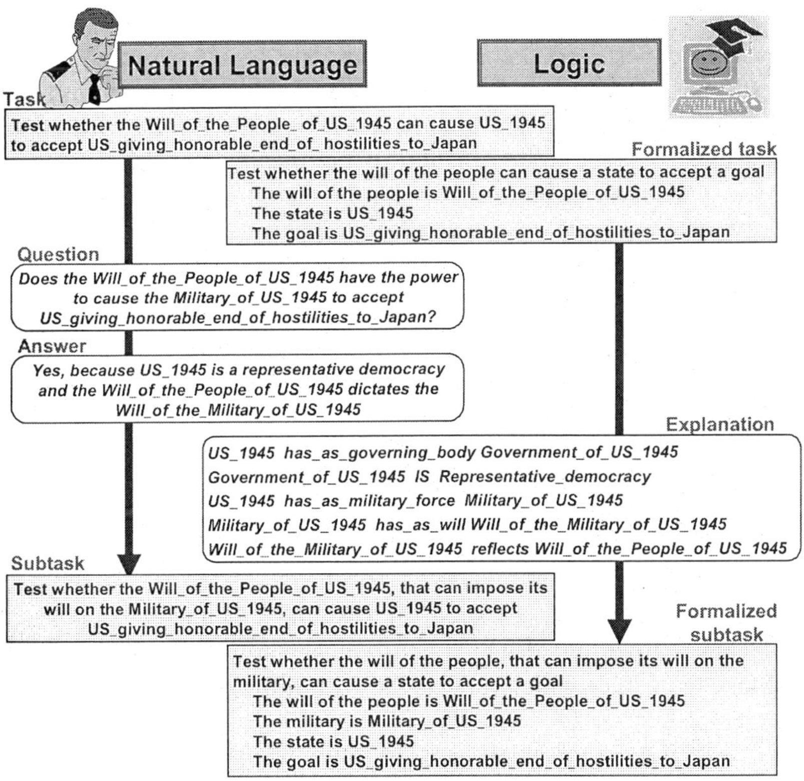

Figure 9. Mixed-initiative language to logic translation.

First the natural language expression of each task is structured into an abstract phrase, called the task name, and several specific phrases, called the task's features. The task name should not contain any instance (such as "US_1945"), but only general concepts (such as "state"). The task's features are specific phrases (such as "The state is US_1945"). Together, the task name and the task features should have an equivalent meaning with the natural language expression of the task. The formalizations of the two tasks from the left had side of Figure 9 are shown in the right hand side of Figure 9. The task formalizations are proposed by the agent and may be modified by the expert.

Next the expert and the agent collaborate to also formalize the question and the answer from the left hand side of Figure 9 into the explanation from the right hand side of Figure 9. This explanation represents the best approximation of the meaning of the question-answer pair that can be formed with the elements of the object ontology. The explanation consists of various relations between certain elements from the agent's ontology (some of which are shown in Figure 6):

- US_1945 has_as_governing_body Government_of_US_1945
- Government_of_US_1945 is representative_democracy
- US_1945 has_as_military_force Military_of_US_1945
- Military_of_US_1945 has_as_will
  Will_of_the_Military_of_US_1945
- Will_of_the_Military_of_US_1945 reflects
  Will_of_the_People_of_US_1945

These explanation pieces state, in Disciple's language, that US_1943 has a government which is a representative democracy and a military the will of which reflects the will of the people.

An expert can understand these formal expressions because they actually correspond to his question-answer pair. However, he cannot be expected to be able to define them because he is not a knowledge engineer. For one thing, he would need to use the formal language of the agent. But this would not be enough. He would also need to know the names of the potentially many thousands of concepts and features from the agent's ontology (such as "has_as_governing_body").

While defining the formal explanation of this task reduction step is beyond the individual capabilities of the expert and the agent, it is not beyond their joint capabilities. Finding these explanation pieces is a mixed-initiative process involving the expert and the agent. In essence, the agent will use analogical reasoning and help from the expert to identify and propose a set of plausible explanation pieces from which the expert will have to select the correct ones (Tecuci *et al.* 2001).

Based on the formalizations from Figure 9 and the object ontology illustrated in Figures 4 and 6, the Disciple agent learns the general IF-THEN task reduction rule shown in Figure 10. This rule has an informal structure, shown in the upper part of Figure 10, and a formal structure, shown in the lower part of Figure 10.

The informal structure of the rule is obtained by simply replacing the five instances from the expert's example from the left hand side of Figure 9, with five different variables. This form of the rule preserves the natural language of the expert and is used in agent-user communication.

The formal structure of the rule is obtained by using a knowledge-based generalization of the formalized tasks and explanation from the right hand side of Figure 9 (Boicu *et al.* 2000, Tecuci 1998). This is the form that is used in the internal formal reasoning of the agent. This is an IF-THEN structure that indicates the condition under which the task from the IF part can be reduced to the task from the THEN part. However, instead of a single applicability condition, the rule learned by Disciple contains two plausible conditions that bound the exact condition to be learned during the ***Task and rule refinement*** phase. Initially, when the agent has no rules and no tasks, the expert teaches Disciple how to solve problems and Disciple generates partially learned tasks and rules, as indicated above. As Disciple learns from the expert, the interaction between the expert and Disciple evolves from a teacher-student interaction, toward an interaction where both collaborate in solving a problem. During this mixed-initiative ***Problem Solving*** phase, Disciple learns not only from the contributions of the expert, but also from its own successful or unsuccessful problem solving attempts. Indeed, the plausible upper bound condition of the rule in Figure 10 allows the rule to be applicable in many analogous situations, but the result may not be correct. The agent will apply this rule to solve new problems and the feedback received from the expert will be used to further refine the rule. Usually, the plausible upper bound condition is specialized to no longer apply in situations that produce incorrect solutions. Similarly, the plausible lower bound condition is generalized to apply in situations that produce correct solutions. In this way the two conditions will converge toward one another, both approaching the exact applicability condition of the rule. Rule refinement could lead to a complex task reduction rule, with additional Except-When conditions which should not be satisfied in order for the rule to be applicable (Boicu *et al.* 2000).

It is important to stress that the expert does not deal directly with the learned rules, but only with their examples used in problem solving. Therefore, the complex knowledge engineering operations of defining and debugging problem solving rules are replaced in the Disciple approach with the much simpler operations of defining and critiquing specific examples.

After the Disciple agent has been trained, it can be used in the autonomous problem solving mode, to identify and test the strategic COG candidates for a new scenario, as was illustrated in the previous section.

IF: Test whether the ?O1 can cause ?O2 to accept ?O3

Question: Does the ?O1 have the power to cause the ?O4 to accept ?O3?
Answer: Yes, because ?O2 is a representative democracy and the ?O1 dictates the ?O5

THEN: Test whether the ?O1, that can impose its will on the ?O4, can cause ?O2 to accept ?O3

IF: Test whether the will of the people can cause a state to accept a goal
The will of the people is ?O1
The state is ?O2
The goal is ?O3

Explanation
?O2 has_as_governing_body ?O6 is representative_democracy
?O2 has_as_military_force ?O4 has_as_will ?O5 reflects ?O1

Plausible Upper Bound Condition
?O1 is will_of_agent
?O2 is force
has_as_military_force ?O4
has_as_governing_body ?O6
?O3 is force_goal
?O4 is military_force
has_as_will ?O5
?O5 is will_of_agent
reflects ?O1
?O6 is representative_democracy

Plausible Lower Bound Condition
?O1 is Will_of_the_People_of_US_1945
?O2 is US_1945
has_as_military_force ?O4
has_as_governing_body ?O6
?O3 is US_giving_honorable_end_of_hostilities_to_Japan
?O4 is Military_of_US_1945
has_as_will ?O5
?O5 is Will_of_the_Military_of_US_1945
reflects ?O1
?O6 is Government_of_US_1945

THEN: Test whether the will of the people, that can impose its will on the military, can cause a state to accept a goal
The will of the people is ?O1
The military is ?O4
The state is ?O2
The goal is ?O3

Figure 10. Rule learned from the example in Figure 9.

In the last two 3-hour sessions of the Spring-2001 Military Applications of Artificial Intelligence course, the students have participated in a controlled agent development experiment that was videotaped in its entirety. Each of the five 2-student teams was provided with a copy of Disciple-RKF/COG that contained a generic object ontology, but no specific instances and no rules. They received a 7-page report describing the Okinawa scenario (that they have not seen before), and were asked to train their Disciple agent to identify center of gravity candidates, based on that scenario. After each significant phase of agent training and knowledge base development (i.e., scenario specification, (i.e., scenario specification, modeling, rule learning, and rule refinement) a knowledge engineer reviewed their work, and the team then made any necessary corrections under the supervision of the knowledge engineer. Each team used the Scenario Specification tool to populate Disciple's ontology with different instances and features. On average they defined 85.40 instances and 93.80 feature values in 1 hour and 6 minutes. After that, each team taught its Disciple agent to identify COG candidates in the Okinawa scenario. The average number of rules per team was 18.80, and the average time interval was 4 hours and 7 minutes. Although obviously incomplete (both because of the use of a single training scenario, and because of incomplete training for that scenario), the knowledge bases were good enough for identifying correct COG candidates not only for the Okinawa (evaluation) scenario, but also for new scenarios whose inputs were taken from the class projects.

At the end of this final experiment, the students completed a detailed questionnaire, containing questions about the main components of Disciple. One of the most significant results was that 7 out of the 10 experts agreed, 1 expert strongly agreed and 2 experts were neutral with respect to the statement: "I think that a subject matter expert can use Disciple to build an agent, with limited assistance from a knowledge engineer." We consider this experiment to be a very significant success. Indeed, to our knowledge, this is the first time that subject matter experts have trained an agent their own problem solving expertise, with very limited assistance from a knowledge engineer.

3 Course of Action Critiquing

A military Course of Action (COA) is a preliminary outline of a plan for how a military unit might attempt to accomplish a mission. A COA is not a complete plan in that it leaves out many details of the operation such as exact initial locations of friendly and enemy forces.

After receiving orders to plan for a mission, a commander and his staff analyze the mission to conceive alternative COAs, evaluate them, and select the best one. Then they prepare a detailed plans to accomplish the mission based on the selected COA. The general practice is for the staff to generate several COAs for a mission, and then to make a comparison of those COAs based on many factors including the situation, the commander's guidance, the principles of war, and the tenets of army operations. The commander makes the final decision on which COA will be used to generate his or her plan based on the recommendations of the staff and his or her own experience with the same factors considered by the staff (Jones 1999). DARPA has chosen COA critiquing as one of the challenge problems for the High Performance Knowledge Bases (HPKB) program. The goal of the HPKB program was to produce the technology needed to rapidly construct large knowledge-bases that provide comprehensive coverage of topics of interest, are reusable by multiple applications with diverse problem-solving strategies, and are maintainable in rapidly changing environments. The participating organizations were given the challenge of rapidly developing knowledge-based systems for that problem domain, and then modifying their systems quickly to solve further problems in the same domain. The aim of the exercise was to test the claim that, with the latest Artificial Intelligence technology, large knowledge bases can be built quickly and efficiently. The COA challenge problem is described in the next section.

3.1 The Course of Action Critiquing Problem

The COA challenge problem consists of rapidly developing a knowledge-based critiquing agent that can automatically critique COAs for ground force operations, can systematically assess selected aspects of a COA, and can suggest repairs to it. The role of this agent is to act as an assistant to the military commander, helping the commander to choose between several COAs under consideration for a certain mission. The agent could also help students to learn to develop courses of action.

The input to the COA critiquing agent consists of the description of a COA that includes the following three aspects:

1. The COA sketch, such as the one in the top part of Figure 11, is a graphical depiction of the preliminary plan being considered. It includes enough of the high level structure and maneuver aspects of the plan to show how the actions of each unit fit together to accomplish the overall purpose, while omitting much of the execution detail that will be included in the eventual operational plan. The three primary elements included in a COA sketch are: control measures which limit and control interactions between units; unit graphics that depict known, initial locations and make up of friendly and enemy units; and mission graphics that depict actions and tasks assigned to friendly units. The COA sketch is drawn using a palette-based sketching utility.
2. The COA statement, such as the partial one shown in the bottom part of Figure 11, clearly explains what the units in a course of action will do to accomplish the assigned mission. This text includes a description of the mission and the desired end state, as well as standard elements that describe purposes, operations, tasks, forms of maneuver, units, and resources to be used in the COA. The COA statement is expressed in a restricted but expressive subset of English.
3. Selected products of mission analysis, such as the areas of operations of the units, avenues of approach, key terrain, unit combat power, and enemy COAs.

Based on this input, the critiquing agent has to assess various aspects of the COA, such as its viability (suitability, feasibility, acceptability and completeness), its correctness (array of forces, scheme of maneuver, command and control), and its strengths and weaknesses with respect to the principles of war and the tenets of army operations, to justify the assessments made and to propose improvements to the COA.

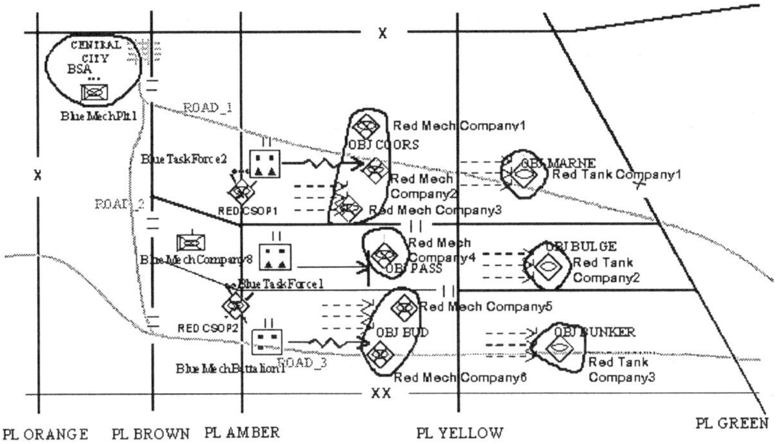

Mission: BLUE-BRIGADE2 attacks to penetrate RED-MECH-REGIMENT2 at 130600 Aug in order to enable the completion of seize OBJ-SLAM by BLUE-ARMOR-BRIGADE1.

Close: BLUE-TASK-FORCE1, a balanced task force (MAIN-EFFORT) attacks to penetrate RED-MECH-COMPANY4, then clears RED-TANK-COMPANY2 in order to enable the completion of seize OBJ-SLAM by BLUE-ARMOR-BRIGADE1.
BLUE-TASK-FORCE2, a balanced task force (SUPPORTING-EFFORT1) attacks to fix RED-MECH-COMPANY1 and RED-MECH-COMPANY2 and RED-MECH-COMPANY3 in order to prevent RED-MECH-COMPANY1 and RED-MECH-COMPANY2 and RED-MECH-COMPANY3 from interfering with conducts of the MAIN-EFFORT1, then clears RED-MECH-COMPANY1 and RED-MECH-COMPANY2 and RED-MECH-COMPANY3 and RED-TANK-COMPANY1. ...

Reserve: The reserve, BLUE-MECH-COMPANY8, a mechanized infantry company, follows Main Effort (MAIN-EFFORT1), and is prepared to reinforce MAIN-EFFORT1.

...

Figure 11. A sample of a COA sketch and a fragment of a COA statement.

The COA challenge problem was solved by developing an integrated system composed of several critiquers, each built by a different team, to solve a part of the overall problem. The teams were Teknowledge-Cycorp, the Expect team from the Information Science Institute of the University of Southern California (ISI/Expect), the Loom group from the Information Science Institute (ISI/Loom), and Disciple group from George Mason University (GMU/Disciple). All these teams shared an input ontology and used the same internal representation of the input generated by Teknowledge, Artificial Intelligence Applications Institute of the University of Edinburgh, and Northwestern University, from COA descriptions provided by Alphatech.

3.2 The Disciple-COA Critiquer

GMU has developed a COA critiquer, called Disciple-COA, that identifies the strengths and the weaknesses of a course of action with respect to the principles of war and the tenets of army operations (Field Manual 100-5 1993).

There are nine principles of war: objective, offensive, mass, economy of force, maneuver, unity of command, security, surprise, and simplicity. They provide general guidance for the conduct of war at the strategic, operational and tactical levels. The tenets of army operations describe the characteristics of successful operations. They are: initiative, agility, depth, synchronization and versatility. Figure 12, for instance, shows some of the strengths of the COA from Figure 11 with respect to the Principle of Mass, identified by Disciple-COA.

In addition to generating answers in natural language, Disciple also provides the reference material based on which the answers are generated, as shown in the bottom of Figure 12. Also, the Disciple-COA agent can provide justifications for the generated answers at three levels of detail, from a very abstract one that shows the general line of reasoning followed, to a very detailed one that indicates each of the knowledge pieces used in generating the answer.

The goal of Disciple is to learn to reason in a similar way with the military expert, combining by-the-book doctrine and tactics with the lessons learned from him. The Disciple agent would not replace the commander or staff, but with no loss of capacity due to stress or other environmental factors, it would be a useful knowledge-based assistant, providing concise, relevant and explainable considerations that the commander can take into account when making decisions.

Assess COA411 with respect to the Principle of Mass

There is a major strength in COA411 with respect to mass because BLUE-TASK-FORCE1 is the MAIN-EFFORT1 and it acts on the decisive point of the COA (RED-MECH-COMPANY4) with a force ratio of 10.6, which exceeds a recommended force ratio of 3.0. Additionally, the main effort is assisted by supporting action SUPPRESS-MILITARY-TASK1 which also acts on the decisive point. This is good evidence of the allocation of significantly more than minimum combat power required at the decisive point and is indicative of the proper application of the principle of mass.

There is a strength in COA411 with respect to mass because BLUE-TASK-FORCE1 is the main effort of the COA and it has been allocated 33% of available combat power but this is considered just a medium level weighting of the main effort.

There is a strength in COA411 with respect to mass because BLUE-MECH-COMPANY8 is a COMPANY-UNIT-DESIGNATION level maneuver unit assigned to be the reserve. This is considered a strong reserve for a BRIGADE-UNIT-DESIGNATION level COA and would be available to continue the operation or exploit success.

Reference: FM 100-5 pg 2-4, KF 113.1, KF 113.2, KF 113.3, KF 113.4, KF 113.5 - To mass is to synchronize the effects of all elements of combat power at the proper point and time to achieve decisive results. Observance of the Principle of Mass may be evidenced by allocation to the main effort of significantly greater combat power than the minimum required throughout its mission, accounting for expected losses. Mass is evidenced by the allocation of significantly more than minimum combat power required at the decisive point.

Figure 12. Solutions generated by Disciple-COA.

For Disciple-COA, an initial ontology was defined by importing the input ontology built by Teknowledge and Cycorp for the COA challenge problem. The input ontology contains the terms needed to represent the COAs to be critiqued, and was shared by all the developed critiquers. The top level of this ontology includes concepts for representing geographical information, military organizations and equipment, descriptions of specific COAs, military tasks, operations and purposes. The ontology was further developed by using the ontology tools of Disciple.

After the COA ontology has been developed, Disciple-COA was taught to critique COAs with respect to the principles of war and the tenets of army operations. First, the expert formulates the critiquing task to be performed, for instance, *Assess COA411 with re-*

spect to the Principle of Mass. To perform this assessment one needs a certain amount of information that is obtained by asking a series of questions. The answer to each question allows one to reduce the current assessment task to a more specific one, as illustrated in Figure 13. From each task reduction step in Figure 13, Disciple has learned a task reduction rule, in a way that is similar to the process explained in Section 2.3.

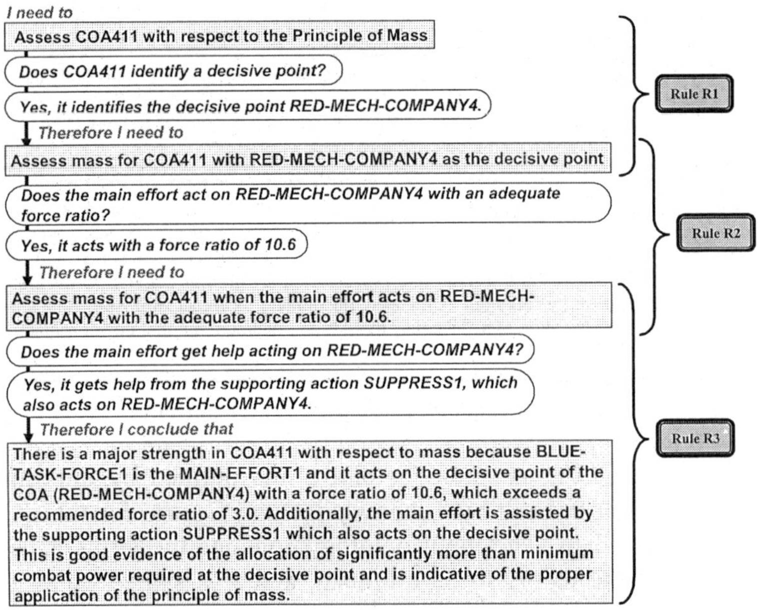

Figure 13. COA critiquing through task reduction.

While the critiquer developed by GMU was based on the Disciple approach, the other three critiquers were based on different approaches. Teknowledge and CYC have developed a critiquer based on the CYC system (Lenat 1995). The ISI/Expect group developed a critiquer based on the Expect system (Kim and Gil 1999), and ISI/Loom group developed a critiquer based on the Loom and PowerLoom systems (MacGregor 1999). All the critiquers were evaluated as part of the HPKB's annual evaluation that took place during the period July 6-16, 1999, and included five evaluation items of increasing difficulty. Each item consisted of descriptions of various COAs and a set of questions to be answered about each of them. Item1 consisted of COAs and questions that were previously provided by DARPA to guide the development of the COA critiquing agents. Item2 included new test questions about the same COAs. Items 3, 4, and 5 consisted of new COAs that were increasingly more complex and required further development of the COA agents in order to properly answer the asked questions. Each of the Items 3, 4 and 5 consisted of two phases. In the first phase each team had to provide initial system responses. Then the evaluator issued the model answers and each team had a limited amount of time to repair its system, to perform further knowledge acquisition, and to generate revised system responses.

Figure 14 compares the recall and the coverage of the developed critiquers for the last three most complex items of the evaluation. For each item, the beginning of each arrow shows the coverage and recall for the initial testing phase, and the end of the arrow shows the same data for the modification phase. The best results are those from the upper-right corner of the graph. This graph shows that all the systems increased their coverage during the evaluation. In particular, the KB of Disciple increased by 46% (from the equivalent of 6229 simple axioms to 9092 simple axioms), which represents a very high rate of knowledge acquisition of 286 simple axioms/day. The final knowledge base contained 801 concepts, 444 object and task features, 360 tasks and 342 task reduction rules. Also, each COA was represented with around 1,500 facts.

4 Workaround Reasoning

Workaround reasoning is another domain used in the DARPA's High Performance Knowledge Bases Program to focus the research and development efforts and measure the effectiveness of alternative technical approaches to knowledge base and agent development. Workaround reasoning consists of estimating enemy's best way of working around damage to an infrastructure, such as a damaged bridge or a cratered road. Workaround reasoning supports air campaign planning in two ways. First it facilitates the evaluation of targeting strategies aimed at disabling enemy infrastructure systems. Second, it provides information about enemy assets that should be preemptively targeted to impede its efforts to work around battle damage. The specific workaround challenge problem addressed is presented in the next section.

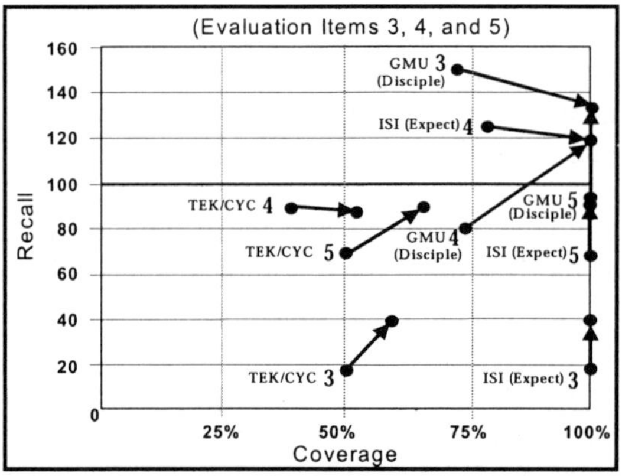

Figure 14. Coverage vs. recall pre- and post-repair.

4.1 The Workaround Reasoning Challenge Problem

The Workaround Reasoning problem consists of rapidly developing a special knowledge-based planning agent that can estimate which is the best way for an enemy unit to workaround some damage in its path (Jones 1998).

The input to the agent includes two elements:

1. a description of the damage (e.g., a bridge is destroyed – see Figure 15), and of the terrain (e.g., the soil type, the slopes of the river's banks, the river's speed, depth and width);
2. a detailed description of the resources in the area that could be used to repair the damage (e.g., a description of the engineering assets of the military unit that has to workaround the damage, as well as the descriptions of other military units in the area that could provide additional resources).

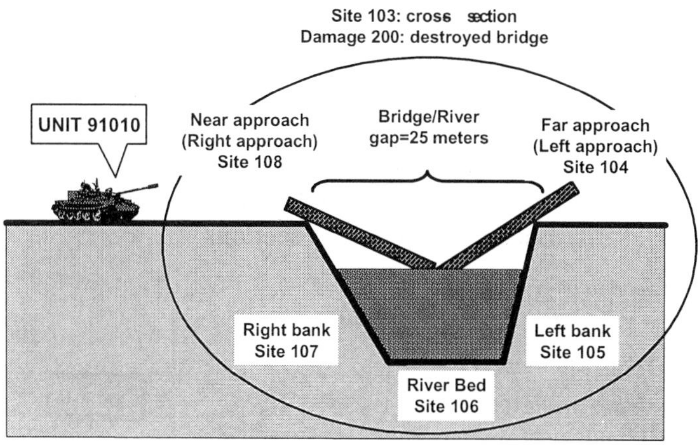

Figure 15. An example of a workaround problem.

The output of the agent consists of the most likely repair strategies, each described in terms of three elements:
1. a reconstitution schedule, giving the transportation capacity of the damaged link (bridge, road or tunnel), as a function of time, including both a minimum time and an expected time;
2. a partially ordered plan of engineering actions to perform the repair, and the minimum as well as the expected time that each of these actions require;
3. a set of required resources for the entire plan and for each action.

To solve this problem we have developed the Disciple-Workaround learning agent (Tecuci *et al.* 1999), as discussed in the next section.

4.2 The Disciple-Workaround Planner

The Disciple-Workaround planning agent was developed very rapidly. First, an initial object ontology was built by representing knowledge from Military Engineering manuals. After that, Disciple was taught based on sample solutions provided by the Alphatech's domain expert, by using an approach similar to that presented in the previous sections. For instance, the teacher defines a specific workaround task (such as, "Workaround obstacle by unit10" in Figure 16). Then he asks a question related to this task, the answer of which leads to the reduction of this task to a simpler one (or, in other cases, to several simpler tasks). This process continues until the top level task is reduced to a partially ordered sequence of elementary tasks that represent the plan to workaround the obstacle. From each task reduction step Disciple learns a general task reduction rule, as discussed in the previous sections.

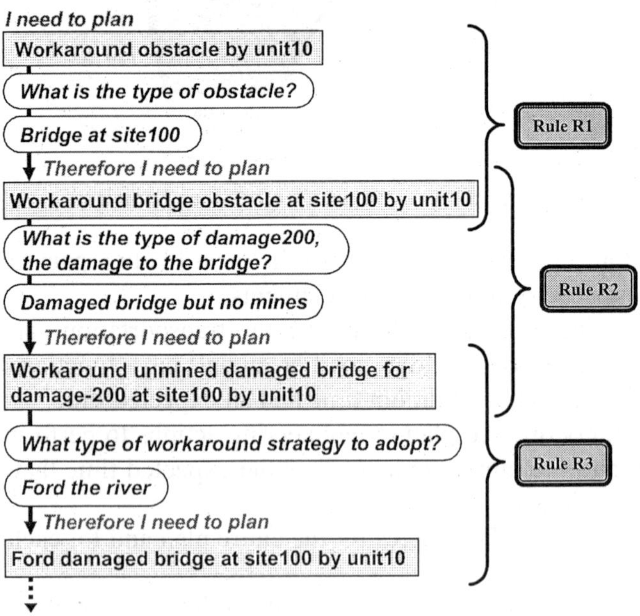

Figure 16. Sample task reduction tree.

Disciple-workaround was evaluated together with three other planners, one developed by the Information Science Institute, one developed by Teknowledge, and one developed the Artificial Intelligence Applications Institute. The planning agents were tested on several sets of problems. Then the model solutions of the problems were provided, and the agents needed to be updated in a limited amount of time, to improve their performance (Cohen *et al.* 1998). The problem illustrated in Figure 15 was one of the testing problems. The damage is a destroyed bridge and UNIT91010 has to work around it, to cross the river. The best solution generated by Disciple is shown in Figure 17, and consists of installing an AVLB bridge over the river gap. It is estimated that this will take a minimum of 11h:4m:58s, the expected duration being 14h:25m:56s. UNIT91010 will need the help of UNIT202, which has AVLB equipment, and of UNIT201, which has a bulldozer. After the bridge is installed, it will allow a traffic rate of 135.13 vehicles/h. The plan consists of 12 elementary actions. UNIT91010 has to obtain operational control of UNIT202 which has the AVLB. Then this unit has to come to the site of the destroyed bridge. Also, UNIT91010 has to obtain operational control of UNIT201 which has a bulldozer. This unit will have to move to the site of the destroyed bridge and then to narrow the river gap from 25m to 17m. These actions can take place in parallel with the actions of bringing UNIT202 to the bridge site. Then the AVLB bridge is emplaced, the bulldozer moves over the bridge and clears the other side of the river to restore the flow of traffic. This plan was generated by successively reducing the WORKAROUND-OBSTACLE task to simpler subtasks, until this task was reduced to the 12 tasks in Figure 17.

During the 17 days of DARPA's evaluation, the knowledge base of Disciple was increased by 72%, from the equivalent of 5,920 simple axioms to 10,162 simple axioms. It was extended with 147 concepts, 104 tasks and 206 task reduction rules, more than any other system. The Disciple-Workaround agent has also achieved the best correctness among all the teams that participated in the workaround challenge problem, and was selected to represent the HPKB program at EFX'98, the Air Force's show case of the most promising technologies.

TASK: WORKAROUND-OBSTACLE
BY UNIT91010

ENGINEERING ACTION:	INSTALL AVLB
MIN-DURATION	11H:4M:58S
EXPECTED-DURATION	14H:25M:56S
RESOURCES REQUIRED	(AVLB-UNIT202 BULLDOZER-UNIT201)
LINK CAPACITY	135.13 VEHICLES/HR

DETAILED PLAN:

S1 OBTAIN-OPERATIONAL-CONTROL-FROM-CORPS	
OF-UNIT	UNIT202
BY-UNIT	UNIT91010
MIN-DURATION	4H:0M:0S
EXPECTED-DURATION	6H:0M:0S
TIME-CONSTRAINTS:	NONE

S2 MOVE-UNIT	
FOR-UNIT	UNIT202
FROM-LOCATION	SITE0
TO-LOCATION	SITE100
MIN-DURATION	1H:8M:14S
EXPECTED-DURATION	1H:8M:14S
TIME-CONSTRAINTS:	AFTER S1

S3 REPORT-OBTAINED-EQUIPMENT	
FOR-EQ-SET	AVLB-UNIT202
MIN-DURATION	0S
EXPECTED-DURATION	0S
TIME-CONSTRAINTS:	AFTER S2

S4 OBTAIN-OPERATIONAL-CONTROL-FROM-CORPS	
OF-UNIT	UNIT201
BY-UNIT	UNIT91010
MIN-DURATION	4H:0M:0S
EXPECTED-DURATION	6H:0M:0S
TIME-CONSTRAINTS:	NONE

S5 MOVE-UNIT	
FOR-UNIT	UNIT201
FROM-LOCATION	SITE0
TO-LOCATION	SITE100
MIN-DURATION	1H:8M:14S
EXPECTED-DURATION	1H:8M:14S
TIME-CONSTRAINTS:	AFTER S4

S6 REPORT-OBTAINED-EQUIPMENT	
FOR-EQ-SET	BULLDOZER-UNIT201
MIN-DURATION	0S
EXPECTED-DURATION	0S
TIME-CONSTRAINTS:	AFTER S5

S7 NARROW-GAP-BY-FILLING-WITH-BANK	
FOR-GAP	SITE103
FOR-BR-DESIGN	AVLB70
MIN-DURATION	5H:19M:44S
EXPECTED-DURATION	6H:7M:42S
RESOURCES REQUIRED	BULLDOZER-UNIT201
TIME-CONSTRAINTS:	AFTER S6

S8 EMPLACE-AVLB	
FOR-BR-DESIGN	AVLB70
MIN-DURATION	5M:0S
EXPECTED-DURATION	10M:0S
RESOURCES-REQUIRED	AVLB-UNIT202
TIME-CONSTRAINTS:	AFTER S3, S7

S9 REPORT-EMPLACED-FIXED-BRIDGE	
FOR-MIL-BRIDGE	FIXED-MILITARY-BRIDGE-EQ
MIN-DURATION	0S
EXPECTED-DURATION	0S
TIME-CONSTRAINTS:	AFTER S8

S10 MOVE-EQUIPMENT-OVER-UNSTABILIZED-MIL-BRIDGE	
FOR-EQ-SET	BULLDOZER-UNIT201
FOR-BR-DESIGN	AVLB70
MIN-DURATION	2M:0S
EXPECTED-DURATION	10M:0S
RESOURCES-REQUIRED	AVLB-UNIT202
TIME-CONSTRAINTS:	AFTER S9

S11 MINOR-BANK-PREPARATION	
OF-BANK	SITE105
MIN-DURATION	30M:0S
EXPECTED-DURATION	50M:0S
RESOURCES-REQUIRED	BULLDOZER-UNIT201
TIME-CONSTRAINTS:	AFTER S10

S12 RESTORE-TRAFFIC-LINK	
FOR-UNIT	UNIT91010
FOR-LINK	AVLB70
LINK-CAPACITY	2.25 VEHICLES/MIN
MIN-DURATION	0S
EXPECTED-DURATION	0S
TIME-CONSTRAINTS:	AFTER S11

Figure 17. Sample workaround plan generated by Disciple-Workaround.

5 Long-Term Research Vision

The Learning Agents Laboratory of George Mason University is developing the Disciple theory, methodology and tools, for building instructable knowledge-based agents. This effort directly addresses the knowledge acquisition bottleneck which we consider to be one of the major barriers in the development and maintenance of Artificial Intelligence applications.

The military applications presented in this chapter have shown that Disciple has reached a significant level of maturity, being usable to rapidly develop complex knowledge based agents. However, these are only initial results of a long-term research the objective of which is to develop the Artificial Intelligence theory and practice that will change the way intelligent agents are built, from "being programmed" by a knowledge engineer to "being taught" by a user that does not have prior knowledge engineering experience. Achieving this objective will allow a normal computer user, who is not a trained knowledge engineer, to build by himself an intelligent assistant as easily as he now uses a word processor to write a paper. It is expected that this research will contribute to a new revolution in the use of computers, probably even more important than the creation of personal computers. Indeed, it will allow every person to no longer be only a user of programs developed by others, but also an agent developer himself.

Acknowledgments

The Disciple approach was started with the support of Mihai Draganescu, at the Research Institute for Informatics and the Romanian Academy. The research reported in this chapter was done in the GMU Learning Agents Laboratory (LALAB) with support from the Defense Advanced Research Projects Agency, Air Force Research Laboratory, Air Force Material Command, USAF, under agreement number F30602-00-2-0546, from the Air Force Office of Scientific Research grant number F49620-97-1-0188 and grant number F49620-00-1-0072, and from the US Army War College, and under the direction of David Gunning, Murray Burke, Robert Herklotz,

Alex Kilpatrick, William Rzepka, Douglas Campbell and David Cammons. Several members of the LALAB have contributed to the development of the versions of Disciple presented in this chapter, especially Michael Bowman, Dorin Marcu, Kathryn Wright, Bogdan Stanescu, and Cristina Boicu (Cascaval). Jerry Comello was the primary subject matter expert for the center of gravity application of Disciple.

References

Boicu, M., Tecuci, G., Marcu, D., Bowman, M., Shyr, P., Ciucu, F., and Levcovici, C. (2000), "Disciple-COA: from agent programming to agent teaching," *Proceedings of the Seventeenth International Conference on Machine Learning*, Stanford, CA, Morgan Kaufmann.

Boicu, M., Tecuci, G., Stanescu, B., Marcu, D., and Cascaval, C. (2001), "Automatic knowledge acquisition from subject matter experts," *Proceedings of the International Conference of Tools with Artificial Intelligence*, Dallas, Texas, November.

Burke, M. (2002), "Rapid Knowledge Formation Program Description," available at http://reliant.teknowledge.com/RKF/projects/Darpa_RKF_PIP.htm, as of July 14, 2002.

Chaudhri, V.K., Farquhar, A., Fikes, R., Park, P.D., and Rice, J.P. (1998), "OKBC: a programmatic foundation for knowledge base interoperability," *Proceedings of the Fifteenth National Conference on Artificial Intelligence*, Menlo Park, California: AAAI Press, pp. 600-607.

Clausewitz, C.V. (1832), *On War*, translated and edited by Howard, M. and Paret, P. (1993), Princeton, NJ: Princeton University Press.

Cohen, P., Schrag, R., Jones, E., Pease, A., Lin, A., Starr, B., Gunning, D., and Burke, M. (1998), "The DARPA High-Performance Knowledge Bases Project," *AI Magazine*, vol. 19, no. 4, pp. 25-49.

Field Manual 100-5 (1993), "Operations," Headquarters, Dept. of the Army, Washington, DC: U.S. Government Printing Office.

Field Manual 3-0 (2001), "Operations," Headquarters, Dept. of the Army, Washington, DC: U.S. Government Printing Office.

Giles, P.K. and Galvin, T.P. (1996), *Center of Gravity: Determination, Analysis and Application*, CSL, U.S. Army War College, PA: Carlisle Barracks.

Gunning, D. and Burke, M. (2002), "High Performance Knowledge Bases (HPKB) Program Description," available at http://reliant.teknowledge.com/HPKB/about/about.html, as of July 14, 2002.

Joint Publication 3-0 (1993), "Doctrine for Joint Operations," Dept. of Defense, Washington, DC: U.S. Government Printing Office.

Jones, E. (1998), "HPKB Year 1 End-to-End Battlespace Challenge Problem Specification," Burlington, MA.

Jones, E. (1999), "HPKB Course of Action Challenge Problem Specification," Alphatech, Inc., Burlington, MA.

Kim, J., and Gil, Y. (1999), "Deriving expectations to guide knowledge base creation," *Proceedings of the Sixteenth National Conference on Artificial Intelligence*, Menlo Park, California: AAAI Press, pp. 235-241.

Lenat, D.B. (1995), "CYC: a large-scale investment in knowledge infrastructure," *Communications of the ACM*, vol. 38, no. 11, pp. 33-38.

MacGregor, R. (1999), "Retrospective on LOOM," available at http://www.isi.edu/isd/LOOM/papers/macgregor/Loom_Retrospective.html .

Strange, J. (1996), *Centers of Gravity & Critical Vulnerabilities: Building on the Clausewitzian Foundation So That We Can All Speak the Same Language*, Quantico, VA: Marine Corps University Foundation.

Tecuci, G. (1998), *Building Intelligent Agents: an Apprenticeship Multistrategy Learning Theory, Methodology, Tool and Case Studies*, Academic Press, London, England.

Tecuci, G., Boicu, M., Wright, K., Lee, S.W., Marcu, D., and Bowman, M. (1999), "An integrated shell and methodology for rapid development of knowledge-based agents," *Proc. AAAI/IAAI*, AAAI Press, Menlo Park CA, pp. 250-257.

Tecuci, G., Boicu, M., Bowman, M., Marcu, D., Shyr, P., and Cascaval, C. (2000), "An experiment in agent teaching by subject matter experts," *International Journal of Human-Computer Studies*, vol. 53, pp. 583-610.

Tecuci, G., Boicu, M., Bowman, M., and Marcu, D., (2001), "An innovative application from the DARPA knowledge bases programs: rapid development of a high performance knowledge base for course of action critiquing," *AI Magazine*, vol. 22, no. 2, pp. 43-61. With a commentary by M. Burke.

Tecuci, G., Boicu, M., Marcu, D., Stanescu, B., Boicu C., Comello, J., Lopez, A., Donlon, J., and Cleckner, W. (2002), "Development and deployment of a disciple agent for center of gravity analysis," *Fourteenth Annual Conference on Innovative Applications of Artificial Intelligence, IAAI-2002*, Edmonton, Alberta, Canada. Deployed Application Award.

Chapter 9

A Parallel Associative Processing Bayesian Algorithm for IR Dim Point Target Detection

R.C. Warren

Methods for detecting anti-ship sea-skimming missiles by staring infrared search and track (IRST) systems will be presented. The aim is to detect targets at long ranges over the ocean. Target detection algorithms will be described which would be effective and efficient when implemented on massively parallel processors with limited processing and memory capacities.

For the case of uncluttered backgrounds a Bayesian track-before-detect (TBD) target detection algorithm is presented, and it is tested on a real target image sequence. The TBD method gives better performance than the standard thresholding method. Small support image filters based on mathematical morphology are used to enhance target detectability and they are suitable for implementation on a massively parallel processor.

None of the standard approaches is able to give sufficient discrimination between the target and background in heavily cloud cluttered backgrounds to allow reliable detection. However, target scintillation is a physical phenomenon which offers the possibility of discrimination between the target from the background in the circumstances of interest. A sufficiently detailed understanding of the scintillation of point targets does not exist, so a phenomenological model of the scintillation is used to generate synthetic sequences of images of point targets in cloud backgrounds.

A new combination of filters is proposed to detect targets by their scintillation. The method is based on taking the absolute differences between adjacent frames in the sequence, but this procedure alone is not sufficient to suppress the clutter because sensor jitter causes spatially correlated clutter to appear on the difference images. It has previously been claimed that it was necessary to register the images before subtraction, and that this would be too computationally demanding. However, two small support filters are able to be used to suppress this clutter. One filter is based on a 2 point mean, and the other is a 2 point morphological opening filter.

The performance of the new filters is evaluated using synthetically generated image sequences. A detection and pre-tracking program has been written which was intended to be implemented on a parallel processor. A series of runs with varying values of the parameters of the detection program was used to determine optimum performance of each filter. The 2 pixel opening filter was found to have the best performance.

The new algorithm had sufficient performance to be considered for IRST applications and was suitable for massively parallel processing.

1 Introduction

Infrared search and track (IRST) systems have been developed for land, air and sea based applications. While these have much in common, each application has specific requirements and operating modes which affect its design and performance. The subject of this work is IRST in the maritime environment, and in particular ship self defense. There exists a requirement to improve the capability of surface ships to detect the approach of anti-ship sea-skimming missiles (ASMs) when a quiet electromagnetic environment must be maintained, or when radar is inadequate.

Current generation IRST systems are based on scanning sensors where the whole sensor head, which includes the optics and possibly the detectors, rotates on a mount on the mast. The update rates are of the order of 1 Hz. The scanning method limits the time spent at any one pixel to the order of microseconds, which gives a low signal level and a poor signal to noise ratio (SNR). A number of summaries of IRST system technology have been published. De Jong (1995) reviewed most of the factors relevant to IRST systems. A discussion of the IRST system principles is given in the IR/EO Systems Handbook (Accetta 1993), and by Blackman and Popoli (1999).

An IRST system which has had a long development time and is in service is the Vampir MB from SAGEM. Missirian and Ducruet (1997) discussed the use of IRST for ASM defense, and also some of the experience gained from the first generation Vampir systems in service. Technical details of the system were given by Missirian and Ducruet and by Scott and Lok (1996). The sensor head scans at a rate of 90 rpm with a 5° vertical field of view (FOV). The long wave IR (LWIR, 3-5 μm) and mid wave IR (MWIR, 8-12 μm) sensors are Sofradir 288x4 focal plane arrays (FPAs) with time delay integration. Scott and Lok state that up to 100 targets can be tracked simultaneously, and that it is predicted that subsonic missiles could be tracked at 16 km, fighter aircraft at 18 km, and supersonic missiles at 27 km.

A 'low cost' IRST system has been reported by Gaussorgues and Campos (2000), based on the SOFRADIR LWIR 288x4 FPA. It has an instantaneous field of view (IFOV) of 0.23 mrad by 0.25 mrad, and a vertical FOV of 4.2°. The revisit rate is 1 per second.

It is being realized that IRST systems can offer capabilities for horizon surveillance that that are not able to be provided by radar (Horman *et al.* 1999). ASMs are being developed with very low radar cross sections and the ability to fly very close to the water.

Most radar beams have diameters greater than 1°, so when they are used to search the horizon their field of view will include a large amount of ocean which would return a strong clutter response. The much finer resolution of the IRST would allow only the sky above the horizon to be scanned, where the clutter is much reduced. The best use of radar would be to confirm possible targets provided by the IRST, and a greater proportion of the radar energy and time could be spent interrogating a small number of possible targets. Because IRST false alarms could be eliminated by the radar, the sensitivity of the IRST could increased to improve detection performance. A combination of IRST and radar would give the weapons officer the accurate azimuth and elevation of the target, and the range and range rate, which would enable the optimum engagement scenario to be determined. A further discussion of the use of an IRST system and its interaction with radar has been given by Missirian and Ducruet (1997).

A target of concern for IRST is the stealthy subsonic turbojet ASM, which would have a very low IR signature and be very difficult to detect. However its subsonic speed means that detection range can be shorter and still give sufficient time for a response. Subsonic missiles can also be rocket powered. Another target is the supersonic rocket powered missile. The detection range needs to be longer in this case, but the missile signature is much higher due to the greater surface heating and exhaust gas emission. Aircraft can also be a threat, but they are often maneuvering or traveling with a tangential velocity, and are easier to detect.

There are various sources of noise and background clutter which interfere with the detection of ASMs. There is photon noise caused by the limited capacity of the pixel photodetectors, quantization noise in the detectors, and electronic noise. There is clutter from solar scattering from low clouds and possible distant land masses. Even in a clear sky there is the variation of solar radiance with azimuth and elevation from solar scattering by atmospheric molecules and aerosols. High levels of background radiance can lead to a reversal of contrast of the target.

In addition to the background there are false targets such as sunglints from cloud edges, buoys and birds. For locations near land, birds can be a serious source of false target detections.

It is necessary to detect ASMs at maximum possible range, ideally as soon as they appear over the horizon, and the extreme range of the missile means that the projected target image would be subpixel in size and its angular motion would be very small, and it will be assumed to be zero. The problem is essentially the detection of ultra-dim, stationary, point-like targets in clutter in sequences of IR images.

The next generation of IRST systems will be based on staring sensors operating at frame rates of 25-50 frames/s. The main problem in designing a staring IRST for anti-ship missile defense (ASMD) is the mapping of the thin ribbon of the horizon field of view (FOV) to the nominally square FPA. Continuous scanning systems achieve this mapping easily at the cost of very short exposure time at any point (pixel). The staring systems will have much longer integration times and higher update rates, and so will have a considerable increase in detection and tracking capability. The higher frame rate would also allow the temporal properties of targets and backgrounds to be used for target detection. It must be recognized that in comparing scanning and staring systems the staring systems would use considerably more sensor hardware and processing resources than current scanning systems.

An approach to the mapping problem for staring systems is to split the horizon FOV into segments and have a dedicated sensor for each segment. A second generation IRST will possibly be based on a focal plane array (FPA) of 2000×2000 pixels. Using an anamorphic lens with a ratio of 3 to 1, a field of view 60°×20° can be mapped onto the FPA, with each pixel representing a horizontal rectangle with aspect ratio 3×1. This would give a resolution of 100 pixels/° in the vertical direction and 33 pixels/° in the horizontal direction. To obtain an estimate of the processing power required for ASM detection, assume that the mapping to the IRST FPA is 100 pixels/°, that a target has to be detectable from the horizon to 10 km, and that the maximum height of interest is 20 m. This situation corresponds to a vertical angular range of 2 mrad, or 12 pixels. If 6 sensors are used to cover the whole horizon, the horizontal angular range for each sensor is 60°, or 1980 pixels. The total number of pixels for each sensor to be analyzed is then 24,000.

This number of pixels makes it feasible to use a massively parallel processor, and to associate a single processing element with each pixel. In this case each pixel-processing element becomes an individual sensor. The use of parallel processing for this problem necessitates a different view of processing methods, and different algorithms may become more efficient than those used in the serial processing case. Algorithms will be considered which will use spatial filters to reduce the magnitude of clutter and temporally decorrelate the residual clutter, and a Bayesian track-before-detect process for target detection. The algorithms were either directly implemented on a parallel processor, or their parallel implementation was emulated in the C language on a PC.

The natural processing methods for massively parallel processors are the associative processing paradigm and mathematical morphology, and these are discussed in Sections 2.1 and 2.2. Elements of image filtering and target detection theory are discussed in Section 2.3. A detection method which is particularly suited to staring IRST and parallel processing is a Bayesian method, which is discussed in Section 2.4. The performance of detection methods for targets in low clutter backgrounds is assessed in Section 3, and methods for cluttered backgrounds are discussed in Section 4. Overall conclusions are given in Section 5.

2 Background Theory

2.1 Associative Processing

Image processing lends itself naturally to parallel processing. An image to be processed can be considered to consist of an array of nominally identical read-only memory elements, and the data in the elements constitutes the pixel values of the image. Image processing consists of transforming these data into different data in a separate read/write memory with the same topological structure as the initial memory. This conceptual link between pixels and memory makes it natural to establish a one-to-one link between the pixels of an image and the processing element (PE) memories of a massively parallel processor (MPP).

An objection often raised to parallel processing is Amdahl's law, which relates the speed up factor for a combined serial and parallel process to the number of processors and the proportion of time spent in the serial mode (Amdahl 1967). The general result is that only a small amount of time spent in the serial mode negates the effect of a large number of parallel processors. However, it is envisaged that the filtering and detection will be done in the parallel mode, so Amdahl's Law does not apply.

The history and types of MPPs were briefly reviewed by Lea (1990). Early MPPs did not perform to expectations for a number of reasons, but two reasons were the mismatch of processor communication to the application, and the limitations caused by processor communication complexity. There are three types of processor connectivity commonly used; 4 and 8 mesh, and string connectivity. Four or 8 mesh connectivity would provide natural communication between PEs for pixel based applications, and this would promote simple programming. However, for large scale processors the mesh connectivity becomes very complex, and scaling up becomes difficult and expensive. The string connectivity has the advantage of easy scalability as it is just necessary to add more units to the chain. String processors can use sophisticated memory transfers and inter-PE communication to overcome some of the disadvantages of simpler connectivity.

Many processing operations are determined by the values of the data in the individual PE and not on their location in the image. This concept is equivalent to associative processing, which is sometimes referred to as content-address processing. Up to the mid-1990s there was much research on associative processing, in particular by Potter and his group at Kent State University, USA (1994). They developed an associative computing paradigm designated ASC, which was a total parallel processing paradigm in opposition to the von Neumann type serial processing paradigm (Potter 1992). While this paradigm appeared to be intriguing for research into associative programming, it was too general and diverse for the IRST application, which only requires simple operations.

The processor used in this work was the System V Associative String Processor (ASP) from Aspex Microsystems, High Wycombe, Buckinghamshire, UK. The system used had 14,372 active PEs. The ASP was chosen for investigation because of its relatively low unit cost, inherent scalability and software support. As its name indicates, the ASP is string connected. The Aspex processor consists of 1bit adders and 70 bit wide comparators as processing elements, and 70 bits of memory per processor. A general background to the type of processing used by this type of processor is given by Krikelis and Weems (1994), and specific computer vision applications for the ASP are given by Krikalis (1991).

While processing in the ASP does not rely explicitly on the pixel address, the address of individual pixels can be tracked by the mapping of the pixel array to string processors when loading and dumping the data to and from the ASP.

2.2 Mathematical Morphology

2.2.1 Background

Mathematical morphology is reported by Ritter *et al.* (1990) to have originated from the work of Minkowski (1903) and Hadwiger (1957) on measure theory, through the work of Matheron (1975) and Serra (1982). The initial application was to binary images, and mathematical morphology was first extended to grayscale images by Sternberg (1986). Grayscale morphology can also be defined in terms of binary morphology of the threshold decomposition of the grayscale function, which may be useful in some applications (Shih and Mitchell 1989). Subsequently there has been a large amount of work extending the theory and applications of the subject, and some of this work is listed below.

Much work on the theoretical basis of mathematical morphology has been carried out in Brazil by the group associated with Barrera and Salas (1996), and a toolbox was developed for the KHOROS image processing environment (Barrera *et al.* 1994).

Mathematical morphology can be used to generate size distributions of both binary and grayscale objects and the process is called granulometry (Dougherty *et al.* 1989, 1992, Vincent 1994). Matheron conceived granulometries as sieves where particles in a image are filtered according to sizes (1975). Granulometric distributions can be used as a basis for texture measurement and segmentation.

Morphological gradients are useful in a number of aspects of image processing, including edge detection and segmentation (Rivest *et al.* 1993). They are a non-linear equivalent of differentiation. A non-linear equivalent of normal interpolation is morphological interpolation, which has been described by Salembier *et al.* (1996).

Morphological filtering covers a wide range of operations, from the application of basic operators, to segmentation, texture measurement, noise removal and image coding. The top hat filter was initially introduced as the rolling ball algorithm by Sternberg (1986). If an image is considered as a topographical relief map, the operation is like a small diameter ball rolling beneath the surface. The ball is not able to enter small diameter upward protrusions from the surface, so the surface generated by the motion of the ball center would correspond to a smoothed version of the original. The top hat version is when a flat element is pushed up beneath the surface and moved over the whole surface, and the output is given by the difference between the original surface and the surface swept out by the origin of the element. The results would depend on the size and shape of the element. Methods for implementing the algorithm will be given in Section 3.1.2.

Image algebra was a response to a call by the US Air Force Armament Laboratory and the Defense Advanced Research Projects Agency for a systematic common processing language to describe all image processing. Image algebra was developed by Ritter and his group at the University of Florida. It was claimed that "mathematical morphology, as used in actual image processing, is a special substructure of the mathematical structure represented by the image algebra" (Ritter *et al.* 1990).

The image algebra was naturally suited for implementation on massively parallel processors and there were a number of papers describing implementation on single instruction multiple data (SIMD) mesh connected computers for binary images (Shi *et al.* 1997), and grayscale images (Ritter and Gader 1985), but it was stated that for efficient implementation it would require far more than 128 bits per memory cell. Despite the apparent advantages of image algebra, and a number of papers from Ritter's group, it has failed to be adopted by the general image processing community.

Mathematical morphology has been used in the military context as part of automatic target recognition (ATR) algorithms. Braga-Neto and Goutsias (1999) used mathematical morphology filtering for background removal and binarization of targets for tracking targets in forward-looking infrared (FLIR) images. Morphology and constant false alarm rate operations were combined for ATR in FLIR images by Pham *et al.* (1997). Peli *et al.* (1993) used the watershed algorithm for segmentation of targets and backgrounds in FLIR images. Schmalz and Ritter (1994) reported on the use of image algebra for ATR of small targets in multispectral image sequences. These applications have involved segmenting extended targets from the background, and are not directly applicable to ASM detection by IRST systems.

2.2.2 Mathematical Morphology for ASM Detection by IRST

A brief discussion is given below of the basics of mathematical morphology used in this work, based on the overview by Sternberg (1986). Mathematical morphology emphasizes the shape aspects of image analysis, and was originally applied to binary images. The foundation of binary mathematical morphology is set theory, and some required definitions are given below.

A binary image will be represented by a 2D set X in the discrete plane Z^2, where Z is the set of integers. The translation of set X by a distance b is denoted

$$X_b = \{b + x : x \in X\},$$

and the reflection of the set X about the origin is denoted

$$X' = \{-x : x \in X\}.$$

The complement of the set X is given by

$$X^c = \{x \in Z^2 : x \notin X\}.$$

The fundamental operations in mathematical morphology are dilation and erosion. The dilation of the set X by another small set B, called a structuring element, is defined as

$$X \oplus B \equiv \bigcup_{b \in B} X_b = \{x + b : x \in X, b \in B\}.$$

The erosion of set X by another small set B, is defined as

$$X \Theta B \equiv \bigcap_{b \in B} X_{-b} = \bigcap_{b \in B'} X_b = \{p : b + p \in X, b \in B\}.$$

Each of these operations involves some loss of information which cannot be recovered, and they are not inverse but dual, i.e.,

$$X \oplus B = (X^c \Theta B)^c.$$

Two further basic operations are opening and closing, which are defined in terms of erosion and dilation. Opening is defined as an erosion followed by a dilation with the same structuring element:

$$X \circ B \equiv (X \Theta B) \oplus B,$$

and closing is defined as

$$X \bullet B \equiv (X \oplus B) \Theta B.$$

The preceding operations have been generalized to cover grayscale images. A grayscale image will be represented by a 2D function with domain in Z^2 and range in R, where R is the set of real numbers.

Grayscale dilation of function $f(x)$ with grayscale structuring element $g(x)$ is defined as:

$$f \oplus g(x) \equiv \max\{f(x - y) + g(y) : y \in Z^2\}$$

and erosion is defined as:

$$f\Theta g(x) \equiv \min\{f(x+y) - g(-y) : y \in Z^2\}$$

Opening and closing are defined in the same way as for the binary image case.

If the structuring element $g(x)$ is flat, the definitions can be simplified as

$$f \oplus g(x) = \max\{f(x-y) : y \in G\}, \text{ and}$$

$$f\Theta g(x) = \min\{f(x+y) : y \in G\},$$

where G is the support of $g(x)$.

The opening operation corresponds to pushing the structuring element $g(x)$ up against the lower surface of the function $f(x)$, and the surface swept out by the origin of the structuring element is the opening. The top hat filter is then given by:

$$TH = f(x) - f \circ g(x).$$

A top hat filter based on a union of structuring elements has been used for detection of dim point targets in cloud backgrounds (Horak and Mattioli (1996). The characteristics of a background image were learned by opening with a series of small, 5×5, structuring elements, and choosing the elements which minimized the difference between the original and opened image. The union of elements was used as the basis for a top hat filter for other images which may have contained a target.

2.3 Filtering and Detection

There have been many approaches to the problem of detection of dim point targets. Linear spatial filters are often used in airborne and maritime IRST systems. Spatial 2D filters are applied to each frame separately to increase the signal to clutter ratio, and the output is thresholded to indicate the presence of possible targets. The aim of linear matched filtering is to generate an array of predicted pixel values made up of linear combinations of surrounding pixels, on the assumption that the surrounding pixels are correlated with the pixel of interest. A comparison of the predicted image with the original image highlights possible targets. An optimal filter is one that maximizes some performance parameter. The most useful performance parameter is the detection probability for a fixed false alarm rate, but unless the image pixel values follow a Gaussian distribution, the calculation of linear filter weights is not tractable. A more useable parameter is the signal to noise ratio (SNR) gain, or signal to clutter ratio (SCR) ratio gain, obtained from the filter.

A filter is often developed by assuming that the target signal is known, but the clutter is unknown. In the IRST case the target may only cover a few pixels, and so it is not possible to obtain much discrimination on target shape, and filters are more useful for clutter suppression.

Clutter is defined by its covariance matrix. For images where the clutter statistics are stationary (i.e., do not vary across the image), it is possible to design a filter based on the spatial correlation of the whole image. Where the clutter is not stationary, but the statistics vary sufficiently slowly so the clutter can be considered to be quasi-stationary, an adaptive filter can be designed to respond to the locally changing clutter. When the clutter is non-stationary the covariance is estimated at each pixel from the portion of the image in a window centered at (or in some cases adjacent to) the current pixel. However in many images there will be regions of rapid change in intensity which the usual type of adaptive filter may not be able to handle. In these cases it may be advantageous to segment the image into regions of different statistics and apply locally optimal filters in each region. In some cases a mean removal operator can be applied to reduce dependence on intensity and increase dependence on shape.

There are two commonly used bases for constructing linear filters, non-parametric and parametric. Parametric methods involve descriptions of the clutter statistics using models with a number of parameters, and Markov models are the most common. For non-stationary clutter, it is necessary to segment the image into a number of regions of different model parameters. Non-parametric methods do not use models of the clutter and calculate the statistical properties of the particular image directly.

There are a number of linear adaptive spatial filters such as that of Soni *et al.* (1992), but they are computationally intensive for non-stationary clutter. Three dimensional spatial-temporal adaptive filters have been described by Melendez and Modestino (1995), and these are more computationally intensive. A method involving adaptive thresholding of point intensity contrast from surrounding 8 pixels and accumulating results over 95 frames was proposed by New *et al.* (1999). Stationary targets were not considered. Developments of this method were given by Ronda *et al.* (1999, 2000). Li and Shen (1994) used a frame summing after registration to detect moving targets. A method for clutter suppression using simultaneous adaptive spatial-temporal filtering and jitter compensation has been described by Tartakovsky and Blazek (2000), but it appears to be computationally intensive.

A common method of target detection based on small size non-linear spatial filters works by predicting the background by some type of smoothing operation, then subtracting the predicted background from the original image. The aim is to fit the predicted background as closely as possible in the original image without diminishing the target signal.

Non-linear spatial filters based on median or mathematical morphology operators are candidate filters for detection of dim point targets in IR image sequences. Both median subtraction filters and filters based on erosions and dilations have been used (Barnett

1989, Barnett *et al.* 1993, Deshpande *et al.* 1999, Li *et al.* 1997, Sang *et al.* 1996, 1998, Tom *et al.* 1993). These filters are related, as median filtering can be achieved by a combination of erosions. Median filtering in 1D is well understood, but the generalization to 2D is not straight forward. Arce and McLoughlin (1987) showed that the standard square window which is often used gave a greater loss of resolution of small features than the max/median filter, which consisted of the maximum of a series of linear filters through the point of interest. The max/median filter used here is the maximum of the 3 element linear median filters through the point in the horizontal, vertical and the 2 diagonal directions.

Spatial filtering based on horizontal and vertical morphological opening was used by Zhu *et al.* (1999) for small target detection. Max/mean and max/median filters have been used by Deshpande *et al.* (1999) for moving point target detection.

The detection of ultra-dim targets poses considerably greater difficulties than the detection of brighter targets. The problem with all spatial filters for detection of ultra-dim point targets in strong clutter with single pixel features is that there is no spatial information in a single pixel to allow discrimination from clutter containing point-like features. If the target is moving, filters based on the temporal profile of the target as it moves through a pixel can be used. Most examples of temporal filtering in the literature have assumed that the target has a significant angular velocity with respect to the sensor. One such filter is described by Tzannes and Brooks (1999). However, the assumption is that the angular motion of targets at the horizon is negligible, so these types of filter are not useable. The average brightness of the target is expected to increase over time as it approaches the sensor, and in the short time frame there will be considerable variation in brightness due to scintillation. These temporal behaviors may be able to be exploited for target detection.

None of the filters considered above appear to be ideal for the detection of ultra-dim scintillating targets in heavy clutter at video frame rates. Spatial and temporal filters separately are not able to provide sufficient detection capability, but a suitable combination of spatial and temporal filters will be shown to provide acceptable performance.

If the frames in an image sequence are highly accurately registered then taking the absolute value of frame differences alone would provide a possible method of detection of scintillating targets. Frame differencing was used as a clutter suppression filter by Nishiguchi *et al.* (1995), but they did not consider the possibility of sensor jitter. Yang *et al.* (1994) used a combination of positive frame differencing and temporal averaging of the frame difference to detect moving targets with velocities less than 0.3 pixels/frame. It has been reported that "for frame-to-frame-differencing type temporal processing algorithms, (frame registration) accuracies to the order of 1% of a pixel are necessary ..." (Accetta 1993 p. 312). Registration to this accuracy would be very computationally expensive.

The sensor would be mounted on a gimbaled, stabilized platform, but the sensor is still likely to be subjected to a variety of motions including random vibrations and ship motion. The stabilization is unlikely to be perfect and some image movement would occur between frames. When taking frame differences this would cause a non-zero response to occur in regions of high brightness slope. These responses could appear as false targets. However, the spatial variation of the frame differences of the background is correlated, since the image moves as a whole. It is possible to efficiently minimize the magnitude of the motion induced clutter in differenced images by filtering with small support spatial filters. The spatial correlation only extends to adjacent pixels so an opening operation with a 2 element structuring element aligned in the direction of motion should closely follow the intensity distribution of the blurring caused by the motion.

Since the temporal variations of target scintillations and background are temporally uncorrelated, the spatial-temporal operations should leave the scintillating single point target in much reduced clutter.

The simplest 2 element opening filter which would be effective regardless of the direction of the image shift would be a max/opening filter analogous to the max/median filter. The top hat filter obtained by subtracting the max/opening from the original image will be designated Orl2 filter. The image sequence obtained from taking absolute frame differences of the original sequence will be designated AFD. The AFD operated on by an Orl2 filter should remove most of the effect of motion. Another spatial filter which was considered for reducing the motion clutter was a 2 point max/mean filter. This filter was obtained by taking the maximum of the means of the 2 opposite adjacent points from the point of interest for the horizontal, vertical and diagonal directions. The top hat filter obtained by subtracting the max/mean from the original image will be designated the Mmn2 filter.

The AFD, Orl2 and Mmn2 filters were implemented on the Aspex parallel computer. Since the filters involved only simple integer arithmetic and comparisons, they ran very quickly, and demonstrated that the Aspex machine would be suitable for other IRST applications which used only the same operations.

2.4 Track before Detect Method

Bayesian track-before-detect (TBD) methods offer efficient means of enhancing target detectability by using the history of the signal intensity at each pixel to aid discrimination between targets and clutter. An efficient Bayesian TBD technique has been developed by Arnold *et al.* (1993). A detailed description of Bayesian methods for target detection and tracking has been given by Stone *et al.* (1999). Bayesian TBD methods have been applied to IRST systems by Branlund *et al.* (1996) and Merlo *et al.* (1997). The application was to tracking as well as to detection, as the IRST systems were scanning types with an update time of 0.5 to 1 second. At this low rate the target could move significantly between frames and so sophisticated tracking is necessary, but the time available for computation would also be longer and so the more complex methods would be practical. In the current work it is assumed that the target has very small motion and that the frame rate is high, so that tracking of the target is not difficult, but target detection is the major problem. In these circumstances it is possible to drastically reduce the complexity of the Bayesian likelihood calculation.

The aim is to detect the weakest possible target signal at maximum range. A target will have very small angular motion, and will be assumed to be confined to 1 pixel. Scintillation or pixelation effects may spread the target over at most 3 pixels, but usually only one pixel is bright. Scintillation also causes the target signal to fluctuate wildly, and a detection method must be robust against missed detection. Each pixel will be considered to be an individual sensor. Only one target would be present to a sensor at any time. Detection will be made by calculating the likelihood of target presence at a pixel location by a Bayesian procedure described by Stone *et al.* (1999).

Let S be the state space of the target, and $X(t)$ be the target state at time t. Prior information about the target is represented by the stochastic process $\{X(t);\, t \leq 0\}$.

Assume the presence of observations obtained at times $0 \leq t_1 \leq .. \leq t_k$. Let Y_k be a random variable corresponding to the set of observations obtained at time t_k. Let y_k denote a value of the random variable Y_k. Assume that the likelihood function for the target present can be calculated as

$$L_k(y_k \mid s) = \Pr\{Y_k = y_k \mid X(t_k) = s\}.$$

Let $Y(t) = (Y_1, \ldots\ldots\ldots, Y_K)$ and $\mathrm{y} = (y_1, \ldots\ldots\ldots, y_K)$

Define $L(\mathrm{y} \mid s_1, \ldots\ldots, s_K) = \Pr\{\mathrm{Y}(t) = \mathrm{y} \mid X(t_1) = s_1, \ldots\ldots, X(t_K) = s_K\}$

which is the probability that the observations Y(t) = y are obtained given that the target has passed through states $s_1, \ldots, s_K$.

The prior probability that the process {X(t); t ≤ 0} passes through states $s_1, \ldots, s_K$ at times $t_1, \ldots, t_K$ is defined as

$$q(s_1, \ldots\ldots, s_K) = \Pr\{X(t_1) = s_1, \ldots\ldots, X(t_K) = s_K\}.$$

The posterior probability, given the observations Y(t) = y, is

$$p(t_K, s_K) = \Pr\{X(t_K) = s_K \mid \mathrm{Y}(t_K) = \mathrm{y}\}.$$

Using Bayes' Theorem

$$p(t_K, s_K) = \frac{\Pr\{\mathrm{Y}(t_K) = \mathrm{y\_}and\_X(t_K) = s_K\}}{\Pr\{\mathrm{Y}(t_K) = \mathrm{y}\}}$$

$$= \frac{\int L(\mathrm{y} \mid s_1, \ldots\ldots, s_K) q(s_1, \ldots\ldots, s_K) ds_1 \ldots\ldots ds_{K-1}}{\int L(\mathrm{y} \mid s_1, \ldots\ldots, s_K) q(s_1, \ldots\ldots, s_K) ds_1 \ldots\ldots ds_K}.$$

Two assumptions about L and q allow a recursive evaluation of $p(t_K, s_K)$:- the stochastic process {X(t); t ≤ 0} must be Markovian, and the observations at different times must be independent of each other given $(X(t_1) = s_1, \ldots\ldots, X(t_K) = s_K)$. The likelihood function can then be written

$$L(\mathrm{y} \mid s_1, \ldots\ldots, s_K) = \prod_{k=1}^{K} L_k(y_k \mid s_k).$$

This means that the observations at a particular time depend only on the state at that time.

Define a transition function

$$q_k(s_k \mid s_{k-1}) = \Pr\{X(t_k) = s_k \mid X(t_{k-1}) = s_{k-1}\}$$

for $k \geq 1$, and q_0 is the probability density function for X(0).

Since the stochastic process $\{X(t);\ t \leq 0\}$ is Markovian,

$$q(s_1, \ldots\ldots, s_K) = \int_S \prod_{k=1}^{K} q_k(s_k \mid s_{k-1}) q_0(s_0) ds_0 .$$

Substituting,

$$p(t_K, s_K) = \frac{1}{C} L_K(y_K \mid s_K) \int_S q_K(s_K \mid s_{K-1}) p(t_{K-1}, s_{K-1}) ds_{K-1} .$$

The recursion relations become

$$p(t_0, s) = q_0(s) \text{ for } s \in S$$

for $k \geq 1$ and $s \in S$.

Motion update

$$p^-(t_k, s_k) = \int_{S+} q(s_k \mid s_{k-1}) p(t_{k-1}, s_{k-1}) ds_{k-1} .$$

Information update

$$p(t_k, s) = \frac{1}{C} L_k(y_k \mid s) p^-(t_k, s).$$

To allow for the target not existing in any of the states of S, augment the target space to S+ with the null state ϕ. Assuming at most one target in S,

$$p(\phi) + \int_{s \in S} p(s) ds = 1 .$$

The formalism then becomes

$$p^-(t_k, s_k) = \int_{S+} q(s_k \mid s_{k-1}) p(t_{k-1}, s_{k-1}) ds_{k-1} \text{ for } s \in S+$$

$$p(t_k, s) = \frac{p^-(t_k, s) L_k(y_k \mid s)}{C(k)} \text{ for } s \in S$$

$$p(t_k, \phi) = \frac{p^-(t_k, \phi) L_k(y_k \mid \phi)}{C(k)}$$

$$C(k) = p^-(t_k, \phi) L_k(y_k \mid \phi) + \int_{s \in S} p^-(t_k, s) L_k(y_k \mid s) ds .$$

Define the target likelihood ratio, Λ(s):

$$\Lambda(s) = \frac{p(s)}{p(\phi)} \text{ for } s \in S$$

$$\Lambda(t, s) = \frac{p(t, s)}{p(t, \phi)}$$

$$\Lambda^-(t, s) = \frac{p^-(t, s)}{p^-(t, \phi)}$$

and the measurement likelihood ratio

$$\mathrm{L}_k(y \mid s) = \frac{L_k(y \mid s)}{L_k(y \mid \phi)} \text{ for } y \in H_k,\ s \in S.$$

If it can be assumed or organized that

$$p^-(t_k, \phi) = q_k(\phi \mid \phi) p(t_{k-1}, \phi) + \int_S q_k(\phi \mid s) p(t_{k-1}, s) ds = p(t_{k-1}, \phi)$$

$$\text{then } \Lambda^-(t_k, s_k) = q_k(s_k \mid \phi) + \int_S q_k(s_k, s) \Lambda(t_{k-1}, s) ds \text{ for } s \in S.$$

A likelihood recursion ratio recursion can then be written

$$\Lambda(t_0, s) = \frac{p(t_0, s)}{p(t_0, \phi)} \quad \text{for } s \in S.$$

Motion update

$$\Lambda^-(t_k, s_k) = q_k(s_k \mid \phi) + \int_S q_k(s_k, s_{k-1}) \Lambda(t_{k-1}, s) ds_{k-1}.$$

Information update

$$\Lambda(t_k, s) = L_k(y_k \mid s) \Lambda^-(t_k, s)$$

Taking natural logarithms

$$\ln \Lambda(t_k, s) = \ln \Lambda^-(t_k, s) + \ln L_k(y_k \mid s) \quad \text{for } s \in S.$$

The aim is to detect the weakest possible target signal at maximum range. A target will have very small angular motion, and will be assumed to be confined to 1 pixel. Scintillation or pixelation effects may spread the target over at most 3 pixels, but this effect will be ignored here. Each pixel will be considered to be an individual sensor. Only one target would be present at any time, so for each pixel there are only 2 states, target absent or target present, X(t) = ϕ or s.

Initially it will be assumed that the chance of a state changing due to "motion" factors will be very small, and the values of $q(s_k, s_{k-1})$ will be given by

$$q(s, s) = 1 - \delta$$

$$q(s, \varphi) = \delta$$

$$q(\varphi, s) = \delta$$

$$q(\varphi, \varphi) = 1 - \delta$$

where δ is a small fraction.

The integrals in the function definitions now become summations over 2 terms:

$$p^-(t_k, s_k) = q(s_k, s)p(t_{k-1}, s) + q(s_k, \varphi)p(t_{k-1}, \varphi)$$

$$p^-(t_k, s) = q(s, s)p(t_{k-1}, s) + q(s, \varphi)p(t_{k-1}, \varphi) = (1-\delta)p(t_{k-1}, s) + \delta p(t_{k-1}, \varphi)$$

$$p^-(t_k, \varphi) = q(\varphi, s)p(t_{k-1}, s) + q(\varphi, \varphi)p(t_{k-1}, \varphi) = \delta p(t_{k-1}, s) + (1-\delta)p(t_{k-1}, \varphi)$$

The posterior probabilities are then

$$p(t_k, s) = \frac{L_k(y_k \mid s)p^-(t_k, s)}{C(k)} = \frac{L_k(y_k \mid s)[(1-\delta)p(t_{k-1}, s) + \delta p(t_{k-1}, \varphi)]}{C(k)}$$

$$p(t_k, \varphi) = \frac{L_k(y_k \mid \varphi)p^-(t_k, \phi)}{C(k)} = \frac{L_k(y_k \mid \varphi)[\delta p(t_{k-1}, s)] + (1-\delta)p(t_{k-1}, \varphi)}{C(k)}$$

Recalling

$$\Lambda(t, s) = \frac{p(t, s)}{p(t, \varphi)} \text{ and } \Lambda^-(t, s) = \frac{p^-(t, s)}{p^-(t, \varphi)}$$

$$\Lambda(t, s) = \frac{L_k(y_k \mid s)[(1-\delta)p(t_{k-1}, s) + \delta p(t_{k-1}, \varphi)]}{L_k(y_k \mid \varphi)[\delta p(t_{k-1}, s) + (1-\delta)p(t_{k-1}, \varphi)]}$$

Dividing by $p(t_{k-1}, \varphi)$ gives

$$\Lambda(t, s) = \frac{L_k(y_k \mid s)[(1-\delta)\Lambda^-(t_{k-1}, s) + \delta]}{L_k(y_k \mid \varphi)[\delta\Lambda^-(t_{k-1}, s) + (1-\delta)]}$$

Since δ is assumed to be very small, approximate $\Lambda(t, s)$ by setting $\delta = 0$.

$$\Lambda(t, s) \approx \frac{L_k(y_k \mid s)\Lambda^-(t_{k-1}, s)}{L_k(y_k \mid \varphi)} \text{ and}$$

$$\ln \Lambda(t, s) \approx \ln L_k(y_k \mid s) - \ln L_k(y_k \mid \varphi) + \Lambda^-(t_{k-1}, s)$$

Assume the background noise has a zero mean Gaussian distribution with standard deviation σ, and the target signal has a Gaussian distribution with mean r and standard deviation σ.

$$\ln L_k(y_k \mid \varphi) = \ln\left[\frac{1}{K}\exp\left(\frac{-y_k^2}{2\sigma^2}\right)\right] = \frac{-y_k^2}{2\sigma^2} - K \text{ and}$$

$$\ln L_k(y_k \mid s) = \ln\left[\frac{1}{K}\exp\left(\frac{-(y_k - r)^2}{2\sigma^2}\right)\right] = \frac{-(y_k - r)^2}{2\sigma^2} - K$$

therefore

$$\ln \Lambda(t_k, s) = \frac{r y_k}{\sigma^2} - \frac{r^2}{2\sigma^2} + \ln \Lambda(t_{k-1}, s)$$

A target could be declared when the log likelihood ratio exceeds a given threshold, and declared lost when the ratio drops below a threshold.

Rescaling the log ratio

$$\ln \Lambda^*(t_k, s) = \sigma^2 \ln \Lambda(t_k, s)$$

$$\ln \Lambda^*(t_k, s) = r y_k - \frac{r^2}{2} + \ln \Lambda^*(t_{k-1}, s)$$

If the threshold level is rescaled, detections can be declared in the usual manner.

The variable y_k is the current iteration pixel intensity. The mean target signal may not be known, so the parameter "*r*" describing the sensitivity of detection is usually taken as a multiple of the standard deviation of the background. The time at iteration number k is t_k. It is assumed that the data has been rescaled so that the background has zero mean. The method also assumes that the background and target intensity distributions are Gaussian, and this is usually true for the background. However, because of the effects of scintillation the target intensity distribution is log Gaussian (Hufnagel 1985).

This difference will be ignored here, but its effect would be to increase the sensitivity of the method.

If the calculation of likelihood is allowed to proceed unchecked as the image sequence is processed, the likelihood in background regions would drop to very low values, and when a target appeared it would take an inordinate amount of time for the likelihood to build up to detectable levels. Hence it is necessary to set a minimum allowable value for the likelihood. Similarly strong, brief false alarms could build up sufficient likelihood to be detected as a target and persist for long times, so a maximum allowable value of likelihood needs to be set. These maximum and minimum valucs arc cxtra parameters which need to be specified.

The resulting procedure is as follows. The background standard deviation is determined or estimated and the value of r calculated, and it is used to calculate the maximum and minimum values. The initial likelihood is set to 0, and the image sequence processed. A detection is declared when the likelihood exceeds 0 and is dropped when the likelihood drops below 0.

The method described above was designed for the simplest operation. More complex situations can be handled by using different approximations at the appropriate places in the derivation.

3 Algorithm Assessment Using Measured Target Data in an Uncluttered Background

Methods for detecting targets in low clutter backgrounds in experimentally measured image sequences will be considered in this section. Simple spatial filters will be used to enhance the targets and suppress the background, and the operation of a Bayesian TBD algorithm will be demonstrated.

Detection algorithm performance was assessed using a sequence of IR images of an aircraft flying over Gulf St Vincent off Adelaide at an altitude of approximately 30 m. The effective target was the surface of the exhaust pipes of the aircraft. The image sequence was obtained using an Amber Galileo 3-5μm infrared camera. The Galileo focal plane array is 256×256 pixels, each pixel 30 μm on side. The camera was fitted with a 300 mm catadioptric lens and operated at 25 frames/s. The 300 mm lens gives a pixel size, or instantaneous field of view (IFOV), of 0.1 mrad. The field of view (FOV) was 1.467°. The sequence involved an approaching target in a benign background. The final frame of the sequence is shown in Figure 1.

Figure 1. Last frame of image sequence.

The Gabriel software (Aspect Computing, Adelaide, Australia) was used to display image sequences, and also to measure pixel intensities and means and standard deviations of image areas.

3.1 Spatial Filters

Preliminary work showed that the most effective simple morphological filters were an opening filter, and a closing filter followed by an opening filter. The median subtraction filter will be referred to as the median filter. The same elements as used for the median filters will be used for morphological structuring elements.

In the present case only a single pixel target is of interest, and since low computational load is a requirement, filters with only a few elements will be considered. Spatial filters with three and five pixel horizontal and vertical linear elements and a 3×3 pixel square element will be evaluated. The filters chosen for study were:-

- Morphological, Opening, 3 pixel horizontal element – Op3.
- Morphological, Close-open, 3 pixel horizontal element – Co3h.
- Median, 3 pixel horizontal element – Dm3h.
- Morphological, Close-open, 3 pixel vertical element – Co3v.
- Median, 3 pixel vertical element – Dm3v.
- Morphological, Close-open, 5 pixel horizontal element – Co5h.
- Median, 5 pixel horizontal element – Dm5h.
- Morphological, Close-open, 5 pixel vertical element – Co5v.
- Median, 5 pixel vertical element – Dm5v.
- Morphological, Close-open, 3×3 pixel element Co9.

In the cases of 3 and 5 pixel linear elements the results of the horizontal and vertical filtering were combined. Only detections occurring in both filter outputs were counted.

3.1.1 Results

The target aircraft carried a GPS unit and the actual trajectory was recorded at one second intervals. There was significant drift in bearing during the flight. The background sky consisted of only very light distant high level cloud cover 56° in azimuth from the sun.

The resolution of the original image sequence was 175 pixels/°, which was higher than the assumed IRST being simulated. The resolution was reduced by resampling with the average of 2×2 pixels of the original image. This procedure ignores the effect of the lens optical transfer function on the image, but the result was considered to be suitable for the current analysis because the point spread was expected to be less than the enlarged pixel size.

Because of the drift in bearing during the flight the target image moved over a number of pixels, which is contrary to the assumption made for a target in a real IRST. An apparently stationary target location was produced using the following procedure. The target was manually tracked during the periods when it was visible. The values of the target pixel and a small surrounding area were copied to a fixed location using knowledge of the true track. The local mean around the target location and the local mean at the fixed location were calculated, and the copied values were adjusted to allow for the difference in means. The intensity at the old target location was replaced with the average local intensity.

The last frame of the sequence showing the target is given in Figure 1. As well as the target there is a grain silo on the horizon on the other side of the body of water. The silo has a width of 1 pixel, which would be expected to pass the horizontal filters. The bright spot near the center is a bird, and the target is barely visible about 15 mm above and to the right of the bird. The bright area in the foreground is a stone breakwater. The lower intensity of the sky in the center is a result of drift in the calibration of the camera.

The target intensity vs frame number is illustrated in Figure 2. The target first appeared in frame 725. The dimness of the target can be judged from the average intensity at the target brightest is less than 20 counts above a background of 3630.

The area of interest for detection of sea skimming missiles is from 2 pixels below the horizon to 8 pixels above the horizon. This is the area that was used to determine the detection vs false alarm rate efficiency of the various spatial filters. Since the target was expected to be seen against a sky background, the background statistics were obtained from the sky region. A similar process could be carried out for the sea if the target was below the horizon, although this would not occur in most engagement geometries.

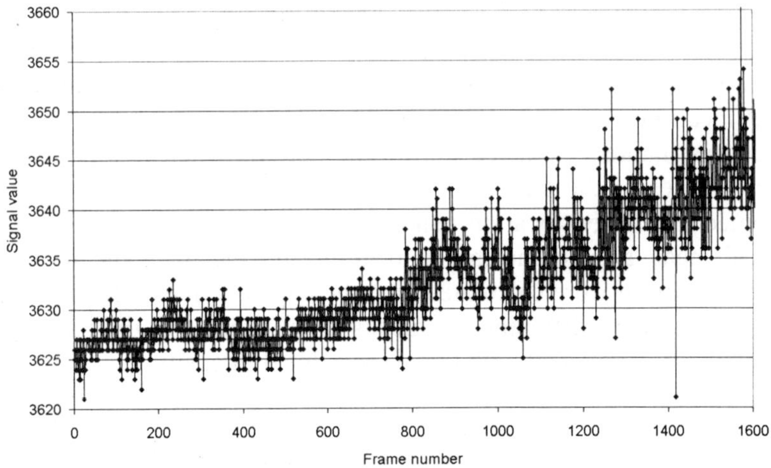

Figure 2. Target intensity vs frame number.

It was decided to select 1 second as the time to allow likelihood to build up for detection, and this corresponded to 25 frames. The minimum allowed likelihood value was set to the value that the background would achieve after 25 frames, and the maximum value was set to the value that a target value of "r" in the likelihood calculation would achieve in 25 frames. Once a detection was declared, a variable denoting the strength of the declaration was increased by 1 for each frame in which a detection occurred to a maximum excess of 12. For each frame where the likelihood dropped below the detection level the variable was reduced by 1. This ensures that if the target disappears, the detection would lapse after 0.5 seconds, but would still make the algorithm robust against the effects of target scintillation.

The filter parameters and results are summarized in Table 1. The value of the parameter "r" was chosen to be twice the standard deviation of the output of the appropriate filter in the background area. The "c" designated the combined output of the 2 linear filters. More details of target detection are given in Table 2. There is not a significant difference in the first detection ability of most of the filters, and their ability to maintain detection is reflected in the number of detections given in Table 1.

Table 1. Filter parameters, target detections and false alarms for the spatial filters.

Filter	Background mean	Background std. dev.	Min	Max	Targets	False alarms	False alarm / frame
op3h	0.6	1.2	-108	36	744	54457	33.51
op3v	0.83	1.4	-156	40	697	40856	25.14
op3c					680	6516	4.01
co3h	-0.49	1.32	- 55	119	733	31861	19.61
co3v	-0.28	1.6	-106	150	753	54665	33.64
co3c					730	5904	3.63
dm3h	0.0	1.43	-102	102	705	30684	18.88
dm3v	0.1	1.41	-106	92	734	22152	13.63
dm3c					679	4398	2.71
co5h	-0.87	1.55	- 53	188	728	19560	12.04
co5v	-0.59	1.86	-118	228	737	35643	21.93
co5c					723	6323	3.89
dm5h	0.0	1.55	-120	120	736	46603	28.68
dm5v	0.24	1.76	-176	134	749	47054	28.96
dmc					731	7341	4.52
co9	-0.74	1.82	- 98	233	741	78987	48.61

Table 2. Frame numbers for detection and loss of targets.

Filter	Detect	Lose	Detect	Lose	Detect	Lose	Detect
op3c	870	952	993	1056	1082	1232	1244
co3c	882	966	980	-	-	-	-
dm3c	888	964	988	1054	1087	1235	1242
co5c	884	966	987	-	-	-	-
dm5c	876	966	978	1064	1072	-	-
co9	882	969	974	-	-	-	-

To show the working of the method, the 3 pixel median case will be discussed in detail. A plot of the filtered intensity of the target pixel is given in Figure 3. From the figure, it can be seen that the mean output for the background was approximately zero. The target appears at about frame 770, and briefly disappears near frames 1050, 1230 and 1400 due to atmospheric refraction effects with a longer timescale than scintillation. This sequence provides a difficult test for the TBD process, as it must track the intensity variations in a balanced way. It must not drop a target which disappears for only a few frames, and it must not maintain a detection for unduly long after it has really disappeared. It is worth noting that several birds flew across the field of view, but these were not detected as false alarms because they were at each pixel for less than the integration time. The output images of the detections from the horizontal and vertical three element linear median filters, and the combined output of both, are given in Figure 4. Of the three detections in the sky region in Figure 4, the leftmost is the target, the center is a defective pixel in the camera, and the rightmost is the 1 pixel wide silo which is a constant in all frames.

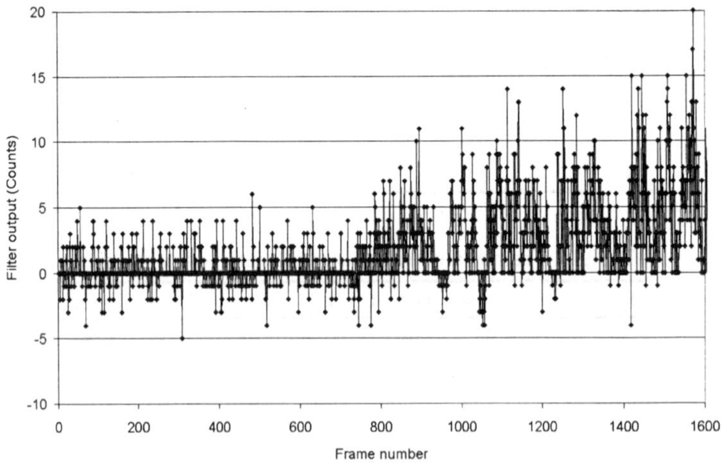

Figure 3. Plot of intensity of target pixel after 3 pixel horizontal median subtraction filter.

(a)

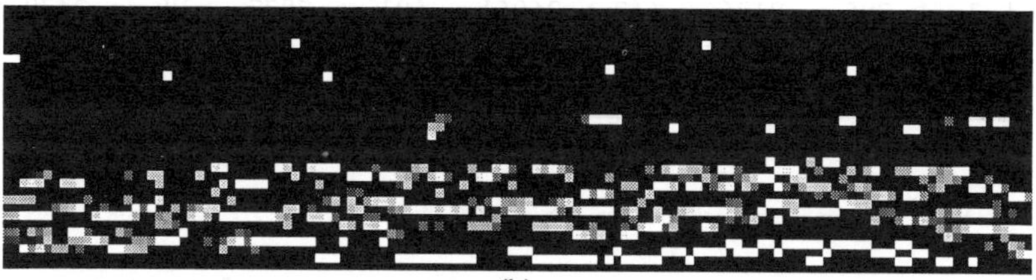

(b)

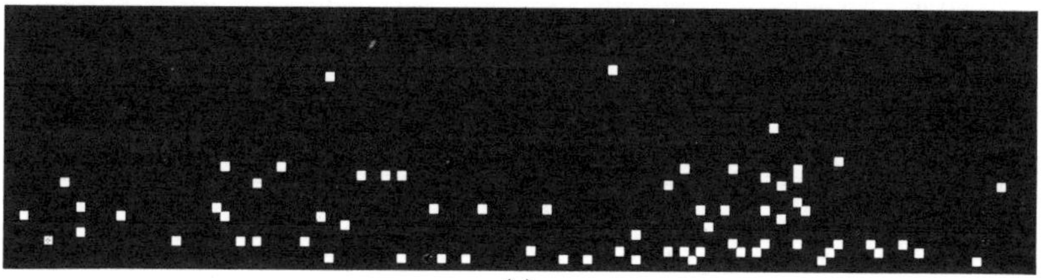

(c)

Figure 4. Outputs from the 3 element median filters for frame 1625.
(a) Horizontal, (b) Vertical, and (c) Both horizontal and vertical.

To show the improvement in performance obtained with the TBD method, a standard threshold exceedance calculation was made. The outputs from the various filters were thresholded at values of 1 to 4 counts, and the results from the horizontal and vertical filters were combined to give detections. The results are given in Table 3 and Figure 5. The values near 900 are likely to be affected by false alarms at the target pixel location, as visual inspection showed that there were frames where detections could not be made. If the threshold detections near 800 are genuine, they are obtained with a much higher number of false alarms than the TBD detections, and hence the TBD method shows clear superiority.

Table 3. Target detections and false alarms calculated from threshold exceedences.

	Threshold = 1		Threshold = 2		Threshold = 3		Threshold = 4	
Filter	Targets	False alarms	Targets	False alarms	Targets	False alarms	Targets	False alarms
op3c	-	-	872	61732	773	20033	630	7730
co3c	699	18468	610	7513	476	4522	-	-
dm3c	795	81766	662	26051	503	7877	396	2479
co5c	656	10433	506	5746	381	4090	-	-
dm5c	902	129124	787	41366	655	13633	518	6058

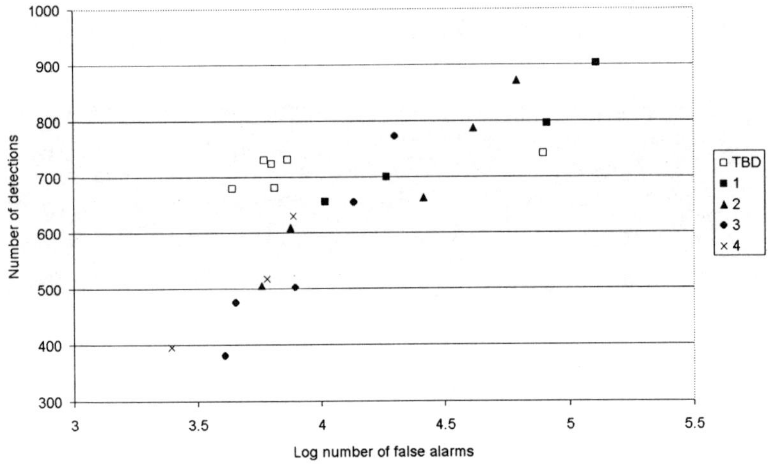

Figure 5. Number of detections vs log false alarms for TBD and threshold methods. Threshold levels given in the legend.

Because the mean output from the background for each filter is not an integer, it is not possible to use the same relative value of threshold for each filter. The threshold values in Table 3 are the absolute values. It can be seen that in comparison with Table 1 the performance of thresholding is significantly poorer than the TBD method. The false alarm rate is much higher for any level of target detection.

The performance of the thresholding method using the combination of the 3 element linear median filter is illustrated in Figure 6. The average number of false alarms per frame is about 5. The rate of detection is very intermittent until about frame 1100, and is not consistent until about frame 1300.

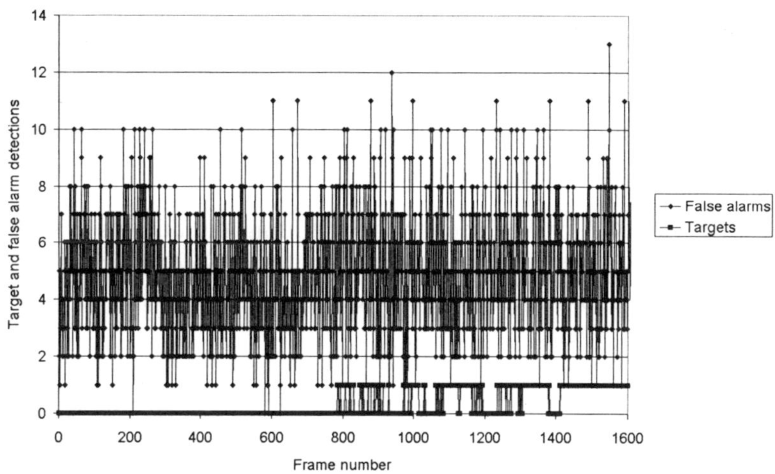

Figure 6. Target detections and false alarms from thresholded output of 3 element median filters, threshold = 3.

3.1.2 Discussion

It can be seen from Table 1 that the best detection efficiency for a reasonable number of false alarms is given by the linear 3-element close-open filter. The 3-element median filter gave a smaller number of detections, but also a significantly smaller number of false alarms, and with tweaking of the parameters it may give a response comparable to the close-open filter. The median filter also has other advantages. The mean value of the filtered background is approximately zero, and the distribution of filtered background intensity values is near to Gaussian. This adds legitimacy to the use of the likelihood processing. All the other morphological filters gave non-zero values for the mean filtered background.

The detection range was not very sensitive to filter type, so the type of filter used can be chosen for computational efficiency. Because of the refraction effects in the atmosphere, it is not certain if the first detection range was the maximum possible range under the circumstances. The intensity of the target would have been relatively high as the exhaust pipes would be hot. A stealthy missile would have a lower intensity and so its detection range would be more sensitive to processing parameters. Measurements of constant intensity sources at a range of distances and under a range of environmental conditions are needed to obtain definitive maximum detection ranges.

From Figure 4 it can be seen that combining the outputs from the horizontal and vertical filters considerably reduced the number of detections in both the sky and sea regions. Similar behavior is seen with the other filters.

The detection efficiency of the TBD process was considerably better than a simple thresholding process. Extra processing would have to be done in the tracking stage on the detections from the simple thresholding process to reduce the number of false alarms.

Also from Figure 4 it can be seen that the little cloud clutter that was present has been suppressed from the sky region. The drift in calibration of the camera also did not affect the output of the filters. Cloud clutter is relatively benign in this case and simple morphological filters were able to suppress it. It is expected that most cloud clutter on the horizon would be at very long range and would appear to be diffuse, so simple filters should suffice in most cases. However, there may be problems in directions near a low sun.

In the sea region there is a greater number of false alarms, but these are often short lived. The detection algorithm was optimized on the standard deviation of the sky region, and so was not optimal for the sea region. If the standard deviation of the sea intensity fluctuations was used in the algorithm the false alarms in the sea region would be much reduced, but the sensitivity to targets would also be reduced. False alarms in the sea may be better handled by using longer time constants in the TBD algorithm.

In an operational system extra intelligence would be required to facilitate the correct assignment of detections to real targets or false targets. The temporal behavior of possible targets could be compared to expected behavior of target intensity. Detections with diminishing intensity would not be targets. Constant intensity pixels would be marked as defective pixels or fixed environmental clutter, and eliminated from consideration. These operations could be carried out in the serial track processor.

3.2 Temporal-Spatial Filtering

The performance of combined spatial and temporal filtering was investigated using absolute frame differencing, AFD, followed by the 2 point max/mean subtraction filter Mmn2. The AFD and the Mmn2 filters were applied to the aircraft sequence and the output for the pixel containing the target and the result is given in Figure 7. It can be seen that there are progressively larger deviations from the mean after the target appears at about frame 770.

The Bayesian TBD algorithm was used for target detection. In order to calculate a likelihood ratio for the Bayesian algorithm, the distribution function of the background values has to be known, and the process is much simpler if that function is a Gaussian. An estimate of the background statistics was obtained from pixel (128,4), which appears to be clear sky for the whole of the sequence. The histograms of the AFD and filtered values for the whole sequence are given in Figure 8. The results from the spatial filters Orl2, Mmn2 and a mean derived from the 8 surrounding pixels, are given in the figure. The AFD values are all non-negative, by definition, and the histogram peaks at 1 count. The Orl2 curve is very asymmetrical and peaks at zero, (the zero value is suppressed because it is far off scale). The mean curves are more symmetrical and bell shaped. A good fit to a Gaussian curve was obtained for the max/mean curve. The mean was −1.41 and the standard deviation was 1.89, with a correlation coefficient of 0.991.

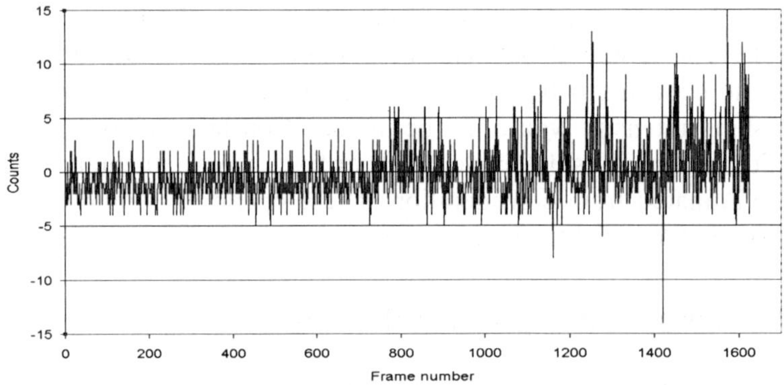

Figure 7. Output of frame differencing plus max/mean filter for the target pixel 40,8 for the Sequence 1 data.

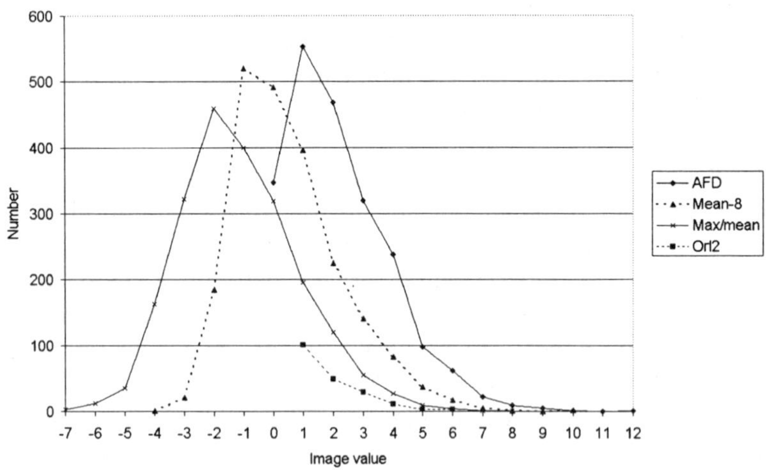

Figure 8. Histograms of values of pixel 128,4 of Sequence 2 for the absolute frame difference, and filters Orl2, mean and max/mean.

The output of the absolute frame differencing plus the max/mean filter for the target pixel is given in Figure 9. There are areas of increased value after frame 770 when the target is present. The filtering produced very little change in the values from the AFD values. There is a small but significant increase in filter output at the target position when the target appears, but the difference is so small that the extra sensitivity of the Bayesian algorithm is required to reliably detect it. The algorithm was set up for the max/mean filter because the background statistics were Gaussian, which gave a simple log likelihood ratio. Optimization of the parameters in the algorithm would probably reduce the false alarm rate.

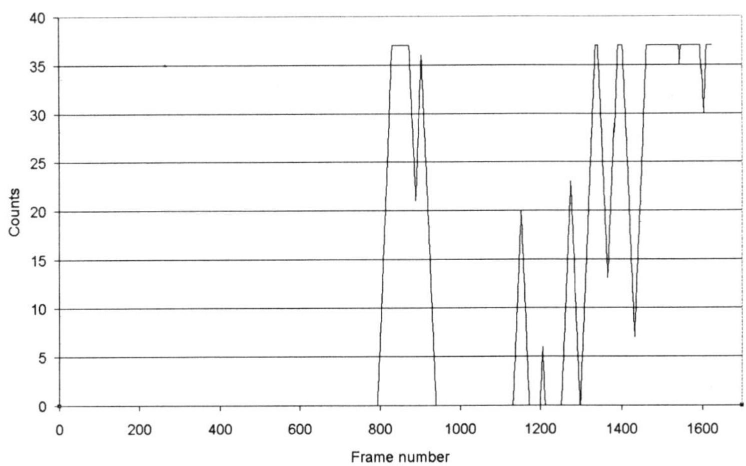

Figure 9. Output from detection algorithm for target pixel 40,8.

Because of the loss of information in the calculating the AFD and applying the temporal/spatial filtering, this procedure would not normally be the most effective detection algorithm for bright targets or benign backgrounds. In practice it might be necessary to segment an image and apply the AFD and temporal/spatial filter to cluttered regions, and more standard spatial and/or temporal filters to uncluttered regions.

3.3 Conclusions

A simple Bayesian track-before-detect algorithm shows promise for detecting ASMs at ranges near the horizon. The computational load of the basic algorithm is low, and the algorithm is suitable for implementation on a massively parallel processor using associative computing methods. The performance is superior to simply thresholding the output of spatial filters.

Morphological filters based on short linear elements were found to effectively suppress light cloud clutter and enhance dim point target detectability.

4 Algorithm Assessment Using Synthetically Generated Target in a Cluttered Background

4.1 Introduction

The previous section covered detection of targets in benign backgrounds where spatial filters were effective in clutter suppression. Target detection is much more difficult in cloud clutter backgrounds. As an example, a frame from a sequence of images of an aircraft flying away from the camera is given in Figure 10. The camera location was the same as for the benign background sequence used in the previous section. The plane had a small but significant angular velocity, and the target appeared to go in and out of clear and cloud regions. The resolution of the image sequence was 0.1 mrad. The isolated black and white points are defective pixels. The target is not visible in this figure, so an enlarged portion containing the target is given in Figure 11. The target is the brightest occurrence in a sequence of about 10 frames around the frame shown. The difficulty of detecting the target is obvious.

Figure 10. Frame 1788 of cloud sequence.

Figure 11. Frame 1788, area containing the target.

Measurement of real targets in cluttered backgrounds and obtaining ground truth to sub-pixel accuracy is extremely difficult. A more practical method of obtaining image sequences for algorithm assessment is to generate synthetic targets images and inject them into image sequences of real backgrounds. When generating synthetic targets there are 2 effects which must be modeled in addition to target intensity; scintillation and image motion from sensor jitter.

The propagation environment near the sea surface is very complex. Random fluctuations in the refractive index caused by turbulence lead to scintillation of point sources. The next generation of staring IRSTs will operate at frame rates high enough to be able to exploit these effects. Unfortunately the effect of scintillation due to atmospheric turbulence is not sufficiently understood to allow accurate target modeling from first principles. An alternative approach is to use a phenomenological model where parameters describing the scintillation are specified. The intrinsic parameters are the magnitude of radiance, blurring effects, image motion, and the correlation times and fluctuations in these quantities. A model has been developed (Warren 2002), and this was used to generate a representative sequence of target images. Care was taken in the target injection process to allow for background occlusion and atmospheric path radiance between the target and camera.

It was expected that an actual IRST in service could not be perfectly stabilized, and the images produced by the system would undergo random motion. The generated synthetic image sequences would have to include this motion. The motions in the row and column directions were assumed to be independent, and were modeled as single step Markov process.

The section near the horizon of the background image in Figure 10 was chosen as representing severe cloud clutter, and a sequence of synthetic background images was generated from it. Image motion from sensor jitter was modeled assuming the standard deviations of the motions in the row and column direction were 0.15 pixels, and the correlation time was 0.05 seconds. An image sequence of 10,000 frames was generated for testing. Synthetic targets were injected into the sequence at two locations corresponding to clear sky and cloud edge where the jitter effect were greatest. Three synthetic targets were used; a subsonic missile at 10 km, a supersonic missile at 30 km, and a target twice the intensity of the subsonic missile. The targets were designated Subsonic, Supersonic and Subsonic-x2 respectively.

4.2 Detection and Tracking Method

The AFD-Orl2 and AFD-Mmn2 filters were used to suppress the cloud and jitter clutter to increase the target signal to noise ratio. For the purpose of comparing the relative effectiveness of the two filters a detector was required that was transparent in operation to make the effect of the filters more apparent. A two stage TBD algorithm was used, where the initial detection process was separate from the tracking process. Detection was by a frame accumulation process, where detections in a neighborhood in the previous frame were carried over into the new frame, and a target declared after a specified number of detections had occurred. This process had some similarity to the Recursive Max Filter used by Nishiguchi *et al.* (1995). In order to determine whether the true target had been detected, and to calculate the number of false alarms, it was necessary to cluster contiguous detections into tracks and determine if any of the tracks included the true target location. The same frame accumulation process that was used for detection was used to link target declarations into target tracks. A similar two stage process, only using Bayesian trackers, was described by Merlo *et al.* (1997).

The detection process was carried out on a pixel by pixel basis over the whole frame in a manner which would lead naturally to a massively parallel implementation on the Aspex ASP. The basis of the processing was to associate a processing element (APE) with each pixel, and program each element simultaneously. To mimic this in the C programming language, an array of structures was declared with a form analogous to the memory in an APE on the ASP. The variables which would be used in the ASP were included in the structure. The serial processing of the structure variables in C would be equivalent to the parallel processing of the APEs on the ASP. The AFD-Mmn2 and AFD-Orl2 filters were coded on the ASP, but the detection program was the C emulation.

The detection method proceeded as follows. A detection at a pixel location was declared when the pixel value was greater than a threshold designated THRESH, and a flag POSSIBLE was set on the appropriate APEs. (Words in small capitals correspond to variables in the program.) A variable, designated DETEC, was defined for each pixel. It recorded the number of POSSIBLE detections that had occurred in the recent past, and its initial value was zero. The DETEC variable would correspond to the likelihood function in a Bayesian algorithm, and this part of the program could be replaced by a Bayesian calculation.

Target scintillation and sensor motion would cause the target to move sporadically to different pixels as the sequence progressed. To account for this circumstance, the values of DETEC in the 8 neighboring pixels were examined as well as the value for the current pixel. If the current pixel had the flag POSSIBLE set, its value of DETEC was set to the maximum DETEC value of the 9 pixels centered on the current pixel. At the end of processing the frame, for each pixel where the flag POSSIBLE was set, the value of DETEC was incremented by 1. If DETEC exceeded a maximum value, DECHI, then DETEC was rest to DECHI. For each pixel where POSSIBLE was not set, DETEC was decremented by 1 to a minimum value of zero.

A target was declared at a pixel if the value of DETEC exceeded a detection threshold, DECLO, and a flag designated TARGET was set for the appropriate pixels. This threshold reduced the number of false alarms, as a possible detection had to persist for the designated time to be declared as a target. Allowing the value of DETEC to exist in a range between DECLO and DECHI made the process robust against missed detections from scintillation, as several missed detections when DETEC was initially well above DECLO may not have reduced the value of DETEC below DECLO. Limiting the value of DETEC to less than DECHI limits the strength of the target, so that if the target disappeared, only a short time would be required for DETEC to drop below DECLO, and the target to be declared dead.

The first stages of the tracking algorithm were similar to the detection algorithm. The program clustered all contiguous pixels where TARGET was set, and assigned to each cluster a unique identifier. The program formed tracks from clusters that overlapped on adjacent frames, and if a pixel was on a track it was given an identifier TRACKNO, which was the number of the track. Non-zero values of TRACKNO corresponded to tracks, and pixels with a TRACKNO value of zero were not on tracks. The assignment of track numbers was to enable true tracks containing the known location of the true target to be separated from false alarms. In an IRST system the higher level tracker would process tracks.

Similarly to detection, each tracking step was carried out on a pixel by pixel basis over the whole frame. For each pixel where the flag TARGET was set, the corresponding pixel on the previous frame, and its 8 neighbors, were tested to determine if TARGET was set for them. If any of these pixels had TARGET set, the pixel on the new frame was given the lowest non-zero TRACKNO of the 9 pixels in the neighborhood on the previous frame. Usually all the TRACKNOs in the neighborhood on the previous frame would be the same, but at times two or more tracks would merge, and so the lowest TRACKNO was given to the pixel on the new frame, and the TRACKNO from the other track was returned to the pool to be available for reuse.

If TARGET was set on a pixel in the current frame, and there was no pixel with TARGET set in the neighborhood on the previous frame, the pixel was assigned the lowest value of TRACKNO from the pool of unused numbers. At a pixel where DETEC had dropped below DECLO, the flag TARGET was reset.

After all the pixels in the frame were processed, a number of steps were taken. The active TRACKNOs were scanned, and any for which there were no pixels with TARGET set were returned to the unused pool. The values of TRACKNO for pixels in contiguous clusters with TARGET set were examined. If more than one TRACKNO was in a cluster, all the pixels were set to the lowest value, and the other TRACKNOs were returned to the pool to be available to be reused. Alternatively, a track might have split into 2 or more different tracks on the new frame, so a search was made for separate clusters of duplicate TRACKNOs, and if found, the extra track(s) would be given new TRACKNO(s).

The track, if any, which included the true target location was designated as a true detection, and all other tracks were counted as false alarms. The program was also set up to calculate the number of pixels in each cluster. This number could be used as the basis of a filter to reduce the number of false alarms, as a target would only cover a few pixels, so large clusters would probably be false alarms. The program output the probabilities of detection and false alarm for clusters of all size, and for clusters less than 6 pixels in size.

4.3 Results

One way of characterizing the detection performance is the receiver operating characteristic (ROC) curve, which is a plot of probability of detection (P_D) vs probability of false alarm (P_{FA}). An ROC curve is obtained by varying the parameters of the detector, such as threshold, and determining P_D and P_{FA}. It is often the case that an acceptable level of false alarms is specified, then the probability of detection is determined under a range of conditions.

The performance of the AFD, AFD-Mmn2 and AFD-Orl2 filters was investigated using the 10,000 frame synthetically generated image sequences. Different values of the detection and tracking thresholds were used to obtain the probability of detection as a function of the false alarm rate. To quantify the effects of the filters, it was decided to take the detection probabilities at an average false alarm rate of 1 per frame. This false alarm rate would probably be acceptable in an IRST application. Only an average false alarm rate can be used because the false alarms often occurred in batches with a number of false alarms in a frame, interspersed with periods of no false alarms. The probability of detection for the nearest false alarm rate to an average of 1 per frame is given in Figure 12 for the clear background case. The values for the raw AFD sequence were very poor, with the AFD-Orl2 filter giving greatly increased detections. The performance of the AFD-Mmn2 filter was intermediate between the two cases. The values for the cloud background are given in Figure 13. The performances of the AFD and AFD-Mmn2 filter cases were similar, with the performance of the AFD-Orl2 filter lower.

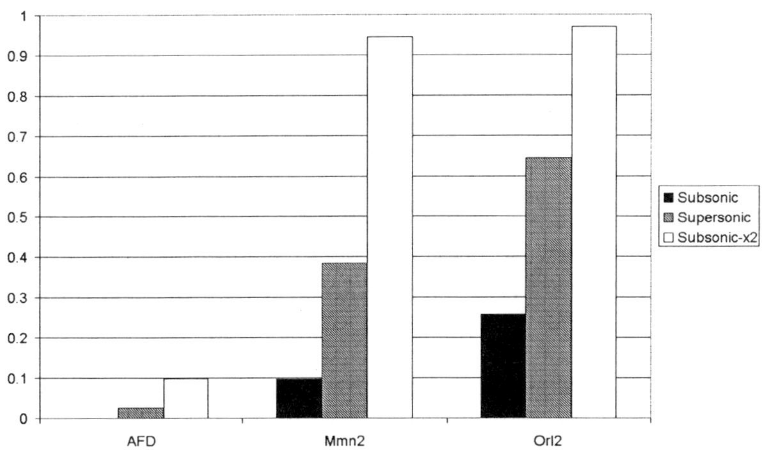

Figure 12. Probabilities of detection for an approximate false alarm rate of 1 per frame, clear background case.

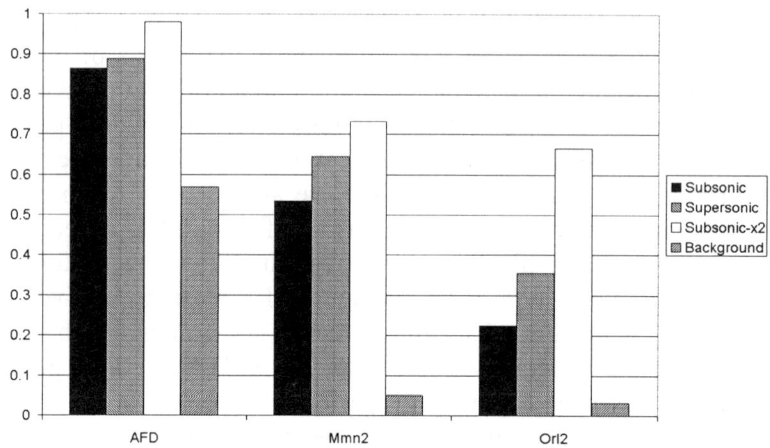

Figure 13. Probabilities of detection for an approximate false alarm rate of 1 per frame, cloud background case for total detections.

The performance values calculated so far are based on the total numbers of target and false alarm tracks, regardless of the number of pixels in each track in each frame. The detection step, which involved setting the detection value to the highest in the neighborhood, had a strong tendency to spread the detection over many pixels. Such diffuse detections would have a lower accuracy for locating possible targets. Figure 14 shows the detection probabilities for the different cases with the cloud background when the target size was limited to 5 pixels or less. In this situation the AFD performed very badly, whereas the AFD-Orl2 filter was little affected, and again the AFD-Mmn2 filter was reduced.

The plots so far have shown the optimum results, with the detection parameters varying in different cases. In an IRST application it may be necessary to have a constant set of parameters over the whole frame. The results for less than 6 detections and the optimum parameters for the clear case were chosen for standardization. The plots are shown in Figure 15. It can be seen that under these conditions that the AFD-Orl2 filter is obviously superior.

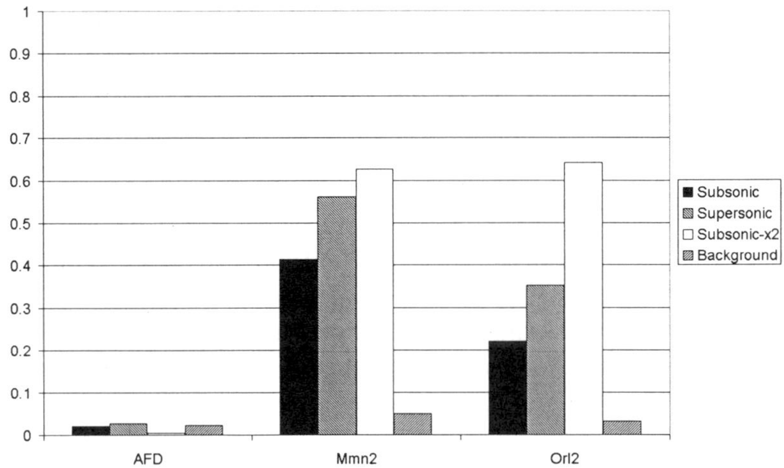

Figure 14. Probabilities of detection for an approximate false alarm rate of 1 per frame, cloud background case for detections less than 6 pixels in size.

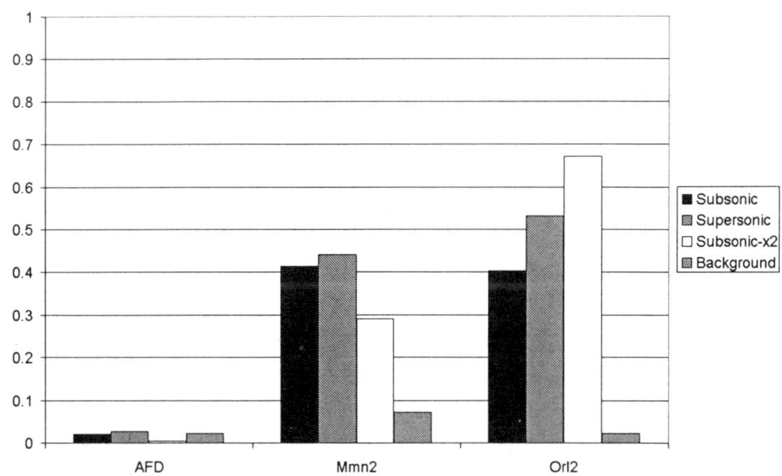

Figure 15. Probabilities of detection for an approximate false alarm rate of 1 per frame, cloud background case for detections less than 6 pixels in size, standardized parameters.

The stability of the detection parameters for the AFD-Orl2 filter, and its superior performance in nearly all cases suggests that it would be suitable for improving target detectability in AFD sequences.

4.4 Conclusions

A morphological top hat filter was able to efficiently minimize clutter induced by sensor jitter in sequences of absolute frame differences. Acceptable target detection performance was obtained by frame differencing followed by top hat filtering, with a detection algorithm based on multiframe accumulation.

A detection and tracking program based on associative processing methods was capable of efficiently quantifying the probability of detection and false alarm for sequences of synthetic targets and backgrounds. The program should execute very rapidly on the parallel processor.

5 Conclusions

Algorithms have been developed for detection of dim, stationary point targets by staring infrared search and track systems. The algorithms showed satisfactory detection performance in cases where more usual algorithms would fail.

The clutter minimization algorithms were based on associative processing methods and mathematical morphology, and detection algorithms were based on Bayesian methods. All the algorithms are intended for implementation on a massively parallel processor. The results indicate that this approach is very promising for the infrared search and track application.

Acknowledgments

The author is pleased to acknowledge Stephane Collignon for his assistance with the sensor system integration and Greg Teague for assistance with the planning and execution of the trials.

References

Accetta, J.S. (1993), "Infrared search and track systems," in Campana, S.B. (ed.), *Passive Electro-optical Systems*, vol. 5 of *The Infrared & Electro-optical Systems Handbook*, Environmental Research Institute Michigan and SPIE – The International Society for Optical Engineering: Ann Arbor, MI, pp. 209-344.

Amdahl, G.M. (1967), "Validity of the single-processor approach to achieving large scale computing capabilities," *AFIPS*, Atlantic City, N.J.: AFIPS Press, Reston, Va.

Arce, G.R. and McLoughlin, M.P. (1987), "Theoretical analysis of the max/median filter," *IEEE Transactions on Acoustics, Speech and Signal Processing*, vol. 35, no. 1, pp. 60-69.

Arnold, J., Shaw, S., and Pasternack, H. (1993), "Efficient target tracking using dynamic programming," *IEEE Transactions on Aerospace and Electronic Systems*, vol. 29, no. 1, pp. 44-56.

Barnett, J.T. (1989), "Statistical analysis of median subtraction filtering with application to point target detection in infrared backgrounds," *Infrared Systems and Components III*, SPIE – The International Society for Optical Engineering.

Barnett, J.T., Billard, B.D., and Lee, C. (1993), "Nonlinear morphological processors for point-target detection versus an adaptive linear spatial filter: a performance comparison," *Signal and*

Data Processing of Small Targets, SPIE – The International Society for Optical Engineering.

Barrera, J. and Salas, G.P. (1996), "Set operations on closed intervals and their applications to the automatic programming of morphological machines," *Electronic Imaging*, vol. 5, no. 3, pp. 335-352.

Barrera, J., Banon, G.J.F., and Lotufo, R.d.A. (1994), "Mathematical morphology toolbox for the Khoros system," *Image Algebra and Morphological Image Processing V*, SPIE – The International Society for Optical Engineering.

Blackman, S.S. and Popoli, R. (1999), *Design and Analysis of Modern Tracking Systems*, Boston: Artech House, 1230, ch. 2.

Braga-Neto, U. and Goutsias, J. (1999), "Automatic target detection and tracking in forward-looking infrared image sequences using morphological connected operators," *23rd Conference on Information Sciences and Systems*, Baltimore, Maryland, USA.

Branlund, E., Davis, P., and Lindgren, U. (1996), "Infrared detection of low-contrast sea-skimming cruise missiles," *Targets and Backgrounds: Characterization and Representation II*, SPIE – The International Society for Optical Engineering.

de Jong, A. (1995), "IRST and its perspective," *Infrared Technology XXI*, SPIE – The International Society for Optical Engineering.

Deshpande, S.D. *et al.* (1999), "Max-mean and max-median filters for detection of small-targets," *Signal and Data Processing of Small Targets*, SPIE – The International Society for Optical Engineering.

Dougherty, E.R., Pelz, J.B., and Marsh, D.G. (1989), "Morphological granulometric analysis of electrophotographic images," *Applications of Digital Image Processing XII*, SPIE – The International Society for Optical Engineering.

Dougherty, E.R. *et al.* (1992), "Morphological image segmentation by local granulometric size distributions," *Journal of Electronic Imaging*, vol. 1, no. 1, pp. 46-60.

Gaussorgues, G. and Campos, T. (2000), "Low cost shipborne passive infrared search and track," *Infrared Technology and Applications XXVI*, SPIE – The International Society for Optical Engineering.

Hadwiger, H. (1957), *Vorlesungen über Inhalt, Oberfläche, und Isoperimetrie*, Berlin: Springer-Verlag.

Horak, R.J. and Mattioli, J. (1996), "Target detection of very dim objects using grey-level morphologic tophat," *Vision Systems: New Image Processing Techniques*, SPIE – The International Society for Optical Engineering.

Horman, S.R. *et al.* (1999), "Seaborne electro-optical sensors and their technologies," *Infrared Technology and Applications XXV*, SPIE – The International Society for Optical Engineering.

Hufnagel, R.E. (1985), "Propagation through atmospheric turbulence," in Wolfe, W.L. and Zissis, G.J. (eds.), *The Infrared Handbook*, Office of Naval Research: Washington, D.C., ch. 6.

Krikelis, A. (1991), "Computer vision applications with the associative string processor," *J. Parallel and Distributed Computing*, vol. 13, pp. 170-184.

Krikelis, A. and Weems, C.C. (1994), "Associative processing and processors," *Computer*, vol. 27, no. 11, pp. 12-17.

Lea, R.M. (1990), "ASP: a parallel computing technology," *Visual Communications and Image Processing*, SPIE – The International Society for Optical Engineering.

Li, J., Shen, Z., and Biao, L. (1997), "Detection of small moving objects in image sequences," *Automatic Target Recognition VII*, SPIE – The International Society for Optical Engineering.

Li, Z.-Y. and Shen, Z.-K. (1994), "Dim point target detecting from image sequences," *Signal and Data Processing of Small Targets*, SPIE – The International Society for Optical Engineering.

Matheron, G. (1975), *Random Sets and Integral Geometry*, New York: Wiley.

Melendez, K.A. and Modestino, J.W. (1995), "Spatiotemporal multiscan adaptive matched filtering," *Signal and Data Processing of Small Targets*, SPIE – The International Society for Optical Engineering.

Merlo, S.P., Lindgren, R.G., and Davis, P.J. (1997), "Application of Bayesian field track-before-detect and peak likelihood track-after-detect trackers to shipboard infrared search and track," *Signal and Data Processing of Small Targets*, SPIE – The International Society for Optical Engineering.

Minkowski, H. (1903), "Volumen und Oberfläche," *Math. Annalen*, vol. 57, p. 447.

Missirian, J.-M. and Ducruet, L. (1997), "IRST: a key system in modern warfare," *Infrared Technology and Applications XXIII*, SPIE – The International Society for Optical Engineering.

New, W.L. *et al.* (1999), "New method for detection of dim point-targets in infrared images," *Signal and Data Processing of Small Targets*, International Society for Optical Engineering.

Nishiguchi, K., Kobayashi, M., and Ichikawa, A. (1995), "Small target detection from image sequences using recursive max filter," *Signal and Data Processing of Small Targets*, SPIE – The International Society for Optical Engineering.

Peli, T., Vincent, L., and Tom, V. (1993), "Morphology-based algorithms for target detection/segmentation in FLIR imagery," *Architecture, Hardware, and Forward-Looking Infrared Issues in Automatic Object Recognition*, SPIE – The International Society for Optical Engineering.

Pham, Q.H., Brosnan, T.M., and Smith, M.J.T. (1997), "Sequential digital filters for fast detection of targets in FLIR image data," *Automatic Target Recognition VII*, SPIE – The International Society for Optical Engineering.

Potter, J.L. (1992), *Associative Computing: a Programming Paradigm for Massively Parallel Computers*, New York: Plenum Press.

Potter, J. *et al.* (1994), "ASC: an associative-computing paradigm," *Computer*, vol. 27, no. 11, pp. 19-25.

Ritter, G.X. and Gader, P.D. (1985), "Image algebra implementation on cellular array computers," *Computer Architecture for Pattern Analysis and Image Database Management*, Miami Beach, Florida USA: IEEE Computer Society.

Ritter, G.X., Wilson, J.N., and Davidson, J.L. (1990), "Image algebra: an overview," *Computer Vision, Graphics, and Image Processing*, vol. 49, pp. 297-331.

Rivest, J.-F., Soille, P., and Beucher, S. (1993), "Morphological gradients," *Journal of Electronic Imaging*, vol. 2, no. 4, pp. 326-336.

Ronda, V. *et al.* (1999), "Multi-mode algorithm for detection and tracking of point-targets," *Acquisition Tracking and Pointing XIII*, SPIE – The International Society for Optical Engineering.

Ronda, V. *et al.* (2000), "Adaptive threshold based spatio-temporal filtering techniques for detection of small targets," *Signal and Data Processing of Small Targets*, International Society for Optical Engineering.

Salembier, P. *et al.* (1996), "Morphological operators for image and video compression," *IEEE Transactions on Image Processing*, vol. 5, no. 6, pp. 881-898.

Sang, N., Zhang, T., and Wang, G. (1996), "Gray scale morphology for small object detection," *Signal and Data Processing of Small Targets*, SPIE – The International Society for Optical Engineering.

Sang, N., Zhang, T., and Shi, W. (1998), "Characteristics of contrast and application for small target detection," *Signal and Data Processing of Small Targets*, SPIE – The International Society for Optical Engineering.

Schmalz, M.S. and Rittter, G.X. (1994), "Image-algebraic design of multispectral target recognition algorithms," *Image Algebra and Morphological Image Processing V*, SPIE – The International Society for Optical Engineering.

Scott, R. and Lok, J.J. (1996), "IRSTs see their second coming," *Jane's Navy International*, December, pp. 26-28.

Serra, J. (1982), *Image Analysis and Mathematical Morphology*, London: Academic.

Shi, H., Ritter, G.X., and Wilson, J.N. (1997), "A fast general algorithm for extracting image features on SIMD mesh-connected computers," *Pattern Recognition*, vol. 30, no. 7, pp. 1205-1211.

Shih, F.Y.-C. and Mitchell, O.R. (1989), "Threshold decomposition of gray-scale morphology into binary morphology," *IEEE Transactions on Pattern Analysis and Machine Intelligence*, vol. 11, no. 1, pp. 31-42.

Soni, T., Zeidler, J.R., and Ku, W.H. (1992), "Adaptive whitening filters for small target detection," *Signal and Data Processing of Small Targets*, SPIE – The International Society for Optical Engineering.

Sternberg, S.R. (1986), "Grayscale morphology," *Computer Vision, Graphics, and Image Processing*, vol. 35, pp. 333-355.

Stone, L.D., Barlow, C.A., and Corwin, T.L. (1999), *Bayesian Multiple Target Tracking*, Boston, Mass.: Artech House.

Tartakovsky, A. and Blazek R. (2000), "Effective adaptive spatial-temporal technique for clutter rejection in IRST," *Signal and Data Processing of Small Targets*, SPIE – The International Society for Optical Engineering.

Tom, V.T. *et al.* (1993), "Morphology-based algorithm for point target detection infrared backgrounds," *Signal and Data Processing of Small Targets*, SPIE – The International Society for Optical Engineering.

Tzannes, A.P. and Brooks, D.H. (1999), "Point target detection in IR image sequences based on target and clutter temporal profile modeling," *Infrared Technology and Applications XXV*, International Society for Optical Engineering.

Vincent, L. (1994), "Fast opening functions and morphological granulometries," *Image Algebra and Morphological Image Processing V*, SPIE – The International Society for Optical Engineering.

Warren, R.C. (2002), "Detection of distant airborne targets in cluttered backgrounds in infrared image sequences," in *School of Electrical and Information Engineering*, University of South Australia: Adelaide.

Yang, W.-p., Shen, Z.-k., and Li, Z.-y. (1994), "The application of difference method to dim point target detection in infrared images," *IEEE National Aerospace and Electronics Conference*, IEEE.

Zhu, Z. *et al.* (1999), "Detection and acquisition of small targets with low signal-to-clutter ratio," *Signal and Data Processing of Small Targets*, SPIE – The International Society for Optical Engineering.

Chapter 10

Intelligent Monitoring of Complex Systems

D.L. Hall

The rapid evolution in microprocessors, micro-scale sensors, and wireless communications systems enable the development of intelligent monitoring systems for military systems such as aircraft, ships, or weapon systems. These intelligent monitoring systems seek to monitor the health and status of complex systems to detect the onset of failure conditions and to predict the time to failure or remaining useful life. Such intelligent monitoring systems allow the implementation of a condition-based maintenance (CBM) philosophy. CBM is a philosophy of performing maintenance on a mechanical system only when there is objective evidence of the need for such maintenance. By contract, time-based maintenance performs maintenance based on the utilization history or after specified duration of time. Time-based maintenance is unduly expensive for two reasons; (1) maintenance is performed on perfectly good machinery before the need for such maintenance, and (2) the very act of maintenance, even when performed correctly, often induces problems in previously functioning machinery. Thus, time-based maintenance is expensive and may induce mechanical failures. Similarly, failure-based maintenance (the philosophy of performing maintenance after a failure has occurred) can be dangerous and lead to lack of operational readiness for military systems.

This chapter provides an overview of the concept and current research in intelligent monitoring systems to enable condition-based maintenance. Applicable technologies span the spectrum, from advanced sensing and signal processing, to pattern recognition, multi-sensor data fusion, automated reasoning, and modeling and prediction. Many of the technologies from military applications such as automatic target recognition, situation assessment, threat assessment, and target tracking and classification are applicable for developing intelligent monitoring system. The systems model for intelligent monitoring systems mirrors the Joint Directors of Laboratories (JDL) model developed for describing military data fusion systems. This chapter introduces the concept of CBM, provides a systems architecture, summarizes applicable techniques and presents examples of recent research and prototype systems.

1 Introduction

The rapid evolution in microprocessor, micro-scale sensors, and wireless communications systems have provided the opportunity to develop intelligent monitoring systems for military systems such as aircraft, ships, or weapon systems. These intelligent monitoring systems seek to monitor the health and status of complex systems to detect the onset of failure conditions and to predict the time to failure or remaining useful life. In particular, such intelligent monitoring systems allow the implementation of a condition-based maintenance (CBM) philosophy. CBM is a philosophy of performing maintenance on a mechanical system only when there is objective evidence of the need for such maintenance. By contract, time-based maintenance performs maintenance based on the utilization history or after specified duration of time. Such time-based maintenance is unduly expensive for two reasons. First, maintenance is performed on perfectly good machinery before the need for such maintenance. Second, the very act of maintenance, even when performed correctly often induces problems in previously functioning machinery (so-called *iatropic* maintenance problems). Thus, time-based maintenance is unduly expensive and may actually induce mechanical failures. Similarly, failure-based maintenance (viz., the philosophy of performing maintenance after a failure has occurred) can be dangerous and lead to lack of operational readiness for military systems.

This chapter provides an overview of the concept and current research in intelligent monitoring systems to enable condition-based maintenance. The technologies required for developing such systems span a broad spectrum, from advanced sensing and signal processing, to pattern recognition, multi-sensor data fusion, automated reasoning, and modeling and prediction. Interestingly, many of the technologies from military applications such as automatic target recognition, situation assessment, threat assessment, and target tracking and classification are applicable for developing intelligent monitoring system. Indeed, as will be shown in this chapter, the systems model for intelligent monitoring systems is very analogous to the Joint Directors of Laboratories (JDL) model developed for describing military data fusion systems (Hall and Llinas 1997, Hall 1992). In the on-going research in data fusion and intelligent monitoring systems, there are numerous opportunities for bi-directional technology transfer.

2 Motivation for Intelligent Monitoring Systems

The problem of determining the condition of complex mechanical systems and accurately predicting remaining useful life is a challenging, multidisciplinary problem. Scientific issues in CBM range from understanding fundamental failure phenomena in materials to advanced sensing, fusion of multisensor data, automatic pattern recognition, mathematics of complex system evolution, distributed computing and sensing systems, and automated reasoning. The integration of these disciplines to address the CBM problem advances basic knowledge in each of the component scientific disciplines.

Development of technology to enable the implementation of condition-based maintenance for mechanical systems is of vital importance to the department of defense and the navy. Maintenance and overhaul of large platforms and weapon systems such as the CVN-76 comprise over 35 per cent of the projected life-cycle cost. This exceeds the original platform design and development costs. The ability to monitor the health of mechanical equipment and to accurately predict remaining useful life will allow a major reduction in maintenance costs while simultaneously increasing safety and improving mission readiness for key assets.

The importance of accurate health assessment and prediction for electrical and mechanical systems has been cited in numerous national defense plans and studies. For example, the 1996 United States Defense Technology Plan objectives included an 80% reduction in aircraft mechanical mishaps and a 35 % reduction in spare parts inventories. A 1995 OASN study on safety and survivability of aircraft recommended a need to achieve a 30 % reduction in life-cycle maintenance costs. Multiple safety reports have identified the impact of undetected maintenance problems that resulted in the loss of life for military personnel. A 1996 United States Air Force Integrated Product Team report identified 58 preventable mishaps resulting in the loss of 16 lives and an equipment cost of $ 365 M.

Two recent documents cite the need to accelerate the application of embedded diagnostics and condition-based maintenance. The U. S. Army VcoS memorandum, dated April 30 1998, assert that, "… include embedded diagnostics on all new and retrofit equipment…". The memorandum further states, "...We will not field systems or retrofit equipment without embedded diagnostics. If a tradeoff must be made because of funding, Army policy will be to obtain fewer, more capable systems." The US Navy OPNAVINST 4700.16-policy document on Condition-Based Maintenance dated May 6 1998 establishes a navy policy to implement CBM. The document asserts that, "CBM Methodology shall be used to determine maintenance decisions and reduce scheduled maintenance and manpower requirements..". The policy also indicates that, "Chief of Naval Operations (CNO) will fund naval programs, processes, and enabling technologies proven applicable and effective in supporting the maintenance, manning, and cost reduction objectives of this instruction."

CBM has enormous commercial application for the transportation and manufacturing industries. Demonstration of CBM enabling technology will provide the basis for improved automobiles, safer commercial aircraft, improved automation of industrial processes, and many other commercial applications. Companies such as Oceana Sensor Technology, Honeywell, Boeing, Lockheed Martin and many other companies are investing in this technology. Commercial products will include new sensors, health monitoring systems, improved automobiles and aircraft, and new factory automation. Overall, CBM technology improvements will lead to increased safety, increased productivity, and reduced costs of products.

Extensive research has been performed to address the condition-based maintenance problem. Research has included work in advanced sensors, signal processing, multi-sensor data fusion, automated reasoning, mechanical models, and non-linear dynamics. Indeed, similar research was conducted at the Pennsylvania State University funded by the Office of Naval Research under the auspices of a Multi-disciplinary University Research Initiative (MURI) program on Integrated Predictive Diagnostics (IPD). Information on this project may be found at the website http://www.arl.psu.edu /areas/soa/conditionmaint.html. This site contains a summary of the research and list of publications. In addition, research at NASA Ames Research Center has focused on spacecraft applications and monitoring the condition of machinery associated with large-scale wind tunnels. The following website, http://ic-www.arc.nasa.gov/ic /projects/condition-based-maintenance/bibliography.html, contains an extensive bibliography of over 450 research papers on condition-based maintenance, as well as links to web sites for industrial, academic, and government-sponsored research. Various assessments of the state of the art in condition-based maintenance have been conducted. See for example the work by Hall *et al.* (1996) and by Byington and Garga (2001).

3 Concept of Condition Monitoring of Complex Systems

A conceptual model for an intelligent monitoring system is shown in Figure 1. A mechanical system or military platform such as a rotorcraft (shown at the top left-hand side of the figure) is to be monitored for status and mechanical health. Failure mechanisms for such a system may include corrosion, wear, lubricant contamination or degradation, and thermo-mechanical fatigue.

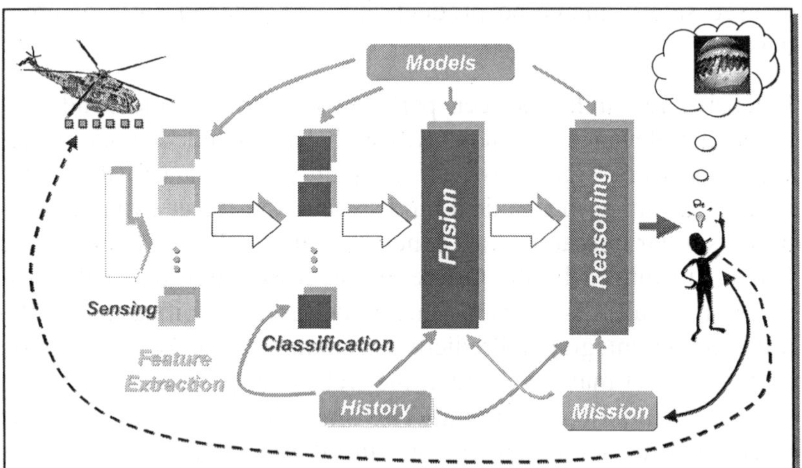

Figure 1. Concept of an intelligent monitoring system (Garga and Hall 1999).

Active and passive sensors monitor the components and subsystems of the rotorcraft. Examples of appropriate sensors include acoustics, infrared sensing for detecting heat, laser monitoring of lubricants, accelerometers and embedded sensors. In addition to these sensors, human inputs may be available to report anomalous operating conditions. In a sense, the problem of determining the health and potential fault conditions of a mechanical system is very analogous to a multi-sensor data fusion system monitoring a battlefield; seeking to determine the current situation and to identify potential threats. The breadth of sensors and utilization of human inputs provide an additional analog to defense-related data fusion problems. These analogies have been further developed by Hall (1994) and by Garga and Hall (1999).

The intelligent monitoring system shown in Figure 1 has multiple components and functions including; (1) active and passive sensors, (2) signal processing and feature extraction, (3) pattern classification, (4) multi-sensor data fusion, (5) automated reasoning, (6) models, (7) historical data input, (8) mission constraints, and (9) human-in-the-loop decision making. Each of these is summarized below.

Active and passive sensing – Many different active and passive sensors are applicable for monitoring a mechanical system. Impending failure conditions cause excessive vibration (Zakrajseh 1989, Brennan *et al.* 1997), noise, heat, debris in lubricants (*LEM* 1987, Richard 1995, Byington *et al.* 1999b), and other symptoms such as acoustic emissions. As might be anticipated a wide variety of sensors have been utilized to monitor such systems. Unfortunately, there are a number of challenges in sensing. First, while mechanical failure is initiated at a materials level, typical sensors observe the results at a subsystem or platform level. Figure 2 illustrates the hierarchy from the material level (where failures are initiated) and the subsystem to platform levels where failure effects are observed.

Thus, we are unable to place sensors where failures actually occur, and must rely on observing the effects of evolving failure conditions propagated to higher levels in the hierarchy. In addition, multiple failure conditions cause very similar observable effects. For example, gear tooth wear, shaft misalignment, and bearing wear all cause excessive vibration and heat. The link between observable conditions and failure causes is many to many. Moreover, the observing conditions are very challenging for sensors. In typical mechanical systems, the sensors fail far more often than the mechanical system that they are designed to monitor.

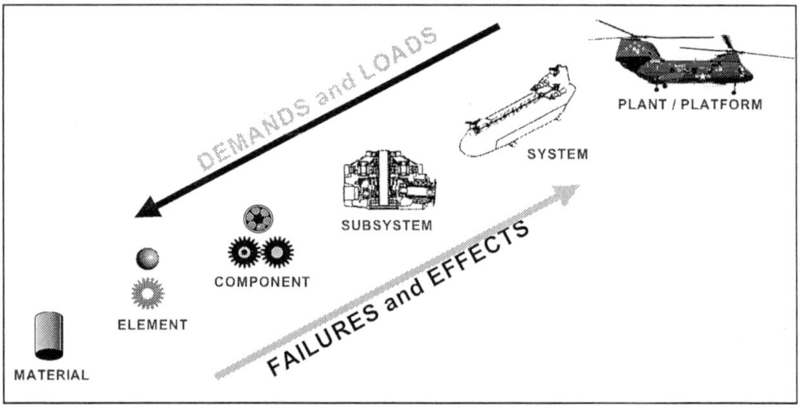

Figure 2. Material to platform hierarchy.

Byington *et al.* (2000b) describe an experiment performed with a simple two-gear mechanical gearbox (called the mechanical diagnostic test bed). In multiple experiments, a two-gear box rated at 10 horsepower was systematically overloaded using a 30 horsepower motor and an applied 70 horsepower load. The gearbox was fitted with approximately 50 sensors including accelerometers, microphones, and torque meters. Information about this experimental test bed can be found at the website http://www.arl.psu.edu/areas/soa /facilitiestools.html#Mechanical.

Under controlled conditions, gearboxes were systematically driven to failure. Even though the load conditions were identical and the sensors were placed in identical locations on the gearboxes, the output from the sensors differed markedly from one transmission to another. Apparently, minor differences in the gearbox housings acted as a non-linear amplifier to result in significantly different readings from the sensors from one test run to another. Hence the idea of using a simple threshold scheme to monitor excess vibration is not a reliable means for detecting failure conditions.

Signal processing and feature extraction – The observational environment for intelligent monitoring systems can very challenging. For example the ambient vibration on a rotorcraft is nearly 1000 g_s. The differential vibration associated with an incipient gear tooth crack may only be ¼ to 1 g. Hence, the signal to be detected has a 1000 to 1 noise to signal ratio. Moreover, the vibrations associated with a rotorcraft transmission involve the fundamental and harmonic frequencies of tens of gear meshes. In general the signals observed by monitoring sensors are very complex and corrupted by noise and interfering signals (Groover *et al.* 2000, Maynard 1999, Wang and McFadden 1993). Hence, extensive signal processing may be performed to reduce noise, focus on frequencies or segments of interest, and perform feature extraction. Numerous features have been utilized for condition monitoring including features in the time domain, frequency domain, and the time-frequency domain.

Pattern classification – After the sensor data are conditioned and features are extracted, then pattern recognition techniques such as neural networks may be used to perform fault detection and identification. While numerous pattern recognition methods have been applied to monitor mechanical systems, the key to success involves selection of robust features. One problem in such selection is that features useful in so-called seeded fault experiments may be brittle or fragile in failure transition experiments (McClintic *et al.* 1999, Garga *et al.* 1999, Byington *et al.* 2000a,b, Garga *et al.* 2000, Hall 2000). Seeded fault experiments involve comparing the results of a normal or no-fault machine with the observations from a machine in which a fault condition has been deliberately planted (or seeded). By contrast, transition experiments attempt to observe the behavior of a feature vector as a machine transitions from normal operation to a failure condition (e.g., induced by overstressing the machine). In general, fault classification based on information from a single sensor tends to be ambiguous with the potential for a high false alarm rate.

The concept of feature extraction and pattern recognition for fault classification is illustrated in Figure 3 below.

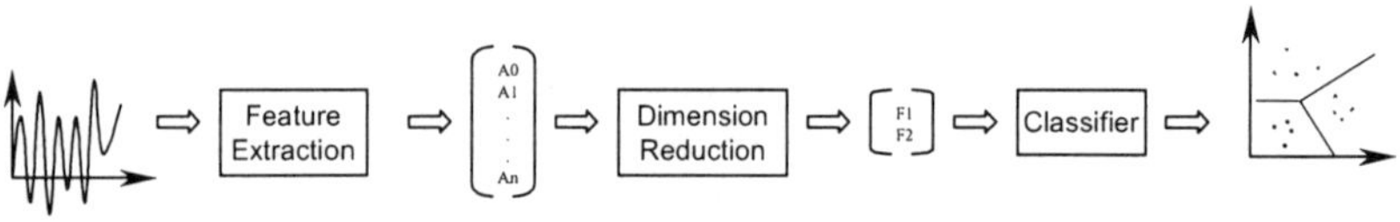

Figure 3. Concept of signal processing and pattern classification for fault identification.

Data from a sensor such as an accelerometer are processed using one or more signal processing algorithms (e.g., extraction of Fourier coefficients, computation of auto-regression (AR) coefficients, wavelet transformations or other methods). Subsequently a feature vector is computed (e.g., a vector of Fourier coefficients). This feature vector is input to a classifier such as a neural net or clustering algorithm. Ideally, the feature vector can be mapped into separate locations in feature space, with each location corresponding to a different fault condition (or to an indication of equipment health). This is illustrated in the right hand side of Figure 3. In practice there are many issues with this approach. Selection of appropriate features, collection and utilization of training data, and choice of classifier are all challenging problems (see, e.g., Garga *et al.* 2000, Hall 2000, Duda and Hart 1973, Huxh and Horne 1993). Typically there is never sufficient training data to develop a robust classifier. Hence, ad hoc substitution methods are used to generate a synthetically large training set.

Multi-sensor fusion – It is intuitive that the use of multiple sensors would improve the ability to determine the health of a mechanical system and to predict time to failure. Improved results should be obtained by expanding the physical baseline of observations and also from combining information from multiple sensors of the same type. Indeed, experiments conducted at The Pennsylvania State University under a Multidisciplinary University Research Initiative grant funded by the Office of Naval Research confirm these anticipated benefits (Byington *et al.* 1999b,c, Garga 1996, Hall and Kasmala 1996, Byington and Garga 2001). Fusion can be performed at a variety of levels from fusion of raw data to feature-level and decision-level fusion. Byington and Garga (2001) provide details on algorithms for fusion at all of these levels. The experiments demonstrate that data fusion can improve the accuracy of fault classification, reduce false alarms, and partially mitigate the effect of poorly performing sensors.

Automated reasoning – Context-based reasoning is important to assist in the monitoring process. Real mechanical systems undergo numerous changes during their operation (e.g., during the course of a mission and during their life span). Symptoms that are meaningful during one part of a mission are meaningless during another part. An example is the vibrations experienced during an aircraft takeoff or landing versus level flight. Hence, the output of sensors and pattern recognition or fusion algorithms cannot simply be taken at face value. Instead the use of automated reasoning assists in interpreting the results (see, e.g., Garga *et al.* 1999). Implicit reasoning techniques such as neural networks or pattern classifiers have been applied for fault detection and classification. In addition, explicit reasoning techniques such as fuzzy-logic rule systems, parametric templates, and case-based reasoning methods have also been applied. Garga *et al.* (1999) have developed hybrid-reasoning methods to improve the performance of both explicit and implicit reasoning methods. In addition, Hall *et al.* (1996) have argued that the use of negative reasoning (viz., reasoning about information that is not observed or conditions that do not appear to be occurring) can improve the diagnosis of mechanical systems. They point out that expert diagnosticians often use negative reasoning.

Modeling – In order to understand the condition of a system and to predict future evolution, models of the system being monitored are required. Ideally, models should include information about the machine construction and makeup, the transformations from platform-induced demands to materials level effects. Models should allow prediction of observations based on hypothesized fault conditions (observation models). Finally, dynamic models are sought to support prediction of the system evolution (viz., state transition models). Unfortunately, such models rarely exist. Much work has been performed to develop materials level models (e.g., crack growth models, stress/strain models, and finite-element models (Dubeck *et al.* 1999a)). However, there are no general models to link materials phenomena to platform results, or to predict observations from hypothesized failure conditions. Finally, system-level failure evolution models do not exist. Hence, it is necessary to combine limited physics-based models with more ad hoc models based on the experience of trained mechanic (Dubeck *et al.* 1999a,b, Begg *et al.* 1999a,b, 2000, Begg and Byington 2000).

Historical data input – Historical maintenance information and information about the previous missions and system use can be used to assist in interpreting the current state and prediction of future conditions. An aircraft that has been utilized for routine flights may have a significantly different condition than an aircraft flown on combat missions – even though they nominally have experienced the same number of flight hours. Moreover, long-term data such as oil debris analysis may be useful to interpret the condition and evolution of a machine. Byington and Garga (2001) point out that for mechanical systems such as gearboxes, the morphology of wear particles in circulating oil is different for benign wear compared to those generated by the active wear associated with pitting, abrasion, scuffing, fracturing and other abnormal conditions. Roylance and Raadnui (1994), for example, correlated particle features (quantity, size, composition and morphology) with wear characteristics (severity, rate, type, and source).

Mission constraints – A key for military systems is mission constraints. While commercial systems may be able to accept recommendations for maintenance, during military operations mission constraints may severely limit the possible maintenance or mitigation actions. Hence, mission constraints and information must be considered when performing context-based reasoning about a system being monitored.

Human-in-the-loop decision-making – Despite advances in intelligent monitoring systems, there remains a role for the human to act as a decision-maker. The pilot, system operator, or maintenance technician may be involved in interpreting the results of the intelligent monitoring system and making final decisions related to system operation or maintenance. The definition of this role is part of the overall systems engineering process in designing a system. For military systems in particular, it is desired for a human operator to be capable of overriding recommendations by an intelligent monitoring system in order to accomplish a mission (even when the act of override may endanger equipment). Byington *et al.* (1999a) provides an excellent discussion of these issues as they relate to flight crews.

4 Challenges for Intelligent Monitoring Systems

Despite many advances in the development of intelligent monitoring systems. There are still remaining challenges. Table 1 provides a summary of some research challenges in the areas of sensing, signal processing and classification, multi-sensor data fusion, automated reasoning, and modeling and prediction. Hansen *et al.* (1994), Hall (1997), and the discussion below provide additional comments on research challenges.

Table 1. Summary of research challenges for intelligent monitoring systems.

Research Area	Challenges & Limitations	Potential Research Tasks
Sensing	Multiple types of sensors exist to measure relevant phenomena; Issues involve - Selection of sensors - Placement of sensors for retrofit to existing machines - Embedded (smart materials) sensors - Environmental effects on sensors - Sensor reliability - Accurate prediction of sensor performance	Development of the following; - Intelligent sensors with embedded processing - Micro-scale and nano-scale sensors - Accurate models for dynamic sensor performance - Self-aware sensors that monitor their own health and performance - Dynamic tasking of sensors to improve health monitoring
Signal Processing & Pattern Classification	Extensive research has been performed on digital signal processing and pattern classification; methods have been developed for the time-domain, frequency-domain, and time-frequency domains	Continued analysis of signal processing methods; - Embedded signal processing in sensors - Automated algorithm selection - Automated feature extraction and feature selection - Context-based processing
Multi-sensor data fusion	Enormous legacy exists from Department of Defense research on data fusion; Initial experiments have been performed to fuse data for intelligent system monitoring	Research is needed to; - Develop fusion methods for non-commensurate sensors - Improve modeling of uncertainty - Dynamic assessment of sensor contributions - Methods to fuse sensors on widely different observation time-scales
Automated Reasoning	Numerous prototype reasoning systems have been developed (e.g., rule-based systems and neural network pattern recognition systems); Challenges exist due to the lack of training data available and lack of transitional failure data;	Advances are required in; - Development of hybrid reasoning systems - New models for uncertainty propagation - Methods to implement negative reasoning - Understanding the scalability of automated reasoning systems

Table 1 (continued).

Research Area	Challenges & Limitations	Potential Research Tasks
Modeling & prediction	Extensive research has been performed to develop models at the materials level (e.g., crack growth and wear models), and system level models (e.g., finite element models).	Additional research is needed to; - Link materials level phenomena with platform level phenomena - Allow prediction of sensor observations given hypothesized fault conditions - Models of the observability of failure phenomena - Models to predict failure evolution and trajectories
Automated reasoning	Development of numerous rule-based expert systems to perform machine diagnosis. Limited success in context-based reasoning and systems level interpretation of sensor data; most systems have very limited success and are brittle in changing contexts	Research is needed to; - Incorporate both implicit and explicit information - Reason with negative information; - Develop a general calculus of uncertainty - Perform robust context-based reasoning

Sensing – As illustrated in Figure 2, failure phenomena occur at the materials level (viz., material corrosion, cracks, fatigue, or other phenomena). In general, sensors are not placed at the microscopic material level. Instead, sensors are used at a subsystem or system level to sense macro-observable quantities such as vibration or heat. As previously indicated this is a challenging sensing environment with large noise to signal ratio. Moreover, sensor systems in these environments tend to be unreliable. Failures often occur in the sensing infrastructure itself (e.g., connectors or power sources). Indeed, when a sensor indicates a failure condition, it is often more likely that the sensor itself has failed. The changing environment associated with machinery causes large changes in ambient vibration and temperature. Under these conditions, sensor calibration becomes very difficult. All of these conditions are exacerbated for situations in which existing machines are retrofit for condition-based maintenance. Potential research areas include the development of intelligent, self-calibrating sensors with embedded processing and wireless communications. The ideal sensor would be self-aware and monitor their performance and health. Sensor control of multi-sensor systems is also an area for research.

Signal Processing – Very extensive research has been conducted in the application of signal processing and pattern classification techniques for fault detection and identification. Signal processing techniques have included time-domain, frequency-domain, and time-frequency domain techniques. The basic concept for signal processing is to condition a signal from a sensor and to characterize the signal via a feature vector. For detection of faults on rotating machinery, a standard practice is to perform time-synchronous averaging, in which a sensor signal is averaged over numerous rotation cycles of the machine. Conceptually this enhances any feature that results from a recurring (cyclical) interaction (e.g., the interaction between a cracked tooth on a gear mesh). Researchers at Penn State University have developed a condition-based maintenance toolbox that incorporates many of these methods. One of the problems in comparing techniques from different researchers is that the algorithms they create and use are often considered to be proprietary. Hence, a researcher may report on the results of a new feature, but not publish the precise feature definition to allow replication of the test results. Key issues in signal processing involve how to automatically select features for effective classification and diagnosis; how robust these features are to changes in operating conditions; how useful the features are for tracking the evolution of a failure condition; and how to use features to predict failure evolution.

Pattern Classification – Given a selected feature vector, one of a number of pattern classification algorithms may be used to transform from feature space into decision space (as illustrated in Figure 3). Examples of algorithms include neural networks, clustering algorithms, decision-trees, and parametric templates. Duda and Hart

(1973) provide an excellent introduction to such techniques. Currently, there are many commercial tool kits for pattern classification (Hall 2002). If a "good" feature vector has been selected, then the pattern recognition process (viz., transforming the feature vector into a declaration of fault condition) is straightforward. However, when there is significant overlap in feature space, it is difficult to unambiguously declare that a particular failure or fault condition is imminent. Even if a robust feature vector can be found, a major challenge is training the pattern classification technique. Hush and Horne (1993) point out that the number of independent training cases to train a classification technique is

$$\text{Number of training cases} = (N_{\text{feature vector}}) \times (N_{\text{classification vector}}) \times (10 - 30).$$

That is, the number of training cases required is equal to the number of components of the feature vector, times the number of components in the classification vector (i.e., the number of fault types), times a number between ten and thirty. It is common for a feature vector to contain ten to fifty coefficients while the number of failure types may range from twenty to fifty. Hence, the number of independent training cases could lie in the thousands. Ideally, these would be collected over a wide variety of operating conditions. Clearly, it is very rare to have such a number of test cases available. Hence the resulting trained pattern classifier is often poorly trained so that the solution is not very general (viz., it is specific only to the test cases presented to date). Garga (1996) has performed some interesting research in which he combines implicit and explicit reasoning. Specifically, he represents explicit information about a machine or failure conditions using fuzzy logic rules (obtained from an expert diagnostician). He subsequently trains a neural network to represent the rule-base. This becomes the baseline network that is in turn trained on sample data. The resulting pattern classifier is much more robust than those trained only from the sample data. Extensive research is required in this area to incorporate a combination of implicit information ("learned" from sample data), explicit knowledge from a domain expert, and model data.

Multi-Sensor Data Fusion – The concept of multi-sensor data fusion has been recently applied to monitoring of complex systems (Byington and Garga 2001) and will be presented in more detail in the next section of this chapter. The concept of data fusion is not new. Humans and animals utilize multiple senses to improve their knowledge of the external world. Very extensive research has been funded by the Department of Defense to automate the fusion of multi-sensor data (Hall 1992, Hall and Llinas 1997, 2001) for applications such as automatic target recognition, threat assessment and situation assessment. A wide variety of techniques are used for this function, ranging from signal processing and statistical estimation methods to automated reasoning and pattern classification methods. For monitoring complex systems, the goal is to improve classification accuracy and to reduce ambiguity. For example, most mechanical faults result in increased vibration. Hence, while accelerometers provide notification of an impending failure condition, they are relatively poor at identifying the specific source of the failure condition. There is a many-to-many mapping between observed phenomena (e.g., vibration, increased temperature) and potential failure conditions. The goal of multi-sensor data fusion is to assist in reducing these ambiguities and improving the accuracy of system characterization.

Modeling & Prediction – Another component of monitoring of complex systems involves modeling and prediction. Conceptually there are two key areas to consider: (1) prediction of the value of observations of macro-phenomena given the hypothesized condition of a machine, and (2) modeling the evolution of a failure condition with continued utilization of a machine. The former predictions are typically performed using finite-element models and geometrical constructs (see for example the discussion by Hansen *et al.* 1994). This is a challenging problem because it requires transforming mechanical conditions at the materials level (e.g., a small crack in a gear tooth) to the resulting large-scale vibration effects or related phenomena. Hence, one must hypothesize the existence of a failure condition and predict macro-observable quantities. These models depend upon the specific configuration and makeup of an individual machine. By the same token, the ability to model the evolution of a fault condition is also difficult. The evolution of a crack or spar on a mechanical element depends upon the details of the originating failure condition, the makeup of the machine, and the loads that are placed on it. Some conditions may evolve quickly, while others may take a very long time to evolve. In experiments conducted at the Pennsylvania State University using a mechanical engineering test bed (Reichard 1995), the time to failure for a simple two-gear transmission varied widely from a few days to more than two months (for nominally identical gearboxes driven under identical overload conditions). While extensive research as been conducted in this area, only limited progress has been made to date. Perhaps the most general approach has been that of Cusumano *et al.* (2000) who have developed a general nonlinear mathematical formulation to describe the evolution of complex systems.

Automated Reasoning – The final component of a monitoring system is the ability to perform automated reasoning to understand the context of the results obtained from models and predictions. Extensive research has focused on development of rule-based expert systems to emulate the diagnostic capability of a mechanic. A survey of these prototype systems is available on the NASA web site previously cited. In general, automated reasoning is a very challenging problem. Hall *et al.* (1996) provide a general review of automated reasoning applied to CBM systems and discuss research challenges and limitations. Briefly, it should be noted that these automated reasoning systems are very limited. Issues related to negative reasoning, non-monotonic reasoning, and representation of uncertainty, knowledge engineering and other factors make the construction of such systems very difficult. While numerous commercial tools exist to assist the prototyping of expert systems, the actual obtaining of expert knowledge and representation of such knowledge is difficult.

5 Multi-sensor Data Fusion

This section provides a brief introduction to multi-sensor data fusion and its application to condition monitoring systems. The basic concept of data fusion is not new. Extensive research has been conducted for military applications over the past twenty years (see, for example, Hall and Llinas 2001). The emergence of new sensors, advanced processing techniques, and improved processing hardware make real-time fusion of data increasingly possible. For department of defense applications, data fusion systems are used extensively for target tracking, automated identification of targets, and limited automated reasoning applications. Spurred by significant expenditures by the Department of Defense, data fusion technology has rapidly advanced from a loose collection of related techniques to an emerging true engineering discipline with standardized terminology, collections of robust mathematical techniques, and established design principles. Software in the area of data fusion applications is becoming increasingly available in the commercial marketplace (Hall 2002).

In principle, fusion of multisensor data provides significant advantages over single source data. In addition to the statistical advantage gained by combining same-source data (e.g., obtaining an improved estimate of a physical phenomena via redundant observations), the use of multiple types of sensors may increase the accuracy with which a quantity can se observed and characterized. For example the ability to characterize the state of a machine may be greatly enhanced if we combine information related to vibration, noise, temperature, oil debris, and pressure. Ambiguity induced by use of a single phenomenon (e.g., vibration) may be resolved by utilizing other physical phenomena.

One of the historical barriers to technology transfer in data fusion has been the lack of a unifying terminology, which crosses application-specific boundaries. Even within military applications, related but different applications such as automatic target recognition (ATR) and identification-friend-foe (IFF) systems have used different definitions for fundamental terms such as correlation and data fusion. In order to improve communications among military researchers and system developers, the Joint Directors of Laboratories (JDL) Data Fusion Working Group, established in 1986, began an effort to codify the terminology related to data fusion. The result was the creation of a process model for data fusion, and a Data Fusion Lexicon (White 1991, Kessler *et al.* 1992). The top level of the JDL data fusion process model is shown in Figure 4.

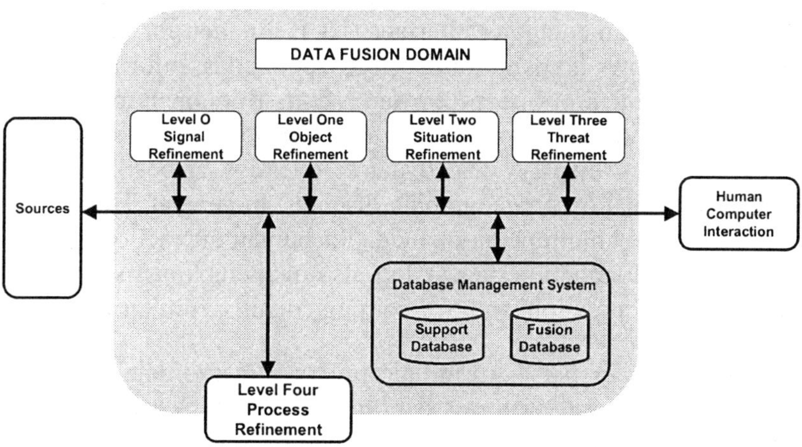

Figure 4. JDL Data Fusion Process Model (Hall 1992).

The JDL process model is a functionally oriented model and is intended to be very general and useful across multiple application areas. The model identifies the processes, functions, categories of techniques and specific techniques applicable to data fusion. The model is a three-layer hierarchy. At the top level shown in Figure 4, the data fusion process is conceptualized by; sources of information, human-computer interaction, source preprocessing, Level 1 processing, Level 2 processing, Level 3 processing, and Level 4 processing. Each of these is summarized below. A summary of data fusion algorithms and techniques is provided in Table 2.

Sources of information – The left side of Figure 4 indicates that a number of sources of information may be available as input including: 1) local sensors associated with a data fusion system (e.g., sensors physically associated with the data fusion system) or organic sensors integrated with a data fusion system platform, 2) distributed sensors linked electronically to a fusion system, and 3) other data such as reference information.

Human-Computer Interaction (HCI) – The right side of Figure 4 shows the human computer interface (HCI) function for fusion systems. HCI allows human input such as commands, information requests, assessment of inferences, and reports from operators. In addition, HCI is the mechanism by which a fusion system communicates via alerts, displays, and dynamic overlays of positional and identity information on geographic displays. In general, HCI incorporates not only multimedia methods for human interaction (graphics, sound and tactic interfaces), but also methods to assist humans in direction of attention, and overcoming cognitive limitations.

Source Preprocessing – An initial process allocates data to appropriate processes and performs data pre-screening. Source preprocessing reduces the data fusion system load by allocating data to appropriate processes (e.g., location and attribute data to Level 1 object refinement). Source preprocessing also directs the data fusion process to concentrate on the data most pertinent to the current situation. Extensive signal processing and detection theory may be required. A special case of source preprocessing is the synthesis of multiple component sensory array data to estimate the location of a target (or fault in the case of a condition monitoring system).

Level 1 Processing (Object Refinement) – This process combines location, parametric, and identity information to achieve refined representations of individual objects. For DoD applications these may include emitters, platforms, weapons, or geographically constrained military units. For condition monitoring systems, these may include the location of fault conditions or anomalies. Level 1 processing performs four key functions; 1) transforms sensor data into a consistent set of units and coordinates, 2) refines and extends in time estimates of an object's position, kinematics, or attributes, 3) assigns data to objects to allow the application of statistical estimation techniques, and 4) refines the estimate of an object's identify of classification.

Level 2 Processing (Situation Refinement) – Level 2 processing develops a description of the current relationships among objects and events in the context of their environment. Distributions of individual objects (defined by Level 1 processing) are examined to aggregate them into operationally meaningful combat units and weapon systems. In addition, situation refinement focuses on relational information (i.e., physical proximity, communications, causal, temporal, and other relations) to determine the meaning of a collection of entities. This analysis is performed in the context of environmental information about the terrain, surrounding media, hydrology, weather, and other factors. Situation refinement addresses the interpretation of data analogous to how a human might interpret the meaning of sensor data. Both formal and heuristic techniques are used to examine, in a conditional sense, the meaning of the Level 1 processing results.

Level 3 Processing (Threat Refinement) – Level 3 processing projects the current situation into the future to draw inferences about enemy threats, friendly and enemy vulnerabilities, and opportunities for operations. Threat assessment is especially difficult because it deals not only with computing possible engagement outcomes, but also assessing an enemy's intent based on knowledge about enemy doctrine, level of training, political environment, and the current situation. Level 3 processing develops alternative hypotheses about an enemy's strategies and the effect of uncertain knowledge about enemy units, tactics, and the environment. Game theoretic techniques are applicable for level 3 processing.

Level 4 Processing (Process Refinement) – Level 4 processing can be considered a meta-process, i.e., a process concerned with other processes. Level 4 processing performs four key functions; 1) monitors the data fusion process performance to provide information about real-time control and long-term performance, 2) identifies what information is needed to improve the multilevel data fusion products (inferences, identities, and locations), 3) determines the source-specific requirements to collect relevant information (i.e., which sensor type, which specific sensor), and 4) allocates and directs the sources to achieve mission goals. This latter function may be outside the domain of specific data fusion functions. Hence, Level 4 processing is shown as partially inside and partially outside the data fusion process (in Figure 4).

Data Management – The most extensive support function required to support data fusion is data management. This collection of functions provides access to, and management of, data fusion databases including data retrieval, storage, archiving, compression, relational queries, and data protection. Database management for data fusion systems is particularly difficult because of the large and varied data managed (e.g., images, signal data, vectors, textual data) and the data rates both for the ingestion of incoming sensor data as well as the need for rapid retrieval.

The application of data fusion techniques to condition monitoring of complex systems would appear to provide advantages for accurately characterizing the state of the system. To date, however, only a few attempts have been made to develop such systems. McGonigal (1997) compared three techniques (neural networks, fuzzy logic, and rule-based reasoning) for decision level fusion of vibration data for the mechanical diagnostic test bed. Koslowski developed a feature-level fusion approach for battery systems that proved to be very robust (see the discussion by Byington and Garga 2001). Erdley (1997) investigated several voting schemes for data fusion for the Penn State mechanical diagnostic test bed, and showed the utility of fusion methods. Finally, Garga (1996) has utilized hybrid-reasoning techniques for feature-level and decision-level fusion.

Table 2. Examples of data fusion algorithms and techniques.

JDL Process	Processing Function	Techniques
Level 1: Object Refinement	Data alignment	• Coordinate transforms • Units adjustment
	Data/object correlation	• Gating techniques (Blackman 1986) • Multiple hypothesis association, probabilistic data association (Bar-Shalom 1989) • Nearest neighbor
	Position, kinematic, and attribute estimation	• Sequential estimation (Gelb 1974) o Kalman filtering o $\alpha\beta$ filter o Multiple Hypothesis (Streit & Luginbuhl 1993) • Batch estimation • Maximum Likelihood (Streit 1994) • Hybrid methods (Poore 1994)
	Object identity estimation	• Physical models • Feature-based techniques o Neural networks (Hush and Horne 1993) o Cluster algorithms (Fukanaga 1990) o Pattern recognition (Fukanaga 1990) • Syntactic models
Level 2: Situation Refinement	Object aggregation Event/activity interpretation Contextual interpretation	• Knowledge-based systems o Rule-based expert systems (Durkin 1994) o Fuzzy logic (Garga *et al.* 1999) o Frame-based KBS • Logical templating (Hall and Linn 1989) • Neural networks • Blackboard systems (Llinas and Antony 1993)

Table 2 (continued).

JDL Process	Processing Function	Techniques
Level 3: Threat Refinement	Aggregate force estimation Intent prediction Multi-perspective assessment	• Knowledge-based systems o Rule-based expert systems o Fuzzy logic o Frame-based KBS • Logical templating • Neural networks • Blackboard systems
Level 4: Process Refinement	Performance evaluation	• Measures of evaluation • Measures of performance • Utility theory
	Process control	• Multi-objective optimization o Linear programming o Goal programming
	Source requirement determination	• Sensor models
	Mission management	• Knowledge-based systems

All of these are promising starts, but much work is still required. In order to assist experiments in this area, Hall and Kasmala (1996) developed a visual programming data fusion tool kit. This tool kit supports rapid prototyping and experimentation. The tool provides a collection of algorithms for signal conditioning, feature extraction, and data fusion. It has been applied to CBM data including gearbox data, data from NASA Ames Research Center wind tunnel equipment, and data collected from a rotorcraft gearbox. The tool provides a "point and click" graphical interface for creation of processing flows, selection of algorithms, and display of output results. The top level graphical interface is shown in Figure 5.

6 Summary

Continuing advances in microprocessors, micro-scale sensors and wireless communications has provided the basis technology for the development of intelligent monitoring systems. Significant research has been performed in the areas of sensing, signal processing and pattern classification, multi-sensor data fusion, automated reasoning, and modeling. These advances are very promising and are anticipated to lead to machines that provide information on their own health (viz., machines that feel their pain) and are capable of predicting the time to failure. When these capabilities are achieved it will have significant impacts on maintenance philosophies, logistics, and military operations. Moreover, such capabilities will migrate to commercial systems to achieve intelligent buildings, intelligent manufacturing systems and intelligent transportation systems. Such facilities and entities will significantly affect our daily lives.

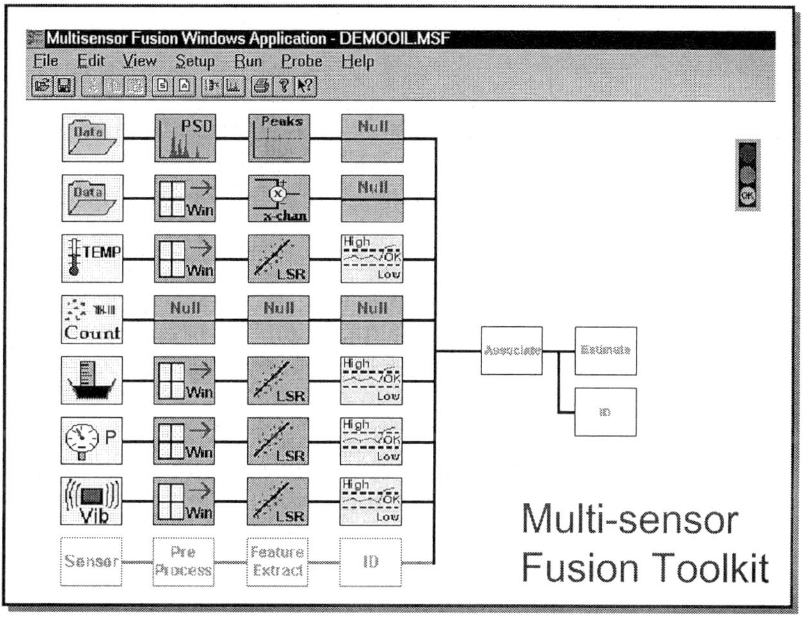

Figure 5. Multi-sensor Data Fusion Tool Kit.

References

Bar-Shalom, Y. (ed.) (1989), *Multi-Target, Multi-Sensor Tracking: Applications and Advances*, Artech House, Inc. Norwood, MA.

Begg, C.D. and Byington, C.S. (2000), "Gear fault transition and ideal torsional structural excitation," *Proceedings of the 13th International Congress and Exhibition for Condition Monitoring and Diagnostic Engineering Management*, Vibration Institute, MFPT Division.

Begg, C.D., Byington, C.S., Maynard, K.P., and Merdes, T.A. (1999a), "Dynamics modeling for mechanical fault diagnostics and prognostics", *Maintenance and Reliability Conference*, 10-12 May.

Begg, C.D., Byington, C.S., Maynard, K.P., and Merdes, T.A. (1999b), "Mechanical system modeling for failure diagnosis and prognosis," *Maintenance and Reliability Conference (MARCON 99)*, May 10-12.

Begg, C.D., Byington, C.S., and Maynard, K.P. (2000), "Dynamic simulation of mechanical fault transition," *Proceedings of the 54th Meeting of the Society for Machinery Failure Prevention Technology*, pp. 203-212.

Blackman, S. (1986), *Multiple Target Tracking with Radar Applications*, Artech House, Inc. Norwood, MA.

Brennan, J., Chen, M.H., and Reynolds, A.G. (1997), "Use of vibration measurements to detect local tooth defects in gears," *Sound and Vibration*, November.

Byington, C.S. and Garga, A.K. (2001), "Data fusion for developing predictive diagnostics for electromechanical systems,"

Chapter 23 in Hall, D. and Llinas, J. (eds.), *Handbook of Multisensor Data Fusion*, CRC Handbook, CRC Press.

Byington, C.S. *et al.* (1999a), "Evaluation of human-machine interfaces for aircrew fault diagnosis and management," *International Conference on Condition Monitoring at Swansea*, 12-15 April.

Byington, C.S., Kozlowski, J.D., and Merdes, T.A. (1999b), "Fusion techniques for vibration and oil debris/quality in gearbox failure testing," *International Conference on Condition Monitoring*, Swansea, 12-15 April.

Byington, C.S., Kozlowski, J.D., and Merdes, T.A. (1999c), "Vibration and oil debris feature fusion in gearbox failures", *53rd Meeting of the Society for MFPT*, April 19-22, 1999.

Byington, C.S., Campbell, R.L., Lebold, M.S., and Scheie, E.O. (2000a), "Generations of HUMS diagnostic estimates using transitional data," *13th International Congress and Exhibition on Condition Monitoring and Diagnostic Engineering Management*, 3-8 December.

Byington, C.S., Campbell, R.L., Lebold, M.S., Maynard, K.P., and McClintic, K.T. (2000b), "Residual and difference feature analysis with transitional gearbox data," *Improving Productivity through Applications of Condition Monitoring*. Presented at the *54th Meeting of the Society of Machinery Failure Prevention Technology*, pp. 635-645, May.

Byington, C.S., Garga, A., Kasmala, G.F., Lang, D.C., and Scheie, E.O. (2000c), "Platform level reasoning for an SH-60B helicopter using intelligent agents," *54th Society of Machinery Failure Prevention Technology Conference*, May 1-4.

Cusumano, J.P., Chatterjee, A., and Chalidze, D. (2000), "Some fundamental issues in damage tracking with application to a low-dimensional model of a gear-pair transmission," *Proceedings of the 54th MFPT Meeting*, April, pp. 507-520.

Dubeck, J., Garga, A.K., and Shirkhodaie, A. (1999a), "Bearing fault diagnostics using physics-based finite element modeling techniques," *Proceedings of the MFPT Society 53rd Meeting*, pp. 483-492, 19 April.

Dubeck, J., Garga, A.K., and Shirkhodaie, A. (1999b), "Physics-based modeling and diagnosis of high performance bearings," *Proceedings of Maintenance and Reliability Conference MARCON '99.*

Duda, R.O. and Hart, P.E. (1973), *Pattern Classification and Scene Analysis*, Wiley, NY.

Durkin, J. (1994), *Expert Systems: Design and Development*, Macmillan, New York.

Erdley, J.D. (1997), "Improved fault detection using multisensor data fusion," M.Sc. Thesis in Electrical Engineering, The Pennsylvania State University.

Fukanaga, K. (1990), *Introduction to Statistical Pattern Recognition*, 2nd ed., Academic Press, New York.

Garga, A.K. (1996), "A hybrid implicit/explicit automated reasoning technique for condition-based maintenance," *Proceedings of the ASME Intelligent Ship Symposium II.*

Garga, A.K. and Hall, D.L. (1999), "A unifying approach for rotorcraft information systems," *Proceedings of the American Helicopter Society 55th Annual Forum*, Montreal, Canada, pp. 25-27, May.

Garga, A.K., Hall, D.L., and Stover, J.S. (1999), "Machinery fault classification – the case for a hybrid fuzzy logic approach," *Proceedings of the MFPT Society 53rd Meeting*, pp. 241-250, 19-22 April.

Garga, A.K., Lang, D.C., and Maynard, K.P. (2000), "Feature extraction strategies and data fusion methodologies: a tutorial," *MFPT Society 54th Meeting*, May 1-4.

Gelb, A. (1974), *Applied Optimal Estimation*, MIT Press, Cambridge, MA.

Groover, C.L., Maynard, K.P., and Threthewey, M. (2000), "Application of double resampling to shaft torsional vibration measurement for the detection of blade natural frequencies," *Proceedings of the 54th Meeting of the Society for Machinery Failure Prevention Technology*, pp. 87-94.

Hall, D.L. (1992), *Mathematical Techniques in Multisensor Data Fusion*, Artech House, Inc., Norwood, MA.

Hall, D.L. (1994), "Application of the JDL data fusion model to condition-based maintenance systems," *Proceedings of the 1994 Data Fusion Systems Conference*, John Hopkins University, Laurel, MD.

Hall, D.L. (1997), "Condition-Based Maintenance Report on Task 6: Application of Prognostic Techniques," project report prepared for the NASA Ames Research Center, (Subcontract No. GFY900240), Oct.

Hall, D.L. (2000), "An overview of pattern recognition techniques," Tutorial given at the *NATO Advanced Study Institute on Multisensor Data Fusion*, Pitlockry, Scotland, June.

Hall, S.A. (2002), “A survey of COTS software for multi-sensor data fusion: what’s new since Hall and Linn?” *Proceedings of the 2002 National Symposium on Sensor and Data Fusion*, August, San Diego, CA.

Hall, D.L. and Kasmala, G. (1996), “A visual programming tool kit for multisensor data fusion,” *Proceedings of the SPIE AeroSense 1996 Symposium*, Orlando, FL, vol. 2764, pp. 181-187, April.

Hall, D.L. and Linn, R.J. (1989), “Comments on the use of templating for multi-sensor data fusion,” *Proceedings of the 1989 Tri-Service Data Fusion Symposium*, vol. 1, pp. 345-354, May.

Hall, D.L. and Llinas, J. (1997), “An introduction to multi-sensor data fusion,” *Proceedings of the IEEE*, vol. 85, no. 1, pp. 6-23.

Hall, D.L. and Llinas, J. (ed.) (2001), *Handbook of Multisensor Data Fusion*, CRC Press.

Hall, D.L., Hansen, R.J., and Lang, D.C. (1996), “The negative information problem of mechanical diagnostics,” *Proceedings of the International Gas Turbine Institute Conference*, Birmingham, England, ASME, June.

Hansen, R.J., Hall, D.L., and Kurtz, S.K. (1994), “A new approach to machinery prognostics,” *International Gas Turbine and Aeroengine Congress and Exposition*, 94-GT-3, June.

Hush, D. and Horne, B. (1993), “Progress in supervised neural networks: what’s new since Lippman?” *IEEE Signal Processing Magazine*, pp. 8-39, January.

Kessler, O. *et al.* (1992), “Functional Description of the Data Fusion Process,” Technical Report, Office of Naval Technology, Naval Air Development Center, Warminster, PA, January.

LEM (1987), "Lubrication fundamentals: fundamental concepts and practices," *Lubrication Engineering Magazine*, September, pp. 22-34.

Llinas, J. and Antony, R. (1993), "Blackboard concepts for data fusion and command and control applications," *International Journal on Pattern Recognition and Artificial Intelligence*, vol. 7, no. 2, April.

Maynard, K.P. (1999), "The application of noise processing to gear fault detection," *Interstitial Processing: Proceedings of the International Conference on Condition Monitoring*, University of Wales Swansea, UK, 12-16 April, pp. 77-86.

McClintic, K.T., Nickerson, G.W., and Reichard, K.M. (1999), "Feature prediction and tracking for monitoring the condition of complex mechanical systems," *Intelligent Ship Symposium*, June.

McGonigal, D.L. (1997), "Comparison of automated reasoning techniques for CBM," M.Sc. Thesis in Electrical Engineering, The Pennsylvania State University, August.

Poore, A. (1994), "Multi-dimensional assignment formulation of data association problems arising from multi-target and multi-sensor tracking," *Computational Optimization Applications*, vol. 3, pp. 27-57.

Reichard, C. (1995), "Oil analysis technique advances with new information technology," *P/PM Technology Magazine*, June, pp. 60-62.

Roylance, B.J. and Raadnui, S. (1994), "The morphological attributes of wear particles – their role in identifying wear mechanisms," *Wear*, 175.

Streit, R.L. (1994), "Maximum likelihood methods for probablistic multi-hypothesis tracking," *SPIE Proceedings*, vol. 2234-5, pp. 394-406.

Streit, R.L. and Luginbuhl, T.E. (1993), "A probablistic multi-hypothesis tracking algorithm without enumeration and pruning,", *Proceedings of the 6th Joint Services Data Fusion Symposium*, John Hopkins University, Laurel, MD, June.

Wang, W.J. and McFadden, P.D. (1993), "Application of the wavelet transform to gearbox vibration analysis," *Structural Dynamics and Vibration Proceedings 16th Annual Energy Sources Technology Conference and Exposition*, Jan 31-Feb 4, American Society of Mechanical Engineering, vol. 52, pp. 13-20.

White, F.E. (1991), *Data Fusion Lexicon*, Data Fusion Subpanel of the Joint Directors of Laboratories Technical Panel for C^3, Code 4202, NOSC, San Diego, CA.

Zakrajseh, J.J. (1989), "An investigation of gear mesh failure prediction techniques," NASA Technical Memorandum 102340 (89-c-005), November.

Index

- A –

abstraction, 260
access control, 10
adding personality, 178
advance constraint, 198
agent,
  architecture, 251, 306
  behaviors, 306
  capabilities, 304
  development, 257, 346
  goals, 301
  objects, 189
  phenotype (self), 296
  pheromone, 322
  technology, 248
agent-enabled sensors, 274
AIP,
  applications, 51
  techniques, 48
air operational analysis, 231
algorithm assessment, 402, 416
AP3-C Orion operating procedure, 245
artificial intelligence, 48,
  paradigms, 49
  technologies, 70, 107
ASM detection, 387
associative processing, 383
attack,
  types, 24
authentication, 6
availability, 11

- B -

battlefield object, 189
BattleModel, 240
Bayesian networks, 76
belief systems, 76
biological systems, 295
biologically inspired agent, 301
blackboard systems, 70

- C -

calculating sensor relationships, 309
carrier sense multiple access (CSMA), 216
case-based reasoning, 72
CDMA, 218
center of gravity analysis, 339, 340
center of gravity problem, 339
centralized control, 277
CGA requirements, 80
CGF systems, 98
classification,
  cognitive, 42
  data processing, 41
  functional, 38
  hierarchical, 40
cluster constraint, 197
cluttered background, 416
code division multiple access (CDMA), 218
cognitive classification, 42
collision avoidance (CA), 216
combat constraint, 197
communication model, 291
communications system, 215
complex systems, 25, 440
computer generated forces, 256
concept of condition monitoring, 440
confidentiality, 4
constraint parameters, 197

control, 183
control management, 46
control v squad mix, 183
coordination management, 47
corporate memory, 322
course of action critiquing, 360,
problem, 361
CSMA/CA, 216

- D -

data processing classification, 41
decentralized control, 277
decision mechanisms, 98
decision modeling, 107
deployment scenario, 211
detection, 389,
method, 419
disband, 321
Disciple agent, 340,
COA critiquer, 364
RKF, 346
workaround planner, 370
display management, 46
distributed problem solving, 52
distributed processing, 261
dMARS, 249
DSTO concept demonstrator, 205

- E -

emergence of form and function, 295
enforced modularity, 261
environment model, 288
evolutionary algorithms, 73
evolutionary computing, 73
expert systems, 71

- F -

F-111 stand-off weapon employment, 241
FHSS, 217
fighter combat, 236
filtering, 389
frequency hopping spread spectrum (FHSS), 217
functional classification, 38
functional requirements, 207
fuzzy logic, 77
fuzzy systems, 77

- G -

genetic algorithms, 73
goal-directed social networking, 313
graphical interfaces, 257, 259

- H -

hierarchical classification, 40
human behavior modeling, 107

- I -

improved interactions, 262
information, 308
integrity, 9
intelligent agents, 234
intelligent monitoring systems, 437, 447
interactions, 262
inter-agent,
communication model, 291
relationships, 315

- J -

JACK, 249

- K -

knowledge-based techniques, 220

- L -

level of machine assistance, 35

- M -

machine assistance, 35
MANA approach, 188
mathematical morphology, 385, 387
military systems, 20
modularity, 261
morphogenesis, 295
motion picture experts group
(MPEG), 219
movement, 193
MPEG, 219
multilevel security, 16
multiple-access techniques, 215
multisensor data fusion, 454
multisensor systems, 277

- N -

nature, 299
neural networks, 78
non-repudiation, 10

- O -

organizational structure, 319

- P -

parameter landscapes, 186
parameter set, 191
perception, 253
personality, 178
pheromone, 322
philosophy, 175
points of attack, 23
presentation management, 46
prototype, 220

- R -

rapid development, 258
real-time video compression, 219
recognition of intention, 253
relationships, 308,
inter-agent, 315
research projects, 83

- S -

scenario, 181
security,
architectures, 12
multi-level, 16
services, 3
self-organizing sensor network, 281
sensor data management, 45
sensors, 274,
networks,
social aspects, 292
relationships, 309
setting goals, 313
simulation environment, 323
skills modeling, 107
social aspects,
of sensor networks, 292
social networking,
goal-directed, 313
social potential, 301
software architectures, 83, 89
software engineering, 254, 262
spatial filters, 403
spectrum, 22
spread spectrum systems, 215
squad mix, 183
squad objects, 189
SWARMM, 237
synthetically generated target, 416
system models, 287, 289
system-high, 12

- T -

target data,
measured, 402
synthetically generated, 416
target model, 290
target observation problem, 270
task coordination, 45
task management, 45
TDMA, 217
team, 316,
formation, 283, 315, 317
forming algorithm, 321
temporal-spatial filtering, 413
time division multiple access (TDMA), 217
track before detect method, 394
tracking method, 419
types of attack, 24

- U -

uncluttered background, 402

- V -

video,
compression, 219
system, 219
virtual air environment, 244
vulnerabilities, 19, 21

- W -

wavelct compression, 220
weapon employment, 241
weight selection, 314
workaround reasoning, 367,
challenge problem, 368

List of Contributors

Sheila B. Banks, Ph.D.
Calculated Insight
Orlando, FL 32828
U.S.A.
sbanks@calculated-insight.com

R.E. Bogner
Department of Electrical and Electronic Engineering
Adelaide University
Adelaide, SA 5005
and
CRC for Sensor Signal and Information Processing
Mawson Lakes Boulevard, Mawson Lakes, SA 5095
Australia

M. Boicu
Learning Agents Laboratory
Computer Science Department, MS 4A5
George Mason University
4400 University Dr, Fairfax, VA 22030
U.S.A.

R.S. Bowyer
Defence Science & Technology Organisation
Electronic Warfare and Radar Division
Edinburgh, SA 5111
and
Department of Electrical and Electronic Engineering
Adelaide University
Adelaide, SA 5005

and
CRC for Sensor Signal and Information Processing
Mawson Lakes Boulevard, Mawson Lakes, SA 5095
Australia

Martin Cross
Air Operations Division
Defence Science and Technology Organisation
Department of Defence
506 Lorimer St., Fishermens Bend, Victoria 3207
Australia

W.M. Daniels
Knowledge Based Intelligent Engineering Systems Centre
University of South Australia
Mawson Lakes, Adelaide, South Australia 5095
Australia

S. Forbes
Weapon Systems Division
Defence Science and Technology Organisation
Salisbury, South Australia 5108
Australia

Michael R. Frater
School of Electrical Engineering
University College
The University of New South Wales
Australian Defence Force Academy
Canberra ACT 2600
Australia

Simon Goss
Air Operations Division
Defence Science and Technology Organisation
Department of Defence
506 Lorimer St., Fishermens Bend, Victoria 3207
Australia

David L. Hall, Ph.D.
School of Information Sciences and Technology
The Pennsylvania State University
Thomas Building, Shortlidge Road, University Park, PA 16802
U.S.A.
dhall@ist.psu.edu

Clint Heinze
Air Operations Division
Defence Science and Technology Organisation
Department of Defence
506 Lorimer St., Fishermens Bend, Victoria 3207
Australia

Lakhmi C. Jain
Knowledge-Based Intelligent Engineering Systems Centre
University of South Australia
Mawson Lakes, Adelaide, South Australia 5095
Australia

Torgny Josefsson
Air Operations Division
Defence Science and Technology Organisation
Department of Defence
506 Lorimer St., Fishermens Bend, Victoria 3207
Australia

M.K. Lauren
Defence Technology Agency
Devonport Naval Base
Private Bag 32901, Auckland
New Zealand

Ian Lloyd
Air Operations Division
Defence Science and Technology Organisation
Department of Defence
506 Lorimer St., Fishermens Bend, Victoria 3207
Australia

Graeme Murray
Air Operations Division
Defence Science and Technology Organisation
Department of Defence
506 Lorimer St., Fishermens Bend, Victoria 3207
Australia

Michael Papasimeon
Air Operations Division
Defence Science and Technology Organisation
Department of Defence
506 Lorimer St., Fishermens Bend, Victoria 3207
Australia

Michael J. Ryan
School of Electrical Engineering
University College
The University of New South Wales
Australian Defence Force Academy
Canberra ACT 2600
Australia

R.T. Stephen
Defence Technology Agency
Devonport Naval Base
Private Bag 32901, Auckland
New Zealand

Martin R. Stytz, Ph.D.
Air Force Research Laboratory
Wright-Patterson AFB, OH 45431
U.S.A.
mstytz@att.net, mstytz@acm.org

G. Tecuci
Learning Agents Laboratory
Computer Science Department, MS 4A5
George Mason University
4400 University Dr, Fairfax, VA 22030
and
Center for Strategic Leadership
US Army War College
650 Wright Ave, Carlisle Barracks, PA 17013
U.S.A.

Michael Turner
Air Operations Division
Defence Science and Technology Organisation
Department of Defence
506 Lorimer St., Fishermens Bend, Victoria 3207
Australia

Pierre Urlings
Air Operations Division
Aeronautical and Maritime Research Laboratory
Defence Science and Technology Organisation
Australia

Robert C. Warren
Electronic Warfare and Radar Division
Systems Sciences Laboratory, DSTO
P.O. Box 1500, Edinburgh, SA 5111
Australia